Hormone Signaling

ENDOCRINE UPDATES

Shlomo Melmed, M.D., Series Editor

---

1. E.R. Levin and J.L. Nadler (eds.): Endocrinology of Cardiovascular Function. 1998. ISBN: 0-7923-8217-X
2. J.A. Fagin (ed.): Thyroid Cancer. 1998. ISBN: 0-7923-8326-5
3. J.S. Adams and B.P. Lukert (eds.): Osteoporosis: Genetics, Prevention and Treatment. 1998. ISBN: 0-7923-8366-4.
4. B.-Å. Bengtsson (ed.): Growth Hormone. 1999. ISBN: 0-7923-8478-4
5. C. Wang (ed.): Male Reproductive Function. 1999. ISBN 0-7923-8520-9
6. B. Rapoport and S.M. McLachlan (eds.): Graves' Disease: Pathogenesis and Treatment. 2000. ISBN: 0-7923-7790-7.
7. W. W. de Herder (ed.): Functional and Morphological Imaging of the Endocrine System. 2000. ISBN 0-7923-7923-9
8. H.G. Burger (ed.): Sex Hormone Replacement Therapy. 2001. ISBN 0-7923-7965-9
9. A. Giustina (ed.): Growth Hormone and the Heart. 2001. ISBN 0-7923-7212-3
10. W.L. Lowe, Jr. (ed.): Genetics of Diabetes Mellitus. 2001. ISBN 0-7923-7252-2
11. J.F. Habener and M.A. Hussain (eds.): Molecular Basis of Pancreas Development and Function. 2001. ISBN 0-7923-7271-9
12. N. Horseman (ed.): Prolactin. 2001 ISBN 0-7923-7290-5
13. M. Castro (ed.): Transgenic Models in Endocrinology. 2001 ISBN 0-7923-7344-8
14. R. Bahn (ed.): Thyroid Eye Disease. 2001 ISBN 0-7923-7380-4
15. M.D. Bronstein (ed.): Pituitary Tumors in Pregnancy ISBN 0-7923-7442-8
16. K. Sandberg and S.E. Mulroney (eds.): RNA Binding Proteins: New Concepts in Gene Regulation. 2001. ISBN 0-7923-7612-9
17. V. Goffin and P. A. Kelly (eds.): Hormone Signaling. 2002 ISBN 0-7923-7660-9

Hormone Signaling

edited by

Vincent Goffin, PhD, and Paul A. Kelly, PhD

Molecular Endocrinology,
Faculty of Medicine Necker,
Paris, France

KLUWER ACADEMIC PUBLISHERS
BOSTON / DORDRECHT / LONDON

Distributors for North, Central and South America:
Kluwer Academic Publishers
101 Philip Drive
Assinippi Park
Norwell, Massachusetts 02061 USA
Telephone (781) 871-6600
Fax (781) 681-9045
E-Mail <kluwer@wkap.com>

Distributors for all other countries:
Kluwer Academic Publishers Group
Distribution Centre
Post Office Box 322
3300 AH Dordrecht, THE NETHERLANDS
Telephone 31 78 6392 392
Fax 31 78 6392 254
E-Mail <services@wkap.nl>

Electronic Services <http://www.wkap.nl>

Library of Congress Cataloging-in-Publication Data

A C.I.P. Catalogue record for this book is available
from the Library of Congress.

Printed on acid-free paper.

Printed in the United States of America

***

List of Contributors

William Bourguet	Institute of Genetics and Molecular and Cellular Biology, University of Strasbourg, France
Marc G. Caron	Howard Hughes Medical Institute Laboratories and Department of Cell Biology and Medicine, Duke University Medical Center, Durham, USA
Guojun Cheng	Department of Medical Nutrition and Department of Bioscience, Karolinska Institute, Huddinge, Sweden
Nicolas de Roux	Inserm Unit 135, Hormone and Molecular Biology Laboratory, Hôpital Bicêtre, Le Kremlin Bicêtre, France
Béatrice Desvergne	Institute of Animal Biology, Faculty of Sciences, University of Lausanne, Switzerland
Beate Doeker	Inserm Unit 135, Hormone and Molecular Biology Laboratory, Hôpital Bicêtre, Le Kremlin Bicêtre, France
Pierre Germain	Institute of Genetics and Molecular and Cellular Biology, University of Strasbourg, France
Vincent Goffin	Inserm Unit 344, Molecular Endocrinology, Faculty of Medicine Necker, Paris, France
Fabrice Gouilleux	Laboratory of Immunology, School of Medicine, University of Picardie, Amiens, France
Hinrich Gronemeyer	Institute of Genetics and Molecular and Cellular Biology, University of Strasbourg, France
Jan-Åke Gustafsson	Department of Medical Nutrition and Department of Bioscience, Karolinska Institute, Huddinge, Sweden
Sébastien Jeay	Inserm Unit 344, Molecular Endocrinology, Faculty of Medicine Necker, Paris, France *Current address*: Boston University, School of Medicine, Department of Biochemistry, Boston, USA
Paul A. Kelly	Inserm Unit 344, Molecular Endocrinology, Faculty of Medicine Necker, Paris, France
Warren J. Leonard	Laboratory of Molecular Immunology, National Heart, Lung, and Blood Institute, NIH, Bethesda, USA
Derek LeRoith	Clinical Endocrinology Branch, National Institute of Diabetes and Digestive and Kidney Diseases, NIH, Bethesda, USA
Liliane Michalik	Institute of Animal Biology, Faculty of Sciences, University of Lausanne, Switzerland
Edwin Milgrom	Inserm Unit 135, Hormone and Molecular Biology Laboratory, Hôpital Bicêtre, Le Kremlin Bicêtre, France

Gastone G. Nussdorfer	Department of Human Anatomy and Physiology, School of Medicine, University of Padua, Italy.
Jacques Pouysségur	Institute of Signaling, Developmental Biology and Cancer Research, Centre Antoine Lacassagne, Nice, France
Michael J. Quon	Cardiology Branch, National Heart, Lung, and Blood Institute, NIH, Bethesda, USA
Gian Paolo Rossi	Department of Clinical and Experimental Medicine, School of Medicine, University of Padua, Italy.
Hideki Sakaguchi	Department of Medical Nutrition and Department of Bioscience, Karolinska Institute, Huddinge, Sweden
Michael A. Shetzline	Howard Hughes Medical Institute Laboratories and Department of Cell Biology and Medicine, Duke University Medical Center, Durham, USA
Véronique Volmat	Institute of Signaling, Developmental Biology and Cancer Research, Centre Antoine Lacassagne, Nice, France
Walter Wahli	Institute of Animal Biology, Faculty of Sciences, University of Lausanne, Switzerland
Ling Wang	Department of Medical Nutrition and Department of Bioscience, Karolinska Institute, Huddinge, Sweden
Margaret Warner	Department of Medical Nutrition and Department of Bioscience, Karolinska Institute, Huddinge, Sweden
Zhang Weihua	Department of Medical Nutrition and Department of Bioscience, Karolinska Institute, Huddinge, Sweden
Alan Wells	Department of Pathology, University of Pittsburgh and VAMC, Pittsburgh, USA
Li-yuan Yu-Lee	Departments of Medicine, Molecular & Cellular Biology, and Immunology, and Cell & Molecular Biology Program, Baylor College of Medicine, Houston, USA
Zhong-Yin Zhang	Department of Molecular Pharmacology, Albert Einstein College of Medicine, New York, USA
Yu Zhao	Department of Molecular Pharmacology, Albert Einstein College of Medicine, New York, USA
Yehiel Zick	Department of Molecular Cell Biology, the Weizmann Institute of Science, Rehovot, Israel

***

Signaling Websites

Below are listed access addresses of a few websites related to intracellular signaling (this list is not exhaustive).

General information on signaling and cell biology

http://www.bio.davidson.edu/people/kabernd/cb/Web.htm
http://stke.sciencemag.org/
http://www.elsevier.com/locate/cellsig
http://www.sugen.com/webpage_templates/sec.php3?page_name=inside_signal
http://www.new-science-press.com/cells-primer.asp
http://www.apnet.com/cytokinereference/
http://web.indstate.edu/thcme/mwking/signal-transduction.html

Signaling, cancer, development, cell motility

http://www.unice.fr/biochimie/
http://path.upmc.edu/divisions/wells-lab/wells.htm
http://researchportfolio.cancer.gov/
http://www.aecom.yu.edu/asb/segall/segall.htm

Sequence abalysis/domain research/protein structure

http://www.ncbi.nlm.nih.gov/Entrez/
http://www.rcsb.org/pdb/
http://smart.embl-heidelberg.de/
http://www.mshri.on.ca/pawson/research1.html
http://saturn.med.nyu.edu/groups/HubbardLab/

Nuclear receptors

http://receptors.ucsf.edu/NR/

G protein-coupled receptors

http://archive.uwcm.ac.uk/uwcm/mg/search/125313.html
http://www.cmbi.kun.nl/7tm/
http://www.gpcr.org/7tm/
http://www.expasy.org/cgi-bin/lists?7tmrlist.txt
http://tinyGRAP.uit.no/
http://www.tumor-gene.org/GPCR/gpcr.html

****

Preface

Multicellular organisms require a means of intracellular communication, in order to first organize and develop the complex body plan that occurs during embryogenesis and then for the cell and organ systems to access and respond to an ever changing milieu. The mediators of this constant exchange of information are growth factors, neurotransmmitters, peptide and protein hormones which bind to cell surface receptors and transduce their signals from the extracellular space to the intracellular compartment. Via multiple signaling pathways, receptors of this general class affect varying processes, including growth, development and differentiation. Other small, hydrophobic signaling molecules, such as steroids and non-steroid hormones, some vitamins and certain metabolic mediators interact with receptors that form a large family known as nuclear receptors. These receptors function as transcription factors affecting gene expression, to regulate the multiple aspects of animal and human physiology, including development, reproduction and homeostasis.

The aim of this book is to cover in five sections various aspects of intracellular signaling involving hormone receptors. The first section is a general overview of the process of hormonal signaling, with a description of the important role of protein phosphorylation and protein-protein interactions. This is followed by a chapter on the function of protein dephosphorylation by tyrosine phosphatases. Since mitogen activated protein kinase (MAPK) pathways play such a key role in physiology and pathology, they are discussed in the last of these introductory chapters, with particular emphasis given to the spatio-temporal context that one should always consider when trying to understand the complex field of hormone signaling.

Section II deals with the specific class of receptor tyrosine kinases (RTK) signaling. This covers the broad category of growth factors in general, and two specific cases are described in two separate chapters: epidermal growth factor (EGF) receptors, and insulin and insulin-like growth factor (IGF) receptors.

The third section is dedicated to the large family of class I cytokine receptors, and is divided into three chapters. The first introduces the Janus kinase/signal transducer and activator of transcription (JAK/STAT) pathway activated by this family of receptors. The second part of this section is devoted to prolactin and growth hormone receptors, while the third covers receptors for erythropoietin, thrombopoietin and leptin.

Section IV covers the broad class of G protein-coupled receptors. The first chapter of this section is a general overview of the discovery and general mechanisms of signaling. This is followed by two examples, the first dealing with glycoprotein (LH, FSH, TSH) receptors, and the next, endothelin receptors.

The last section is devoted to the ever-expanding class of nuclear receptors. One chapter is a general overview with a specific example of receptors of the retinoid family; the second is the example of estrogen receptors (α and β); the final chapter presents the interesting class of sensors for metabolic control that act via members of the nuclear receptor family.

Vincent Goffin and Paul A. Kelly

****

Acknowledgements

We are grateful to Dr Claude Kordon for his valuable help and advice in generating the outline of *Hormone Signaling*. We also thank Claudine Coridun for her excellent secretarial assistance, and Pauline and Valentine Goffin for their precious help during the preparation of this book.

Section I

Fundamental Mechanisms in Signaling

1

Protein Phosphorylation and Protein-Protein Interactions

Vincent Goffin* and Paul A. Kelly

Inserm Unit 344, Molecular Endocrinology, Faculty of Medicine Necker, 156 rue de Vaugirard, 75730 Paris Cedex 15, France

OVERVIEW

Intracellular signaling is the series of events which translates the specific message of circulating ligands into a particular biological response in target cells. The first step in hormone signaling involves the interaction between a ligand and its cognate receptor (membrane or nuclear), which in turn induces stoichiometric and conformational changes on the ligand-receptor complex. In the current state of the art, these modifications appear to be the molecular basis of signal triggering. Whatever the number or the type of proteins involved in signal transduction process within the cell, all known cascades involve one or, in most cases, two basic mechanisms: protein-protein interactions and protein phosphorylation.

Protein-protein interactions are fundamental since signal propagation through the intracellular space requires proteins otherwise not in contact, to interact with each other and thereby, to acquire the ability to transfer the message from one site to another. To maintain signal specificity until the final target is reached (e.g. gene promoters), these non covalent interactions must not occur randomly, but rather in a tightly controlled manner. This means that protein interactions must be restricted to authorized partners, which usually requires some flexibility and subtle conformational changes of biomolecules

* goffin@necker.fr

involved to achieve molecular recognition by "induced fit" [1]. Protein modules (or interaction motifs) are active players in these specific, though not exclusive recognition processes. These modules adopt consensus conformations, which are structurally appropriate to temporarily, but efficiently dock to small target sequences which are present in signalling proteins and exhibit structurally recognizable characteristics. Why are signaling cascades triggered under stimulation only, despite the fact that proteins involved in these cascades are present, but maintained in an inactive state in unstimulated cells? In many instances, it was shown that ligand-induced clustering of receptor chains was sufficient to bring receptor-associated proteins near to each other, thereby allowing their reciprocal activation. A typical example is the transactivation of Janus tyrosine kinases associated with cytokine receptors (see Chapter 6). In a more general fashion, it is believed that the ability of potential partners to interact with each other is driven by conformational/chemical modifications occurring upon receptor activation (oligomerization). These conformational changes involve intrinsic (e.g. proline-rich sequence) or transient, post-translationally added (e.g. phosphotyrosines) structural features, and confer to these amino acid stretches folding characteristics recognized by protein modules. One of the most popular signals to propagate the information along protein cascades is protein phosphorylation. Protein phosphorylation occurs when a protein kinase interacts with tyrosine, threonine or serine residues of its substrate(s), to which it covalently links the γ phosphate of ATP. This chemical modification is "viewed" by cellular proteins as a signal that switches *on* or, in contrast, switches *off* downstream signaling events. For example, if the phosphorylated substrate is a receptor tyrosine kinase (RTK), phosphorylation can turn on its enzymatic activity, which ensures propagation of the message through phosphorylation of downstream effectors. In a more general fashion, and this is true whether the substrate is an enzyme, a receptor or an adapter, phosphotyrosines confer the ability to interact with protein partners containing phosphotyrosine binding motifs.

The following chapters of this book are aimed at deciphering signaling cascades activated by the four various families of receptors: receptor tyrosine kinases (Section II), cytokine receptors (Section III), G protein-coupled receptors (Section IV) and nuclear receptors (Section V). The three chapters of Section I are aimed at providing an overview of some fundamental mechanisms involved in intracellular signaling. Chapter 2 describes protein phosphatases, whose activity is essential to balance kinase activity by catalyzing de-phosphorylation of signaling proteins and participate in the maintenance of cell homeostasis. Chapter 3 describes one of the most important, though often omitted parameters in signaling studies: the spatio-temporal context. In this introductory chapter, we wish to illustrate the two basic mechanisms mentioned above, i.e. protein-protein recognition and protein phosphorylation. With respect to the former, some of the most common motifs involved in protein-protein interactions will be briefly illustrated. With respect to protein phosphorylation, we chose to describe a family of tyrosine kinases involved in signaling cascades activated by several receptors and containing protein modules regulating their enzymatic activity, namely the Src kinase family.

PROTEIN DOMAINS

A class of crucial proteins involved in propagating signals through the cells are proteins termed "adapters". These proteins are devoid of any enzymatic activity, but contain one or more protein interaction motif. Owing to these motifs, adapters link signaling proteins together and, thereby, establish a framework participating in signal specificity. With the identification of new genes, the list of protein motifs and adapters is rapidly growing ([2]; http://smart.embl-heidelberg.de/), therefore it was not possible to exhaustively review all protein modules in the context of this introductory chapter. The following paragraphs describe five of the most classical protein modules which are among the first that were identified and are certainly the best characterized to date (SH2, SH3, PTB, WW and PH). Complementary information on these domains as well as of many others can be found on the website of T. Pawson's laboratory (http://www.mshri.on.ca/pawson/).

General Rules of Protein Motifs

Protein modules/motifs are consensus sequences ranging from 40 to 200 amino acids. Typically, they are independent folding modules (used to retain their intrinsic ability to bind physiological targets when individually expressed using recombinant technology), whose N- and C-termini are usually close to each other in space, while their ligand-binding surface lies on the opposite face of the domain. They are inserted within larger proteins that (sometimes) contain other interaction motifs or exhibit enzymatic activity. Their sequences are characterized by conserved key amino acids and therefore, they can be often (but not always) identified by primary structural analysis. They adopt a consensus fold also referred to as recognition pocket, which interacts with small (3-6 residues) target sequences also called "core" motifs. The specificity of the interaction is defined by two parameters: first, by the most important core amino acid, referred to as residue "0" (e.g. the phosphotyrosine in a sequence recognized by SH2 domains), and second, by its surrounding amino acids designated "-1, -2, -3" or "+1, +2, +3" depending whether they are respectively amino-terminal or carboxy-terminal to central core amino acid. In a very simplified view, one can say that amino acid "0" defines the type of module with which the interaction occurs (phosphotyrosines bind to SH2 and not to SH3 domains), whereas sequences flanking this central residue participate in defining the specificity of the interaction within one domain family (e.g. a phosphotyrosine in a given sequence will bind to SH2 domain of protein A, but not of protein B).

Phosphotyrosine Binding Motifs

SH2 Domain

Src-Homology domain 2 (SH2) is the first consensus protein interaction module that was discovered [3]; as suggested by its name, it was identified as one stretch of highly conserved amino acids within members of the Src tyrosine kinase family (see below). One had to wait for five years after its discovery to identify the function of SH2 domains as modules interacting with phosphorylated tyrosines [4]. Typical SH2 domains contain ~100 residues and fold in two α helices and five β strands forming one β sheet (Fig. 1). The high affinity of the interaction is directed by the attraction between the positively charged pocket of the SH2 domain and the negatively charged phosphate group of the phosphotyrosine (main site). One key arginine present on the second β strand of the pocket is conserved in all SH2 domains and appears to be crucial for interacting with the phosphate oxygens of the phosphotyrosine residue. Target selection by the different SH2 domains is directed by a secondary site interacting with residues C-terminal (in +2 or +3) to the phosphotyrosine (Fig. 1). Extensive literature illustrates the amino acid requirements coding for interaction selectivity [5-7]; optimized phosphopeptide sequences bind to SH2 domains with an affinity of ~50-500 nM (K*d*). SH2 domains are found in a large number of proteins regulating multiple cellular events such as enzyme activity, substrate recruitment and protein localization [5](Fig. 2). In the second part of this chapter, the way SH2 together with SH3 domains control Src kinase enzymatic activity will be illustrated.

PTB Domain

PTB (Phospho-Tyrosine Binding) domains were discovered in Shc protein in 1994 [8] and identified as an alternative to SH2 domains for interacting with phosphorylated tyrosine-containing sequences (dissociation constant in micromolar range). Despite target similarity, PTB and SH2 domains exhibit important differences. First, interaction specificity of PTB domains is governed by residues amino-terminal, and not carboxy-terminal to the phosphotyrosine (Fig 1). Second, and more surprisingly, it was shown that some PTB motifs (e.g.

in proteins such as X11, Dab, Fe65 and Numb; [9]) do not require the tyrosine "0" (central core) to be phosphorylated to achieve a high affinity interaction.

PTB domains contain ~150-200 amino acids folding in 7 β strands and 3 α helices, which form a sandwich of 2 β sheets with one carboxy-terminal helix (Fig. 1). This folding is different from that of SH2 modules and, actually it represents a subclass of pleckstrin homology domains (see below). In contrast to SH2 or SH3 domains, PTB domains exhibit rather low sequence similarity, which prevented the identification of some of them based on the sole parameter of sequence comparison. Binding specificity is directed by the N-P-x sequence N-terminal to the (phospho-)tyrosine. Most of non tyrosine-phosphorylated peptide sequences interacting with PTB domains contain the canonical N-P-x motif, although the latter also appears dispensable for some PTBs to interact with their target [9]. As shown on Fig. 2, PTB domains are frequently found in adapter proteins.

Proline-Rich Binding Motifs

SH3 Domain

Src-Homology 3 (SH3), as SH2 domains, are one of the four conserved motifs found in the Src family [10]; their function was identified a decade ago as a proline-rich binding motif [11]. SH3 domains contain ~60 residues which fold as a compact β-barrel of five/six anti-parallel β-strands. The ligands of SH3 domains are peptide stretches containing two or more proline residues arranged around a minimal consensus sequence P-x-x-P, which adopts a typical conformation referred to as left-handed polyproline II (PPII) helix; the highest affinities of these interactions are in the micromolar range. Two classes of SH3 domains have been defined (Class I and Class II) depending whether they recognize R-x-x-P-x-x-P or P-x-x-P-x-R motifs, respectively [12]. The unique feature of SH3 domains is that the their target sequences can dock within the same binding pocket (composed of three subsites S0, S1 and S2; see Fig.1) in the N→C or C→N orientation of the polyproline peptide. Actually, due to the pseudosymmetrical conformation of PPII helices, the side chains interacting with the SH3 domain are in a virtually identical position irrespective of peptide orientation; directionality is conferred by the interaction of the arginine of the PPII motif with the charged outer face of the SH3 domain. In the "+" orientation, the arginine is N-terminal to the prolines and interacts with S2 site of the SH3 domain, whereas in the "-" orientation, the arginine interacting with S2 site is C-terminal to the prolines [12](Fig.1). Other features of proline-rich sequences binding to various SH3 domains have been widely reviewed elsewhere [5,13-16].

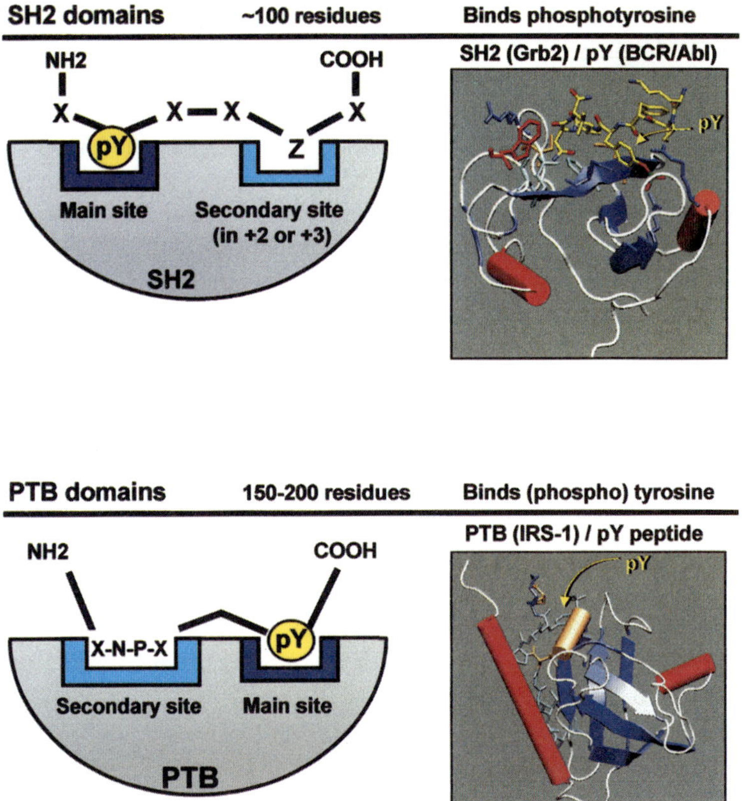

↑→

Figure 1 : Schematic (left) and crystallographic (right) structural representation of the various protein interaction domains described in the text are illustrated. PTB and PH domains exhibit very similar folding despite the fact that their ligands are different (phosphotyrosine and phospholipids, respectively), whereas SH3 and WW domains display no structural homology although both interact with proline-rich sequences (see text for details). The illustrated domains originate from proteins as indicated above the 3D structures. Symbols: pY, phosphotyrosine.

Adapted from Ref. [25]

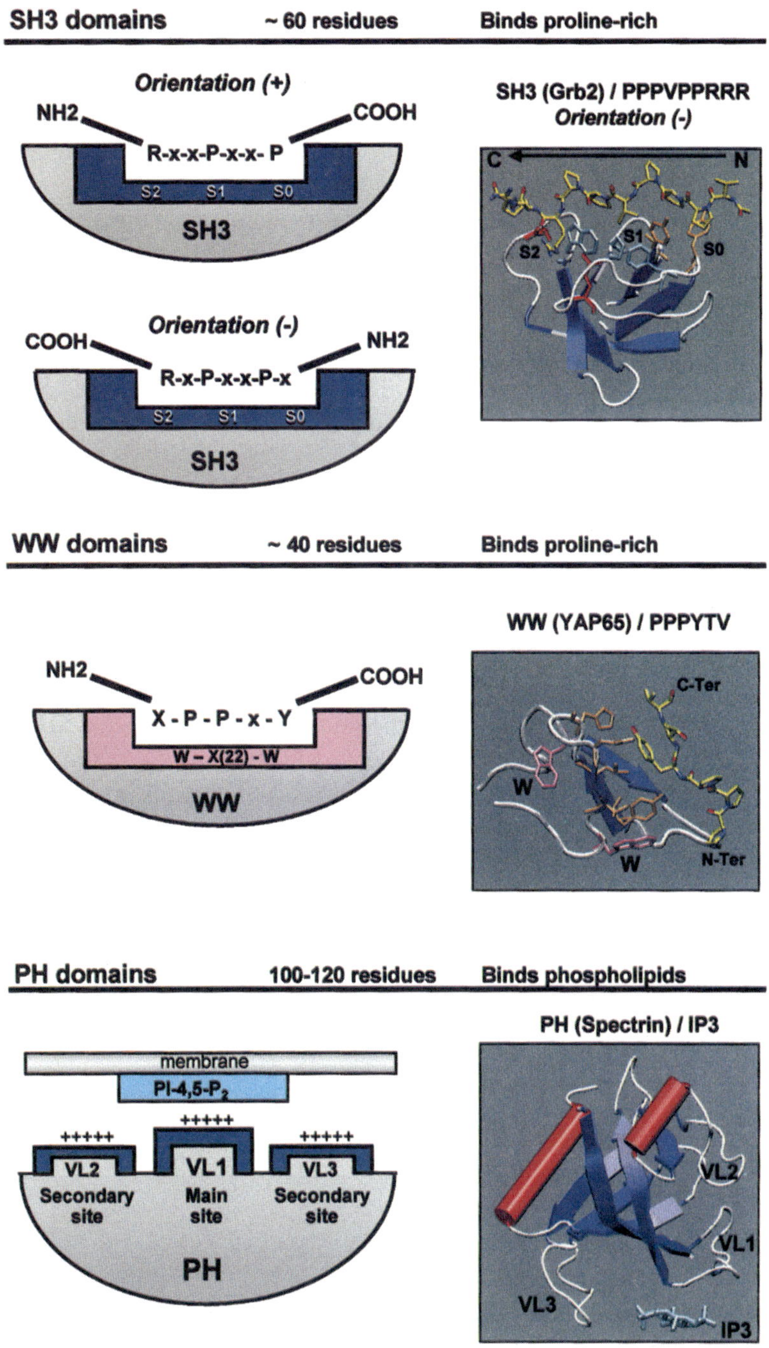

SH3 domains
~ 60 residues
Binds proline-rich
Orientation (+)
NH2
COOH
R-x-x-P-x-x- P
S2 S1 S0
SH3
Orientation (-)
COOH
NH2
R-x-P-x-x-P-x
S2 S1 S0
SH3
SH3 (Grb2) / PPPVPPRRR
Orientation (-)
C
N
S2
S1
S0
WW domains
~ 40 residues
Binds proline-rich
NH2
COOH
X - P - P - x - Y
W – X(22) - W
WW
WW (YAP65) / PPPYTV
C-Ter
W
W
N-Ter
PH domains
100-120 residues
Binds phospholipids
membrane
PI-4,5-P2
+++++
+++++
+++++
VL2
Secondary site
VL1
Main site
VL3
Secondary site
PH
PH (Spectrin) / IP3
VL2
VL1
VL3
IP3

SH3 domains are present in a wide variety of proteins regulating cell-cell communications and signal transduction, of which Grb2 (Growth factor Receptor Bound factor 2) is probably the best known. This 217 amino acid adapter is formed by one SH2 in between two SH3 domains (Fig. 2). The SH2 domain of Grb2 interacts with phosphotyrosine(s) of various activated RTKs (e.g. PDGF receptor), then the SH3 domain recruits Sos, a GTP exchange factor which in turn activates the downstream MAP kinase pathway.

Interestingly, motifs interacting with SH3 domains – the PPII helix - are intrinsic components of target proteins, meaning they allow interactions to occur without ligand-induced post-translational modifications; these interactions are usually more stable than those requiring chemical and/or conformational modification of substrates. One particular example is the intramolecular interaction involving SH3 domain and PPII helix of Src kinases, and this will be illustrated in the second part of this chapter.

WW Domain

After the first two protein recognition motifs were discovered (SH2 and SH3), it was assumed that each sequence identified as a consensus target sequence would interact with a single type of protein module. This predictive model was revised when PTB domains were identified as an alternative to SH2 domains for interacting with tyrosine-containing motifs, and definitely discarded when WW modules were shown to be an alternative to SH3 domains for interaction with proline-rich sequences [17,18]. WW modules are small consensus sequences (40 amino acids) named for the presence of two conserved tryptophane residues spaced by 20-22 residues, the second of which is directly involved in target recognition. Although the 3 β strand folding of this motif is strikingly different from that of SH3 domains (Fig. 1)[19], these two motifs share the ability to interact with proline-rich sequences, adopting a polyproline II structure. However, the canonical P-P-x-Y consensus recognized by WW motifs does not overlap the P-x-x-P motif to which SH3 domains bind, suggesting that these motifs have distinct targets [20]. WW domains bind peptide ligands with dissociation constant ~10 μM.

Phospholipid Binding Motif

PH Domain

The pleckstrin homology (PH) domain was discovered in pleckstrin, the major substrate of protein kinase C [21]. Although sequence similarity between

PH domains is generally low, these 100-120 amino acid motifs adopt a virtually identical conformation containing two anti-parallel β sheets and one α helix (Fig 1). Interestingly, PH domains have a similar structure to PTB domains, despite of the fact that these modules fail to exhibit any significant sequence homology [22] and bind to unrelated targets. Indeed, it is now well established that PH domains mediate protein-lipid interactions and bring their target proteins near the cell membrane. Protein-membrane clustering occurs via interactions involving lipid vesicles and positively charged areas on PH domains (Fig. 1). Some PH domains bind with high affinity (nanomolar range) to specific phosphoinositides, such as phosphatidyl-inositol (PI)-4,5-bisphosphate (PI-4,5-P_2, binding to PLCδ), PI-3,4,5-P_3 (binding to Btk) or PI-3,4-P_2 (binding to Akt/PKB). However, the large majority of PH domains bind weakly and non-specifically to phosphoinositides. A recent review article discusses the possibility that membrane targeting by PH domains exhibiting low affinity for phosphoinositides could be driven by alteration of their oligomeric state, and hence, of the avidity of their membrane binding [23].

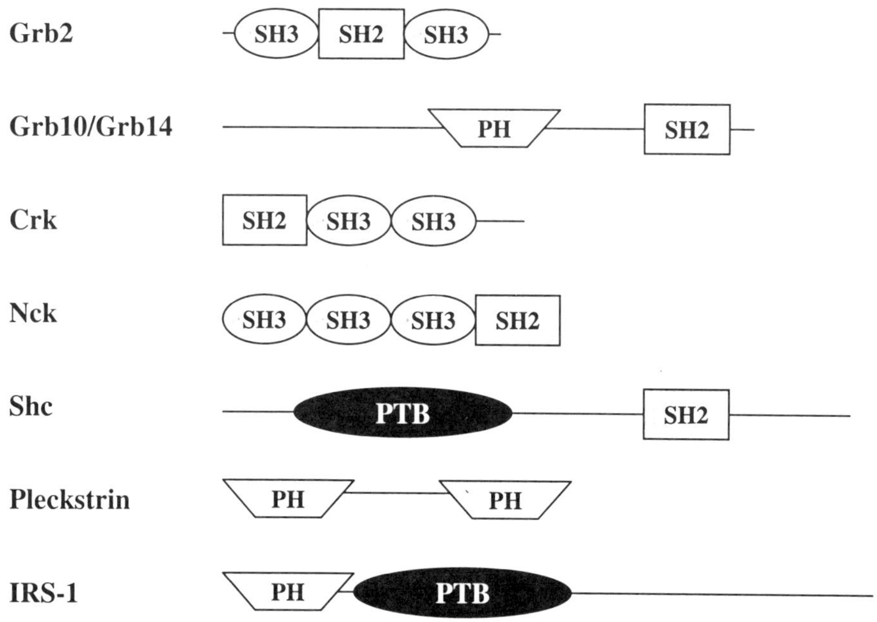

Figure 2. Examples of adapters containing protein interaction motifs. Schematic representation does not strictly respect relative scale of the various proteins. Many other examples can be found in previous reviews [25].

Protein Interaction Motifs : Conclusion

It is interesting to note that protein modules such as PTB, PH or EVH1 (Enabled/Vasp Homology), which bind to phosphotyrosines, membrane phospholipids and proline-rich sequences respectively, share the same global fold. This suggests that they may have evolved from a common ancestor domain functioning as a regulatory binding module. Otherwise, it can be viewed from the opposite angle that this fold may be well suited for protein recognition and hence, reflects convergent evolution [7]. Recently, a comparative analysis of predicted proteomes of yeast (S. *cerevisiae*), fly (D. *melanogaster*) and worm (C. *elegans*) has listed the predicted occurrence of the most frequently occurring protein domains in these organisms [24]. This study highlighted the universal character of protein modules throughout evolution, and also emphasized the correlation between the highest complexity of multidomain proteins and of signaling pathways in mutlicellular (fly, worm) compared to unicellular (yeast) organisms. For example, there are 63 (fly), 55 (worm) and 25 (yeast) candidate proteins containing SH3 domains, and 33, 43 and only 1, respectively, containing SH2 domains.

The large majority of information that regulates cell behavior comes from the extracellular space (hormones, growth factors, extracellular matrix, etc), and is translated inside the cell by signaling cascades to induce a specific biological response. Efficient protein-protein recognition is thus a critical mechanism for directing signaling specificity and to guarantee cell responsiveness as well as cell homeostatis. Although the five protein modules presented above are among the most common found [25], many others have been identified that are involved in mechanisms covering biological areas as different as regulation of apoptosis, interaction with chromatin, proteolysis or protein dimerization. The reader is invited to surf on the website of T. Pawson's laboratory where these motifs are nicely overviewed as individual files, or to consult recent reviews deciphering how protein domains can be integrated into the multiplicity of complex parameters regulating signaling cascades [26-28].

PROTEIN PHOSPHORYLATION, A KEY EVENT IN SIGNALING

Protein Tyrosine Kinases

Protein tyrosine phosphorylation was discovered more than twenty years ago [29,30]. This process involves the transfer of the γ phosphate from ATP to the hydroxyl group of tyrosine phenyl ring and it is likely the most relevant covalent modification occurring in intracellular signaling cascades. This chemical reaction is catalyzed by a family of proteins termed Protein Tyrosine Kinases, or PTKs. Some membrane-anchored proteins combine ligand binding and tyrosine kinase potencies and are therefore referred to as Receptor Tyrosine Kinases, or RTKs. Typical examples of RTKs are receptors for growth factors (EGF, PDGF) or insulin, that are described in Chapters 4 and 5, respectively. In keeping with RTK terminology, cytoplasmic PTKs devoid of ligand binding ability are also referred to as Non Receptor Tyrosine Kinases, or NRTKs [31]. Protein phosphorylation is not restricted to tyrosine, however, since it also occurs on serine and threonine residues. These amino acids are phosphorylated by the same class of kinases, referred to as serine/threonine kinases. In addition to the MAP kinase family (see Chapter 3), the more commonly known serine/threonine kinases are Protein Kinase C (PKC) and cAMP dependent Protein Kinase (PKA). Finally, some kinases such as MAP kinase kinases (MKK) are called dual specificity kinases, since they exhibit both tyrosine and serine/threonine kinase properties (see Chapter 3). The reader can refer to signal transduction websites or to recent reviews for further information on serine/threonine kinases [32].

Eight classes of NRTKs have been identified based on structural and functional parameters, two major of which include Src and Janus (JAK) kinases [31]. JAKs were discovered as kinases associated with cytokine/interferon receptors, and as such, they are described in detail in Chapter 6 (overview of cytokine receptor signaling). To avoid redundancy, this introductory chapter is focused on the largest family of NRTKs, namely Src tyrosine kinases.

Src Tyrosine Kinase Family

The mammalian family of Src kinases contains eight members which among others differ by their tissue distribution: Src, Fyn and Yes are ubiquitously expressed, while Lck, Lyn, Fgr, Hck, and Blk appear to be preferentially expressed in hematopoietic cells. Src, the prototype member of this archetypal NPTK family, was first discovered as the transforming protein in Rous Sarcoma Virus (v-Src protooncogene), then the cellular homolog (c-Src) was identified. Src kinases are involved in signaling pathways activated by many receptor types, including RTKs, integrins, G-protein coupled receptors and cytokine receptors. Src tyrosine kinases are remarkable proteins in that they contain not only an enzymatic site, but also typical protein modules interacting with intramolecular consensus core features, which generates a network of interactions regulating in concert the catalytic activity of the kinase. Following the structural description of protein domains in the first part of this chapter, Src kinases are thus a good example to understand how these modules can function to harmoniously regulate some aspects of intracellular signaling (for reviews, see Ref. [10,33-35]

Primary Structure of Src Kinases

Src family members contain ~530 amino acids (52-62 kDa) and share a common primary structure organization which is divided in four conserved domains referred to as Src-homology (SH) domains (Fig. 3)[10]. The 15-17 N-terminal amino acids (SH4 domain) contain sites of myristoylation (of glycine 2) and palmitoylation (of cysteines within first 10 residues)[36], which ensures anchoring of Src kinases into the cell membrane [37]. Next to the SH4 domain, a unique domain encompassing the first 40 to 70 amino acids does not exhibit significant homology between Src members. The central part of the protein includes two crucial features, the SH2 and SH3 domains. The major part of the C-terminal half of Src kinases (residues 265-516) involves the catalytic domain of the enzyme (also named SH1). This region contains structural features conserved within all family members, such as the tyrosine kinase typical loop (FP^{245}IKWTA) which ensures catalytic selectivity towards tyrosine due to interaction of P^{245} with the phenolic ring of tyrosine. One of the most remarkable conserved features among Src kinases is a tyrosine referred to as Tyr416 (chicken Src numbering; accession number: P00523; [38]), whose phosphorylation stimulates enzymatic activity. Finally, the 20 C-terminal residues also play an essential role in regulating negatively Src kinase activity via the constant Tyr527 (chicken Src numbering).

Regulation of Src Kinase Function

It has been long known that the phosphorylation status of Tyr416 and Tyr527 is a key factor in regulating Src kinase activity [34]. In the unstimulated state, Tyr527 is phosphorylated while Tyr416 is not. The tyrosine kinase involved in phosphorylation of Tyr527 has been identified as Csk (C-terminal Src kinase)[39]. Phosphorylation of Tyr527 inhibits catalytic activity, as emphasized by the viral version of Src (v-Src) whose 19 C-terminal amino acids (including Tyr527) are replaced by a 12 residue tail typical of all Rous Sarcoma Virus and conferring constitutive activity to the kinase domain. Stimulation of Src intrinsic activity involves dephosphorylation of Tyr527 and autophophoryaltion of Tyr 416. The latter lies within a typical structural features shared by RTKs and NRTKs, i.e. the "activation loop" (Fig. 3).

From biochemical studies, it has been suggested for a long time that the negative role of phosphorylated Tyr527 was due to an intramolecular interaction between this residue and the SH2 domain of the kinase (Fig. 3), thereby preventing accessibility to the catalytic site. The SH3 domain was also suspected to play a role in this regulation, yet its mechanism remained unknown.

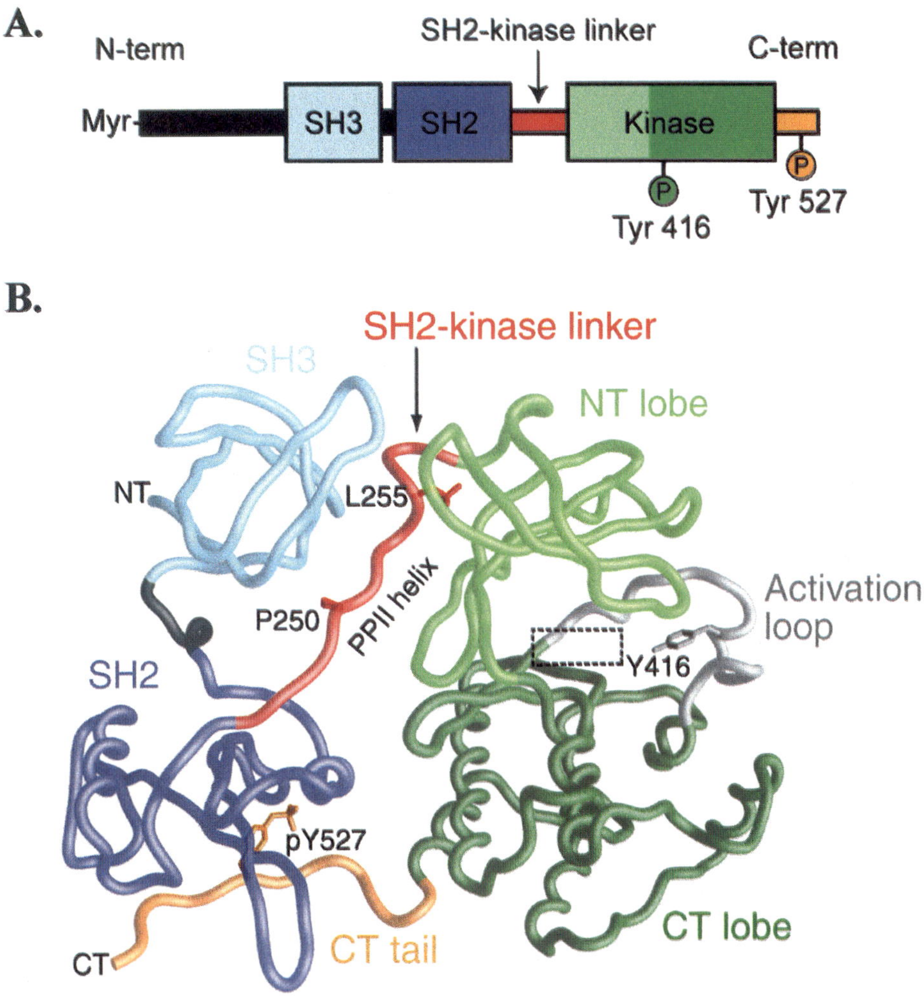

Figure 3. Schematic (A) and 3-D ribbon (B) representation of Src kinases. The conserved features of Src family members are color-coded in both parts of the Figure: the SH3 domain (light blue), SH2 domain (violet), the SH2-kinase linker (red) folding in a poplyproline type II conformation (PPII helix), the enzymatic site (with N-terminal and C-terminal lobes in light and dark green, respectively), the activation loop (gray) with Tyr416 pointing into the active site cleft (dashed box), and the C-terminal tail (orange) with Tyr527 interacting with the SH2 domain.

Reproduced from Ref. [44]

Recent elucidation of crystal structures of Src ([40] and Hck [41]; for reviews, see Ref. [42,43])[44] confirmed the interaction between Tyr527 and the SH2 domain, and furthermore, it brought some insight on the mechanism of kinase regulation by the SH3 domain. This domain interacts with a peptide sequence linking the SH2 and kinase domains, and therefore referred to as the "SH2-kinase linker" (Fig. 3). Interestingly, this linker adopts a PPII helix predicted from former sequence analysis since the minimal SH3 binding motif (P - x - x - P consensus; see above) is only present in Hck or Lck, but is not conserved in other Src family members such as Src itself or Fyn, which only contain the first proline (P^{250}; Fig. 3).

The question is thus to understand how these internal interactions involving SH2 and SH3 domains combined with the shift of phosphorylation site (from Tyr527 to Tyr416) fit into a model leading to Src kinase stimulation. Actually, the mechanism of kinase activation lies on the fact that these internal contacts are not optimal [44,45], therefore, effective competition with phosphotyrosine- (for the SH2) or polyproline- (for the SH3) containing sequences can occur and disrupts internal contacts within the kinase. Substituting intermolecular kinase-substrate interactions for intramolecular kinase interactions modifies the conformation of the enzyme and promotes Tyr416 autophosphorylation, leading to reconfiguration of the activation loop and fully active kinase. Thus, structural determination of various Src family members confirmed the importance of the SH2 domain-mediated interactions, but it also highlighted the unexpected role of the SH3 domain-PPII helix contact, which in some cases has been proposed to be the most relevant event in shifting Src kinase from inactive to active conformation [41,44,46].

CONCLUSION

In multicellular organisms, the information carried by circulating factors such as hormones or growth factors needs to be translated into a language interpreted by living cells as an order to exhibit responses as different as division, secretion, mobility or death. All these events, referred to as intracellular signaling, require a machinery able to rapidly and efficiently translate these messages and direct them towards the right target. This molecular machinery implies the occurrence of transient interactions between protein partners, which among others involve modification of their conformation and/or phosphorylation status, and participation of protein recognition domains. This first chapter has been aimed at providing a very broad Introduction to the extremely complex field of *Hormone Signaling*. The next thirteen chapters will each decipher in details one or another aspect of intracellular signaling by hormone receptors.

Acknowledgements

We are very grateful to Drs A. Ducruix, I. Broutin and S. Hubbard as well as to the Editors of Medecine/Sciences and Nature Structural Biology for permissions to use representations of protein interaction domains (Fig. 1, adapted from [25]) and Src kinase structure (Fig. 3, reproduced without modification from [44]). We also thank A. Ducruix for critical reading of the manuscript. Finally, we would like to acknowledge the participation of Novo Nordisk®, Danemark, for providing a grant for the color Figures.

REFERENCES

1. **Bosshard HR** 2001 Molecular recognition by induced fit: how fit is the concept? News Physiol Sci 16:171-173
2. **Schultz J, Copley RR, Doerks T, Ponting CP, Bork P** 2000 SMART: a web-based tool for the study of genetically mobile domains. Nucleic Acids Res 28:231-234
3. **Sadowski I, Stone JC, Pawson T** 1986 A noncatalytic domain conserved among cytoplasmic protein-tyrosine kinases modifies the kinase function and transforming activity of Fujinami sarcoma virus P130gag-fps. Mol Cell Biol 6:4396-4408
4. **Koch CA, Anderson D, Moran MF, Ellis C, Pawson T** 1991 SH2 and SH3 domains: elements that control interactions of cytoplasmic signaling proteins. Science 252:668-674
5. **Pawson T** 1995 Protein modules and signalling networks. Nature 373:573-580
6. **Songyang Z, Shoelson SE, Chaudhuri M, Gish G, Pawson T, Haser WG, King F, Roberts T, Ratnofsky S, Lechleider RJ, Neel BG, Birge RB, Fajardo JE, Chou MM, Hanafusa H, Schaffhausen B, Cantley LC** 1993 SH2 domains recognize specific phosphopeptides sequences. Cell 72:767-778
7. **Sudol M** 1998 From Src Homology domains to other signaling modules: proposal of the 'protein recognition code'. Oncogene 17:1469-1474
8. **Kavanaugh WM, Williams LT** 1994 An alternative to SH2 domains for binding tyrosine-phosphorylated proteins. Science 266:1862-1865
9. **Forman-Kay JD, Pawson T** 1999 Diversity in protein recognition by PTB domains. Curr Opin Struct Biol 9:690-695
10. **Tatosyan AG, Mizenina OA** 2000 Kinases of the Src family: structure and functions. Biochemistry (Mosc) 65:49-58
11. **Cicchetti P, Mayer BJ, Thiel G, Baltimore D** 1992 Identification of a protein that binds to the SH3 region of Abl and is similar to Bcr and GAP-rho. Science 257:803-806
12. **Feng S, Chen JK, Yu H, Simon JA, Schreiber SL** 1994 Two binding orientations for peptides to the Src SH3 domain: development of a general model for SH3-ligand interactions. Science 266:1241-1247
13. **Sparks AB, Hoffman NG, McConnell SJ, Fowlkes DM, Kay BK** 1996 Cloning of ligand targets: systematic isolation of SH3 domain- containing proteins. Nat Biotechnol 14:741-744
14. **Sparks AB, Rider JE, Hoffman NG, Fowlkes DM, Quillam LA, Kay BK** 1996 Distinct ligand preferences of Src homology 3 domains from Src, Yes, Abl, Cortactin, p53bp2, PLCgamma, Crk, and Grb2. Proc Natl Acad Sci (USA) 93:1540-1544
15. **Koyasu S, Tse AG, Moingeon P, Hussey RE, Mildonian A, Hannisian J, Clayton LK, Reinherz EL** 1994 Delineation of a T-cell activation motif required for binding of protein tyrosine kinases containing tandem SH2 domains. Proc Natl Acad Sci (USA) 91:6693-6697
16. **Pawson T, Schlessinger J**. 1993 SH2 and SH3 domains. Curr.Biol. 3:434-442.

17. **Sudol M, Chen HI, Bougeret C, Einbond A, Bork P** 1995 Characterization of a novel protein-binding module--the WW domain. FEBS Lett 369:67-71
18. **Sudol M** 1996 The WW module competes with the SH3 domains? Trends Biochem Sci 21:161-163
19. **Macias MJ, Hyvonen M, Baraldi E, Schultz J, Sudol M, Saraste M, Oschkinat H** 1996 Structure of the WW domain of a kinase-associated protein complexed with a proline-rich peptide. Nature 382:646-649
20. **Aghazadeh B, Rosen MK** 1999 Ligand recognition by SH3 and WW domains: the role of N-alkylation in PPII helices. Chem Biol 6:R241-R246
21. **Haslam RJ, Koide HB, Hemmings BA** 1993 Pleckstrin domain homology. Nature 363:309-310
22. **Lemmon MA, Ferguson KM, Schlessinger J** 1996 PH domains: diverse sequences with a common fold recruit signaling molecules to cell surface. Cell 85:621-624
23. **Lemmon MA, Ferguson KM** 2000 Signal-dependent membrane targeting by pleckstrin homology (PH) domains. Biochem J 350:1-18
24. **Rubin GM, Yandell MD, Wortman JR, Gabor Miklos GL, Nelson CR, Hariharan IK, Fortini ME, Li PW, Apweiler R, Fleischmann W, Cherry JM, Henikoff S, Skupski MP, Misra S, Ashburner M, Birney E, Boguski MS, Brody T, Brokstein P, Celniker SE, Chervitz SA, Coates D, Cravchik A, Gabrielian A, Galle RF, Gelbart WM, George RA, Goldstein LS, Gong F, Guan P, Harris NL, Hay BA, Hoskins RA, Li J, Li Z, Hynes RO, Jones SJ, Kuehl PM, Lemaitre B, Littleton JT, Morrison DK, Mungall C, O'Farrell PH, Pickeral OK, Shue C, Vosshall LB, Zhang J, Zhao Q, Zheng XH, Lewis S** 2000 Comparative genomics of the eukaryotes. Science 287:2204-2215
25. **Broutin I, Ducruix A.** 2000 Domaines structuraux et signalisation. Médecine/Sciences 16: 611-616.
26. **Pawson T, Nash P** 2000 Protein-protein interactions define specificity in signal transduction. Genes Dev 14:1027-1047
27. **Hunter T** 2000 Signaling--2000 and beyond. Cell 100:113-127
28. **Schlessinger J** 2000 Cell signaling by receptor tyrosine kinases. Cell 103:211-225
29. **Hunter T** 1980 Protein phosphorylated by the RSV transforming function. Cell 22:647-648
30. **Sefton BM, Hunter T, Beemon K, Eckhart W** 1980 Evidence that the phosphorylation of tyrosine is essential for cellular transformation by Rous sarcoma virus. Cell 20:807-816
31. **Taniguchi T** 1995 Cytokine signaling through nonreceptor protein tyrosine kinases. Science 268:251-255
32. **McCubrey JA, May WS, Duronio V, Mufson A** 2000 Serine/threonine phosphorylation in cytokine signal transduction. Leukemia 14:9-21
33. **Abram CL, Courtneidge SA** 2000 Src family tyrosine kinases and growth factor signaling. Exp Cell Res 254:1-13
34. **Superti-Furga G, Courtneidge SA** 1995 Structure-function relationships in Src family and related protein tyrosine kinases. Bioessays 17:321-330
35. **Courtneidge SA, Fumagalli S, Koegl M, Superti-Furga G, Twamley-Stein GM** 1993 The Src family of protein tyrosine kinases: regulation and functions. Dev Suppl57-64
36. **Resh MD** 1994 Myristylation and palmitylation of Src family members: the fats of the matter. Cell 76:411-413
37. **Milligan G, Parenti M, Magee AI** 1995 The dynamic role of palmitoylation in signal transduction. Trends Biochem Sci 20:181-187
38. **Takeya T, Hanafusa H** 1983 Structure and sequence of the cellular gene homologous to the RSV src gene and the mechanism for generating the transforming virus. Cell 32:881-890
39. **Superti-Furga G, Fumagalli S, Koegl M, Courtneidge SA, Draetta G** 1993 Csk inhibition of c-Src activity requires both the SH2 and SH3 domains of Src. EMBO J

12:2625-2634
40. **Xu W, Harrison SC, Eck MJ** 1997 Three-dimensional structure of the tyrosine kinase c-Src. Nature 385:595-602
41. **Sicheri F, Moarefi I, Kuriyan J** 1997 Crystal structure of the Src family tyrosine kinase Hck]. Nature 385:602-609
42. **Williams JC, Wierenga RK, Saraste M** 1998 Insights into Src kinase functions: structural comparisons. Trends Biochem Sci 23:179-184
43. **Hubbard SR, Till JH** 2000 Protein tyrosine kinase structure and function. Annu Rev Biochem 69:373-398
44. **Hubbard SR** 1999 Src autoinhibition: let us count the ways. Nat Struct Biol 6:711-714
45. **Gonfloni S, Weijland A, Kretzschmar J, Superti-Furga G** 2000 Crosstalk between the catalytic and regulatory domains allows bidirectional regulation of Src. Nat Struct Biol 7:281-286
46. **Moarefi I, LaFevre-Bernt M, Sicheri F, Huse M, Lee CH, Kuriyan J, Miller WT** 1997 Activation of the Src-family tyrosine kinase Hck by SH3 domain displacement. Nature 385:650-653

2

Control of Signaling by Tyrosine Phosphatases

Yu Zhao and Zhong-Yin Zhang*

Department of Molecular Pharmacology, Albert Einstein College of Medicine, 1300 Morris Park Avenue, Bronx, New York 10461, USA

INTRODUCTION

Extracellular stimuli, such as peptide hormones, growth factors and cytokines, regulate diverse biological functions, including cell cycle progression, proliferation and differentiation, the immune response, and cytoskeletal reorganization, by altering the levels of protein phosphorylation. The initial and crucial event for most signaling pathways inside the cell is protein tyrosine phosphorylation. Indeed, many transmembrane receptors for peptide hormones and growth factors possess intrinsic protein tyrosine kinase (PTK) activity (see Chapters 4 and 5) . Receptors for cytokines (see Chapters 6 to 8) lack intrinsic kinase activity but associate with non-receptor PTKs inside the cell. Levels of cellular protein tyrosine phosphorylation are regulated by two opposing activities catalyzed by PTKs and protein tyrosine phosphatases (PTPs). PTKs have been studied extensively for the past two decades [1]. However, the importance of PTPs has not been fully appreciated until recently. The addition or removal of a phosphoryl moiety from tyrosine residue(s) in a protein can generate novel recognition motifs that regulate protein-protein interactions, control protein stability, and most importantly, modulate enzyme activity. Phosphorylation is usually regarded as an "on" switch to activate a signal transduction pathway, while dephosphorylation as an "off" switch.

---

* zyzhang@aecom.yu.edu

However, this is not always the case. As will be discussed below, PTPs can exert both positive and negative effects on a signaling pathway (see Chapter 1). Often, the critical factor is the balance of phosphorylation, which dictates the extent and duration of an extracellular signal event. To fully understand the fundamental role of protein tyrosine phosphorylation in complex and critical signal transduction pathways, detailed studies of both PTKs and PTPs are required.

CLASSIFICATION OF PTPs

Based on structure and function, PTPs can be grouped into three subfamilies (Fig. 1): the "classical" PTPs, the dual-specificity phosphatases (DSPs), and the low molecular weight phosphatases (LMW-PTPs) [2]. Although the PTPs within each subfamily share significant sequence homology, little amino acid sequence similarity is apparent among PTPs of different subfamilies. The common feature that defines the PTP superfamily is the active site sequence (H/V)**C**$(X)_5$**R**(S/T) in the catalytic domain, also known as the "PTP signature motif". Amino acid sequence comparisons of the catalytic domains of PTPs with the catalytic subunits of protein Ser/Thr phosphatases have shown no sequence similarity between the two families. This is in marked contrast to the situation with protein kinases, where tyrosine specific and Ser/Thr specific kinases do share sequence similarity in both the catalytic domains and other structural and mechanistic properties. Although the two classes of protein phosphatases catalyze the same chemical reactions, the hydrolysis of phosphate monoesters, they have evolved to employ completely different strategies to accomplish the same task. While the protein Ser/Thr phosphatases are two-metal-ion metalloenzymes that effect catalysis by direct attack of an activated water molecule at the phosphorus atom of the substrate, PTPs are non-metalloenzymes and proceed through a covalent phosphocysteine intermediate during catalytic turnover.

Classical PTPs

The subfamily of the classical PTPs can be further divided into receptor-like and intracellular PTPs based on their cellular localizations. *The receptor-like PTPs (RPTPs)* are type 1 transmembrane proteins containing an extracellular domain, a single transmembrane region and a cytoplasmic domain. They can be further subdivided into at least six distinct subtypes based upon the structural motifs in the extracellular domains. The extracellular structural motifs include domains resembling immunoglobulin, fibronectin III, carboxyl anhydrase, MAM (Meprin-Xenopus A2-Mu) domains or cysteine rich regions, which suggest that RPTPs may function in cell-cell or cell-matrix adhesion. The cytoplasmic region of the receptor PTPs contain one or two PTP domains, of which the membrane proximal PTP domain (D1) is responsible for the majority (if not all) of the phosphatase activity. The membrane distal domain (D2) of some RPTPs also displays weak, intrinsic phosphatase activity while that of others is catalytically inactive. Evidence suggests that the D2 domain may play a regulatory role. However, the biological significance of the second PTP domain of most RPTPs remains largely obscure.

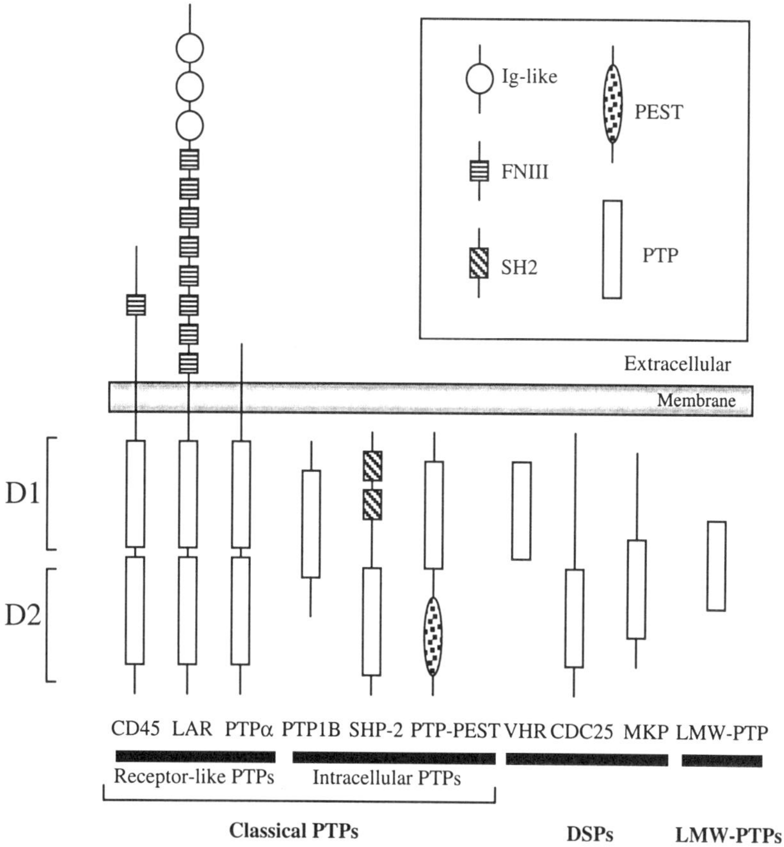

Figure 1. The PTP superfamily. Only representative members of the three subfamilies are shown. Ig-like: immunoglobin-like; FN III: fibronectin III; PEST: Pro, Glu, Ser, Thr-rich motif.

The intracellular PTPs contain a single conserved catalytic domain and highly variable non-catalytic domains located at the amino-terminus, the carboxyl-terminus, or both. The non-catalytic domains of intracellular PTP appear to have regulatory and/or targeting functions, with the net effect of conferring *in vivo* substrate specificity upon PTPs. For example, a carboxyl-terminal hydrophobic region in PTP1B is both necessary and sufficient to target PTP1B to the endoplasmic reticulum. A family of PTPs including PTP-MEG1 and PTPH1 contain a non-catalytic domain that shows significant homology with band 4.1 superfamily of cytoskeletal proteins. The band 4.1 domain may direct these PTPs to the actin filaments near the plasma membrane. SHP-1 and SHP-2 utilize their SH2 domains to target the phosphatases to activated growth factor receptors.

Dual-Specificity Phosphatases

The subfamily of DSPs is capable of catalyzing the dephosphorylation of pTyr, pSer and pThr residues. DSPs can be further divided into three subclasses based on their structures and functions: VHR (VH1-related), MAP kinase phosphatases and Cdc25. This subfamily of enzymes appears to have strong preferences for certain protein substrates. VHR was the first DSP identified in humans and contains a single catalytic domain that is conserved among all DSPs. Cdc25 is a major cell cycle modulator that is capable of removing the inhibitory phosphates from both tyrosine and threonine residues on cyclin-dependent kinases to trigger cell entry into mitosis. Ten mammalian MAP kinase phosphatases (MKPs) have been identified since the isolation of MKP-1/CL100/3CH134 [3]. MKPs are characterized by their high specificity towards MAP kinases (see Chapter 3). They all contain a catalytic domain that is similar in sequence to VHR. What distinguishes them is their highly variable amino-terminal domain that appears to regulate their cellular localization as well as specificity towards different members of MAP kinases. They exhibit distinct substrate specificities toward MAP kinases with different tissue distribution and subcellular localization. Moreover, the expression of many MKPs is subject to induction by different extracellular stimuli.

Low Molecular Weight Phosphatases

The subfamily of low molecular weight phosphatases (LMW-PTPs) consists of small cytosolic enzymes (~18 kDa) with little sequence homology to the other subfamilies of PTPs. This family of enzymes displays substrate specificity primary against pTyr. The LMW-PTPs are expressed ubiquitously and may play an important role in down-regulating receptor PTK signaling. However, the physiological function of these enzymes awaits further investigation.

STRUCTURE AND CATALYTIC MECHANISM

All members of the PTP superfamily contain the PTP signature motif (H/V)**C**$(X)_5$**R**(S/T). Although the signature motif is localized at different positions and the sequences flanking the motif among members of different subfamilies lack general similarity, the three-dimensional structures of the catalytic domains all PTPs that have been solved share strikingly similar features [4]. The PTPs are $\alpha+\beta$ proteins with tertiary folds composed of a highly twisted mixed ß-sheet flanked by α–helices on both sides (Fig. 2). The three dimensional structures of the DSPs and the LMW-PTPs display general folds that resemble the PTP structures, and many of the secondary structural elements of PTPs are also present in the DSPs and LMW-PTPs. The phosphatase active site is located within a crevice (~ 9 Å deep for the PTPs and ~ 6 Å deep for DSPs) on the molecular surface. At the bottom of the active site is the phosphate-binding loop formed by the PTP signature motif. The active site is surrounded by several surface loops, which are important for catalysis and substrate recognition. In PTP1B, one of these loops, called the substrate recognition loop, contains Tyr46 and Arg47. A second loop contains the essential general acid/base Asp181. One important structural feature in PTPs is that ligand binding promotes a major conformational change of the general acid/base containing loop, which brings Asp181 into the catalytic site and forms hydrogen bonds to the phenolic oxygen of pTyr to provide the general acid function (Fig. 2). In the ligand-bound form, the aryl side chains of Tyr46 and Phe182 sandwich the pTyr ring and delineate the boundaries of the pTyr binding pocket in PTPs. The shallower active site in DSPs may be responsible for their ability to hydrolyze both pTyr and pSer/Thr.

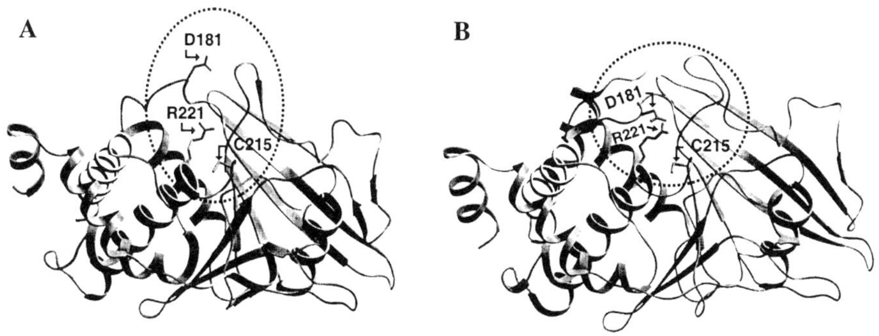

Figure 2. Crystal structures of PTP1B in the open, unliganded from (A), and in the closed, ligand-bound form (B).

Detailed enzymological studies on selected members of each PTP subfamily reveal a common chemical mechanism for the reactions catalyzed by all PTPs [2]. The common catalytic strategy involves a two-step mechanism (Fig. 3). The invariant Cys215 in the PTP signature motif attacks the phosphate ester in the first step, generating a thiol-phosphate intermediate. The invariant Arg221 residue makes bidentate hydrogen bonds with the phosphoryl oxyanions in the substrate through its guanidinium group and plays an important role in both substrate binding and transition state stabilization during hydrolysis. The protonation of the leaving group by the general acid Asp181 greatly enhances the formation of the phosphoenzyme intermediate. In the second step, the dephosphorylation of the phosphoenzyme intermediate occurs by the attack of a nucleophilic water molecule assisted by the general base Asp181, with subsequent release of the free enzyme and inorganic phosphate. As a part of the PTP signature motif, the conserved Ser222 serves to facilitate the breakdown of the phosphoenzyme intermediate by charge stabilization.

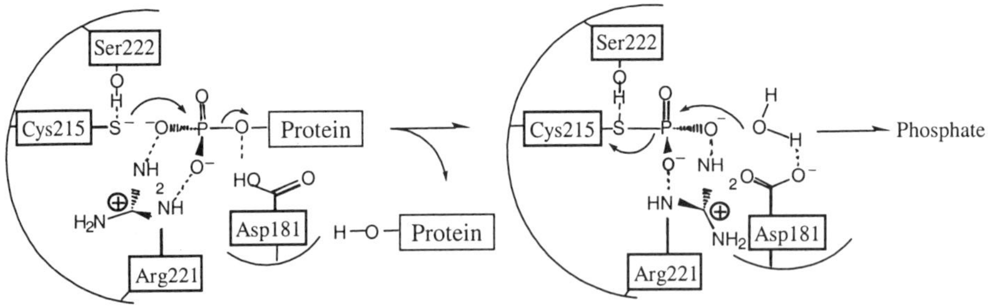

Figure 3. A chemical mechanism for the PTP1B-catalyzed reaction.

PTP SUBSTRATE SPECIFICITY

Understanding the specific functional roles that PTPs play in cellular signaling requires the identification of the physiological substrates for each member of the PTP family. The development and use of "substrate-trapping" mutants have greatly facilitated the discovery of PTP substrates [5]. Two types of substrate-trapping mutants have been used to isolate PTP substrates. In the first, the active site Cys residue is replaced by a Ser while in the second, the general acid Asp residue is substituted by an Ala. Although the Cys to Ser mutant has no measurable phosphatase activity and the catalytic activity of the Asp to Ala mutant is reduced 10^5-fold, both mutants retain the ability to bind substrates [6], enabling their use as "affinity reagents" for the isolation of *in vivo* PTP substrates. Interestingly, in several instances, the Asp to Ala mutant has been found to be a better substrate-trapping reagent than the Cys to Ser mutant [5, 6]. Substrate-trapping mutants have also been employed extensively in structural studies of PTPs. In fact, most of the structural information regarding PTPase-substrate interactions has been obtained not from wild type PTPases, but from substrate complexes with the active site Cys to Ser substrate-trapping mutants [7, 8].

In spite of the remarkable progress in the identification and characterization of new PTPs and in the understanding of PTP catalysis, the molecular basis by which PTPs distinguish and recognize the diverse substrates that they encounter in the cell remains elusive. This is partly because the physiological substrates for most PTPs remain unknown. Furthermore, since the physiological substrates for PTPs are phosphoproteins, even if the identity of the true physiological substrate is known, it remains a major obstacle to obtain quantities of specifically and stoichiometrically tyrosine phosphorylated proteins required for detailed enzymological studies. The current approach for probing PTP substrate specificity relies on the use of synthetic pTyr-containing peptides that correspond to the phosphorylated species found *in vivo*. These studies have shown that PTPs exhibit a broad range of substrate specificity. In one extreme, the MAP kinase phosphatases and cdc25 phosphatases show exquisite specificity, capable of dephosphorylating only their physiological substrates, the MAP kinases and the cyclin dependent kinases, respectively. On the other end of the spectrum, the active site of PTP1B displays extraordinarily plasticity in its ability to accommodate different phosphopeptides [7, 8]. PTP1B is known to down-regulate the activity of receptor PTKs that contain multiple tyrosine autophosphorylation sites. The active site plasticity in PTP1B may be important for efficient termination of receptor PTK signaling.

BIOLOGICAL FUNCTIONS OF PTPs

PTPs play an active role in many diverse biological processes that involve tyrosine phosphorylation [9]. Transgenic mice with targeted mutations in various PTP genes serve as an invaluable tool to provide insight into the physiological functions of PTPs. For example, a defect in T cell development and signaling was observed in CD45-null mice [10], strongly indicating a positive role of CD45 in T cell receptor signaling events. Loss of SHP-1 results in mice with *motheaten* phenotype, suggesting a negative role played by SHP-1 in T cell receptor signaling [11]. Abolishment of SHP-2 [12], PTP-σ [13], or T Cell PTP (TC-PTP) [14], causes various degrees of lethality during mouse development. Recently, results from PTP1B-knockout mice reveal an important role for PTP1B as a negative regulator of the insulin signaling pathway [15, 16].

CD45 is a major receptor-like PTP that is expressed in hematopoietic cells and it acts as an important regulator of Src-family kinases (see Chapter 1). Src kinases have two principle tyrosine phosphorylation sites with opposing regulatory effects. Phosphorylation of the tyrosine residue located in the activation loop of the kinase domain (autophosphorylation site) potentiates Src kinase activity, whereas phosphorylation of the tyrosine residue within the C-terminal tail represses its activity. Both of the regulatory tyrosine sites in Src family kinases are CD45 substrates *in vitro*. Increasing evidence suggests that CD45 is capable of regulating Src kinases both positively and negatively *in vivo*, depending on cell types and state of differentiation or activation. Thomas and Brown [17] proposed a mechanism to explain why CD45 exerts different effects on Src family kinases for integrin-mediated adhesion and antigen receptor signaling. When localized to integrin-mediated adhesion sites, CD45 dephosphorylates the autophosphorylation site on Src that potentiates kinase activity and thus negatively regulates Src kinases. Whereas CD45 acts as a positive regulator of antigen receptor signaling because it is excluded from interacting with Src kinases at sites of engaged receptors. Thus, it can only dephosphorylate the C-terminal inhibitory tyrosine phosphorylation site on Src.

PTP1B is a classical intracellular PTP. There are many reports that implicate PTP1B as a negative regulator of insulin action as well as other signal transduction pathways. Overexpression of PTP1B in rat fibroblast cells reduces insulin-mediated insulin receptor (IR; see Chapter 5) autophosphorylation significantly, whereas an inactive mutant of PTP1B has the opposite effect. Osmotic loading of neutralizing antibodies against PTP1B into rat hepatoma cells increases insulin-stimulated activities. Targeted disruption of the PTP1B gene in mice provides an important clue to its function in the insulin signaling pathway [15, 16]. The PTP1B$^{-/-}$ mice develop normally, are fertile and do not show macroscopic or histological differences compared with wild-type mice (PTP1B$^{+/+}$). Interestingly, the homozygote PTP1B$^{-/-}$ mice show increased IR phosphorylation and enhanced sensitivity to insulin in skeletal muscle and liver. In addition, PTP1B$^{-/-}$ mice have remarkably low adiposity and are protected from diet-induced obesity. Studies of PTP1B-deficient mice provide strong evidence that PTP1B acts directly on the insulin receptor, but its effects on downstream effectors of the insulin signaling pathway cannot be excluded. The tissue-specific effects of PTP1B knockout mice on insulin action cannot be readily explained. Another issue that needs to be addressed is the cellular localization of PTP1B, which has been shown to be localized at the endoplasmic reticulum [18], whereas the IR is located at the plasma membrane and endosomes. It is possible that some

PTP1B is co-localized with the IR. The mechanism of how PTP1B regulates the insulin-signaling pathway requires further investigation.

REGULATIONS OF PTPs

Since the balance of opposing activity of PTKs and PTPs determines cellular levels of tyrosine phosphorylation, understanding the regulation of PTPs is an essential step toward a full appreciation of the mechanism underlying cellular protein phosphorylation. There is no general mechanism for PTP regulation. Although PTPs do not appear to require post-translational modification to maintain their catalytic activity, nature provides diverse means to fine-tune their function in cells. We will discuss some of the potential regulation mechanisms for PTPs.

Phosphorylation on Serine/Threonine Residues

Like other cellular proteins, PTPs can be regulated by phosphorylation on Ser/Thr residues such that Ser/Thr protein kinases can directly modulate the activity and/or cellular localization of PTPs. PTP1B is phosphorylated on serine in HeLa cells [18]. The sites of phosphorylation were identified as Ser-352, Ser-378 and Ser-386 [19, 20]. However, the effect of phosphorylation at these sites on PTP1B enzymatic activity is not clear. Interestingly, PTP1B is also phosphorylated at Ser-50 by the CLK family of dual-specificity protein kinases in human embryonic kidney (HEK) 293 cells [21]. Phosphorylation at Ser-50 activates PTP1B activity by increasing its affinity towards its substrates.

PTP-PEST is a cytosolic PTP that contains PEST sequences which are highly enriched in proline, glutamic acid, serine and threonine residues. PKA and PKC appear to be able to phosphorylate PTP-PEST at Ser-39 *in vivo* and *in vitro* [22]. Phosphorylation at Ser-39 *in vitro* decreases the activity of PTP-PEST by reducing its affinity for substrates. The HePTP (hematopoietic protein tyrosine phosphatase) is a tyrosine specific PTP that is a negative regulator of ERK1/2 and p38 [23]. HePTP binds to MAP kinases through a kinase-interaction motif (KIM) in its amino-terminal non-catalytic domain and dephosphorylates the critical pTyr in the activation loop of MAP kinases [24]. Saxena *et al.* [24] showed that cyclic-AMP-dependent protein kinase (PKA) phosphorylates Ser-23 in the middle of the KIM sequence *in vitro* and *in vivo.* The serine phosphorylation reduces the binding of MAP kinases to HePTP, thus prevents HePTP from inactivating the MAP kinases. Furthermore, they showed that PKA-mediated phosphorylation of HePTP is sufficient to activate

MAP kinases in intact T cells. Therefore, serine phosphorylation provides a mechanism of cross-talk between the cAMP system and the MAP kinase cascades through the regulation of a tyrosine phosphatase. Phosphorylation by PKA, however, does not change the activity of HePTP towards p-nitrophenyl phosphate (pNNP) *in vitro*. This suggests that serine phosphorylation regulates the interaction between HePTP and its physiological substrates by modulating their affinity towards each other.

MKP-1 is the first member of human MAP kinase phosphatase family that has been identified, and is capable of dephosphorylating all three types of MAP kinases. MKP-1 is a labile protein in fibroblast cells and is targeted for degradation by the proteasome complex [25]. Interestingly, MKP-1 itself is also a target of ERK1/2, which phosphorylates MKP-1 on two carboxyl-terminal serine residues. Remarkably, phosphorylation does not appear to have any effect on intrinsic phosphatase activity of MKP-1 towards pNPP or ERK1; instead, it serves to reduce the ubiquitin-dependent degradation of MKP-1. Thus, phosphorylation on serine residues of MKP-1 leads to the stabilization of the protein.

Cdc25 is a major cell cycle regulator that activates the protein kinase cdc2 by dephosphorylating its inhibitory phosphotyrosine and phosphothreonine residues. Cdc25 itself is highly regulated by Ser/Thr phosphorylation. Human Cdc25C is predominantly phosphorylated at Serine-216 during the interphase [26]. Phosphorylation at Ser-216 creates a 14-3-3 protein binding site and association of Cdc25C with 14-3-3 proteins prevents Cdc25C from entering the nucleus by masking its nuclear localization signal as well as precluding its interactions with other proteins, especially cyclins [26]. However, phosphorylation at Ser-216 does not appear to affect the catalytic activity of Cdc25C *in vitro*. Thus, Ser/Thr phosphorylation of PTPs appears to play an important role in regulating the functions of PTPs by modulating their catalytic activity, mediating protein-protein interactions and modifying protein stability.

Phosphorylation on Tyrosine Residues

Tyrosine phosphorylation can also potentially regulate PTPs. However, it does not appear to be a widely employed mechanism of regulation due to the intrinsic catalytic activity and the ability of PTPs to autodephosphorylate. In order to detect phosphorylation on tyrosine in PTPs, it is often necessary to use tyrosine phosphatase inhibitors for the wild-type enzymes or use a catalytically inactive form of PTPs. Another technical issue is that phosphotyrosine on wild-type PTPs, if there is any, is usually far below the stoichiometric level such that it is difficult to assay the effects of tyrosine phosphorylation accurately. Thus, the biological significance of regulation of PTPs by tyrosine phosphorylation is sometimes difficult to evaluate.

In addition to the insulin receptor, PTP1B also interacts with the epidermal growth factor receptor (EGFR; see Chapter 4), which phosphorylates PTP1B on Tyr-66 in A431 cells [27]. The tyrosine phosphorylation is correlated with a three-fold increase in PTP1B phosphatase activity (by increasing its affinity for substrates). So PTP1B potentially can be regulated by both serine and tyrosine phosphorylation. It is interesting to note that the two phosphorylation sites that can modulate PTP1B activity, Ser-50 and Tyr-66, all lie near the substrate-binding pocket of PTP1B. Phosphorylation near the active site may serve to modulate the activity of PTPs by directly affecting its affinity for its substrates.

PTPα is a receptor-like classical PTP with two cytoplasmic PTP domains and a short extracellular domain. PTPα is constitutively phosphorylated on tyrosine and serine in NIH 3T3 cells [28]. Tyr-789 of PTPα, a residue four amino acids away from the C-terminus, appears to be phosphorylated by Src family protein kinases [28]. Tyr-798 phosphorylation generates a Grb2 binding site on PTPα and may modulate Grb2 mediated signaling [29]. PTPα appears to be able to dephosphorylate Src on its inhibitory pTyr527 and activate Src kinase [30]. Ponniah *et al.* [31] showed that Src has increased Tyr-527 phosphorylation and decreased activity in PTP$\alpha^{-/-}$ mice. Mutational results of Tyr-798 suggest that phosphorylation on this residue in PTPα has an important regulatory effect. Phosphorylation at Tyr-798 specifically enables PTPα to dephosphorylate pTyr527 of Src through binding of pTyr798 with Src SH2 domain that displaces the intramolecular pTyr527-SH2 binding [32].

Tyrosine phosphorylation has also been reported to regulate LMW-PTPs. The Src protein kinase phosphorylates the LMW-PTP at two tyrosine residues, Tyr-131 and Tyr-132 [33]. Site-directed mutagenesis suggests that phosphorylation at these two residues has different effects. Tyr-131 phosphorylation results in a strong increase (about 25-fold) of the PTP activity, whereas phosphorylation at Tyr-132 leads to Grb2 recruitment without affecting enzymatic activity [34]. Thus, phosphorylation appears to play an important role in regulating PTP activity, interactions with other proteins and its intracellular localization.

Dimerization

The crystal structure of the membrane-proximal catalytic domain (D1) of PTPα shows that it crystallizes as a dimer [35]. A structurally constrained amino-terminal helix-turn-helix region was found inserted into the active site of the other monomer in the dyad, resulting in blocking of the active site from substrate access. Thus, it was proposed that PTPα dimerization *in vivo* would lead to inhibition of the intrinsic activity of PTPα. Experimental results from artificially induced dimers indicate that dimerization inhibits the *in vivo* biological activity of a CD45-EGFR chimeric protein [36] and PTPα [37]. Recently, Jiang *et al.* [38] provided further evidence that PTPα is capable of homodimerizing on the cell surface leading to its inhibition. These data support the idea that dimerization-mediated negative regulation of PTP activity is physiologically relevant, at least for PTPα.

However, there is also extensive evidence suggesting that dimerization-mediated inhibition is not a general mechanism among receptor-like PTPs. For example, although the D1 domain of receptor PTPμ also crystallizes as a dimer, the orientation of the D1 monomer does not appear to interfere with substrate access to its active site [39]. Also, the crystal structure of the tandem D1 and D2 domains of LAR reveals that the catalytic sites of both domains are readily accessible and there is no indication of dimer formation for LAR, either in the crystal structure or in solution [40].

Substrate/Adapter Protein Binding

Some PTPs may exist in an inactive or low-activity conformation in cells due to autoinhibitory mechanism or improper positioning of critical active site residues. Upon binding to their physiological substrates or other adapter proteins, these PTPs are switched on by either the release of an autoinhibitory sequence from their active site or adoption of an active conformation. One good example is SHP-2, a dual SH2-domain containing cytosolic PTP. SHP-2 potentiates signaling events downstream of hormones, cytokines, and growth factors to control cell proliferation, differentiation, migration and apoptosis. Crystal structure of SHP-2 reveals that the N-SH2 domain binds to the catalytic cleft, resulting in an autoinhibitory state. Indeed, the activity of SHP-2 is low against p-nitrophenyl phosphate *in vitro* but its activity increases dramatically upon engagement of pTyr-containing peptides to its SH2 domains [41]. Thus, the N-SH2 domain functions as an allosteric switch. Recruitment of substrates/adapter proteins to the SH2-domain would induce conformational change, which is likely to destabilize the interaction of the N-SH2 domain with the PTP active site and lead to the attainment of an active conformation in the PTP domain.

An example of a PTP activated by substrate binding is MAP kinase phosphatase-3 (MKP3). MKP3 is a dual specificity phosphatase with high specificity towards ERK1/2. The crystal structure of the catalytic domain of human MKP-3 (Pyst1) shows that it adopts distorted active site geometry in the absence of its substrate [42]. MKP3 undergoes catalytic activation towards pNPP in the presence of ERK2, changing from a low- to high-activity conformation [43]. Detailed kinetic analysis using artificial substrates *in vitro* suggests that upon binding to ERK2, MKP3 switches to an active conformation to ensure its high degree of substrate specificity [44]. Catalytic activation of MKPs through substrate binding may be a general feature of this family of DSPs, and may be an important mechanism for the regulation of MAP kinase activation in mammalian cells. We have discussed several mechanisms of regulation of PTPs, including phosphorylation on serine, threonine and tyrosine residues, dimerization and substrate binding. Other potential mechanisms such as transcriptional/translational regulation and reversible modification by reactive oxygen species are also possible.

CONCLUSIONS AND PERSPECTIVES

All PTPs share a common catalytic mechanism and similar three-dimensional structures. The PTPs catalyze reactions through a two-step, double displacement mechanism involving a transient cysteinyl-phosphate intermediate. A conserved aspartic acid localized on a surface loop acts as a general acid/base during enzyme intermediate formation and breakdown. PTPs play an indispensable role in diverse signal transduction pathways and appear to be regulated through various mechanisms including phosphorylation on serine, threonine and tyrosine residues, dimerization, substrate binding, etc. To better understand the functions of PTPs in diverse signal transduction pathways in mammalian cells, their physiological substrates must be identified and characterized. Regulation mechanisms are best studied in the cellular context involving physiological substrates and enzymes. As the physiological substrates continue to be identified, biochemical studies of PTPs will be performed in more biological relevant settings instead of artificial ones. Genetic approaches will continue to shed light on the *in vivo* function of PTPs. Finally, increased understanding of PTP structure and function will lead to the acquisition of potent and selective PTP inhibitors. These inhibitors will be important reagents to dissect the role of PTPs in signaling and serve as starting points for therapeutic development.

Acknowledgments

This work was supported by National Institutes of Health grants CA69202 and GM55242. Z.-Y. Z is a Sinsheimer Scholar and an Irma T. Hirschl Career Scientist.

REFERENCES

1. **Hunter T.** 2000 Signaling--2000 and beyond. Cell 100:113-127.
2. **Zhang ZY** 1998 Protein-tyrosine phosphatases: biological function, structural characteristics, and mechanism of catalysis. Crit Rev Biochem Mol Biol 33:1-52.
3. **Keyse SM** 1999 The role of protein phosphatases in the regulation of mitogen and stress-activated protein kinases. Free Radic Res 31:341-349.
4. **Barford D** 1999 Structural studies of reversible protein phosphorylation and protein phosphatases. Biochem Soc Trans 27:751-766.
5. **Flint AJ, Tiganis T, Barford D, Tonks NK** 1997 Development of "substrate-trapping" mutants to identify physiological substrates of tyrosine phosphatases. Proc Natl Acad Sci USA 94: 1680-1685.
6. **Zhang YL, Yao ZJ, Sarmiento M, Wu L, Burke TR Jr, Zhang ZY** 2000 Thermodynamic Study of Ligand Binding to Protein-Tyrosine Phosphatase 1B and Its Substrate-Trapping Mutants. J Biol Chem 275: 34205-34212.
7. **Jia Z, Barford D, Flint AJ, Tonks NK** 1995 Structural basis for phosphotyrosine peptide recognition by protein tyrosine phosphatase 1B. Science 268:1754-1758.
8. **Sarmiento M, Puius YA, Vetter SW, Keng YF, Wu L, Zhao Y, Lawrence DS, Almo SC, Zhang ZY** 2000 Structural basis of plasticity in protein tyrosine phosphatase 1B substrate recognition. Biochemistry 39:8171-8179.
9. **Li L, Dixon JE** 2000 Form, function, and regulation of protein tyrosine phosphatases and their involvement in human diseases. Semin Immunol 12:75-84.
10. **Kishihara K, Penninger J, Wallace VA, Kundig TM, Kawai K, Wakeham A, Timms E, Pfeffer K, Ohashi PS, Thomas ML, et al.** 1993 Normal B lymphocyte development but impaired T cell maturation in CD45- exon6 protein tyrosine phosphatase-deficient mice. Cell 74:143-156.
11. **Shultz LD, Schweitzer PA., Rajan TV. Yi T, Ihle JN, Matthews RJ, Thomas ML, Beier DR** 1993 Mutations at the Murine Motheaten Locus Are within the Hematopoietic Cell Protein Tyrosine Phosphatase (*Hcph*) Gene. Cell 73: 1445-1454.
12. **Saxton TM, Henkemeyer M, Gasca S, Shen R, Rossi DJ, Shalaby F, Feng GS, Pawson T** 1997 Abnormal mesoderm patterning in mouse embryos mutant for the SH2 tyrosine phosphatase Shp-2. EMBO J 16:2352-2364.
13. **Elchebly M, Wagner J, Kennedy TE, Lanctot C, Michaliszyn E, Itie A, Drouin J, Tremblay ML** 1999 Neuroendocrine dysplasia in mice lacking protein tyrosine phosphatase sigma. Nat Genet 21:330-333.
14. **You-Ten KE, Muise ES, Itie A, Michaliszyn E, Wagner J, Jothy S, Lapp WS, Tremblay ML** 1997 Impaired bone marrow microenvironment and immune function in T cell protein tyrosine phosphatase-deficient mice. J Exp Med 186:683-693.
15. **Elchebly M, Payette P, Michaliszyn E, Cromlish W, Collins S, Loy AL, Normandin D, Cheng A, Himms-Hagen J, Chan CC, Ramachandran C, Gresser MJ, Tremblay ML, Kennedy BP** 1999 Increased insulin sensitivity and obesity resistance in mice lacking the protein tyrosine phosphatase-1B gene. Science 283:1544-1548.
16. **Klaman LD, Boss O, Peroni OD, Kim JK, Martino JL, Zabolotny JM, Moghal N, Lubkin M, Kim YB, Sharpe AH, Stricker-Krongrad A, Shulman GI, Neel BG, Kahn BB** 2000 Increased energy expenditure, decreased adiposity, and tissue-specific insulin sensitivity in protein-tyrosine phosphatase 1B-deficient mice. Mol Cell Biol 20:5479-5489.
17. **Thomas ML, Brown EJ** 1999 Positive and negative regulation of Src-family membrane kinases by CD45. Immunol Today 20:406-411.

18. **Frangioni JV, Beahm PH, Shifrin V, Jost CA, Neel BG.** 1992 The nontransmembrane tyrosine phosphatase PTP-1B localizes to the endoplasmic reticulum via its 35 amino acid C-terminal sequence. Cell 68:545-560.
19 **Flint AJ, Gebbink MF, Franza BR, Jr., Hill DE, Tonks NK** 1993 Multi-site phosphorylation of the protein tyrosine phosphatase, PTP1B: identification of cell cycle regulated and phorbol ester stimulated sites of phosphorylation. EMBO J 12:1937-1946.
20. **Shifrin VI, Davis RJ, Neel BG** 1997 Phosphorylation of protein-tyrosine phosphatase PTP-1B on identical sites suggests activation of a common signaling pathway during mitosis and stress response in mammalian cells. J Biol Chem 272:2957-2962.
21. **Moeslein FM, Myers MP, Landreth GE** 1999 The CLK family kinases, CLK1 and CLK2, phosphorylate and activate the tyrosine phosphatase, PTP-1B. J Biol Chem 274:26697-26704.
22. **Garton AJ, Tonks NK** 1994 PTP-PEST: a protein tyrosine phosphatase regulated by serine phosphorylation. EMBO J 13:3763-3771.
23. **Saxena M, Mustelin T** 2000 Extracellular signals and scores of phosphatases: all roads lead to MAP kinase. Semin Immunol 12:387-396.
24. **Saxena M, Williams S, Tasken K, Mustelin T** 1999 Crosstalk between cAMP-dependent kinase and MAP kinase through a protein tyrosine phosphatase. Nat Cell Biol 1:305-311.
25. **Brondello JM, Pouyssegur J, McKenzie FR** 1999 Reduced MAP kinase phosphatase-1 degradation after p42/p44MAPK- dependent phosphorylation. Science 286:2514-2517.
26. **Peng CY, Graves PR, Thoma RS, Wu Z, Shaw AS, Piwnica-Worms H** 1997 Mitotic and G2 checkpoint control: regulation of 14-3-3 protein binding by phosphorylation of Cdc25C on serine-216. Science 277:1501-1505.
27. **Liu F, Chernoff J** 1997 Protein tyrosine phosphatase 1B interacts with and is tyrosine phosphorylated by the epidermal growth factor receptor. Biochem J 327:139-145.
28. **den Hertog J, Tracy S, Hunter T** 1994 Phosphorylation of receptor protein-tyrosine phosphatase alpha on Tyr789, a binding site for the SH3-SH2-SH3 adaptor protein GRB-2 in vivo. EMBO J 13:3020-3032.
29. **den Hertog J, Hunter T** 1996 Tight association of GRB2 with receptor protein-tyrosine phosphatase alpha is mediated by the SH2 and C-terminal SH3 domains. EMBO J 15:3016-3027.
30. **Zheng XM, Wang Y, Pallen CJ** 1992 Cell transformation and activation of pp60c-src by overexpression of a protein tyrosine phosphatase. Nature 359:336-339.
31. **Ponniah S, Wang DZ, Lim KL, Pallen CJ** 1999 Targeted disruption of the tyrosine phosphatase PTPalpha leads to constitutive downregulation of the kinases Src and Fyn. Curr Biol 9:535-538.
32. **Zheng XM, Resnick RJ, Shalloway D** 2000 A phosphotyrosine displacement mechanism for activation of Src by PTPalpha. EMBO J 19:964-978.
33. **Tailor P, Gilman J, Williams S, Couture C, Mustelin T** 1997 Regulation of the low molecular weight phosphotyrosine phosphatase by phosphorylation at tyrosines 131 and 132. J Biol Chem 272:5371-5374.
34. **Bucciantini M, Chiarugi P, Cirri P, Taddei L, Stefani M, Raugei G, Nordlund P, Ramponi G** 1999 The low Mr phosphotyrosine protein phosphatase behaves differently when phosphorylated at Tyr131 or Tyr132 by Src kinase. FEBS Lett 456:73-78.
35. **Bilwes AM, den Hertog J, Hunter T, Noel JP** 1996 Structural basis for inhibition of receptor protein-tyrosine phosphatase-alpha by dimerization. Nature 382:555-559.
36. **Majeti R, Bilwes AM, Noel JP, Hunter T, Weiss A** 1998 Dimerization-induced inhibition of receptor protein tyrosine phosphatase function through an inhibitory wedge. Science 279:88-91

37. **Jiang G, den Hertog J, Su J, Noel J, Sap J, Hunter T** 1999 Dimerization inhibits the activity of receptor-like protein-tyrosine phosphatase-alpha. Nature 401:606-610.
38. **Jiang G, den Hertog J, Hunter T** 2000 Receptor-like protein tyrosine phosphatase alpha homodimerizes on the cell surface. Mol Cell Biol 20:5917-5929.
39. **Hoffmann KM, Tonks NK, Barford D** 1997 The crystal structure of domain 1 of receptor protein-tyrosine phosphatase mu. J Biol Chem 272:27505-27508.
40. **Nam HJ, Poy F, Krueger NX, Saito H, Frederick CA** 1999 Crystal structure of the tandem phosphatase domains of RPTP LAR. Cell 97:449-457.
41. **Hof P, Pluskey S, Dhe-Paganon S, Eck MJ, Shoelson SE** 1998 Crystal structure of the tyrosine phosphatase SHP2. Cell 92: 441-450.
42. **Stewart AE, Dowd S, Keyse SM, McDonald NQ** 1999 Crystal structure of the MAPK phosphatase Pystl catalytic domain and implications for regulated activation. Nat Struct Biol 6:174-181.
43. **Camps M, Nichols A, Gillieron C, Antonsson B, Muda M, Chabert C, Boschert U, Arkinstall S** 1998 Catalytic activation of the phosphatase MKP-3 by ERK2 mitogen-activated protein kinase. Science 280:1262-1265.
44. **Zhou B, Zhang ZY** 1999 Mechanism of mitogen-activated protein kinase phosphatase-3 activation by ERK2. J Biol Chem 274:35526-35534.

3

Spatio-Temporal Parameters: The Case of the MAP Kinase Pathway

Véronique Volmat and Jacques Pouysségur*

Institute of Signaling, Developmental Biology and Cancer Research CNRS UMR-6543
Centre Antoine Lacassagne, 33 Avenue Valombrose, 06189 Nice, Cedex 2, France

INTRODUCTION

Eucaryotic cells have developed signal transduction networks that allow them to convey extracellular signals in the nucleus to induce the proper response at the gene level. A common way for the cell to "reprogram" its genetic information is through specific protein phosphorylation. Mitogen Activated Protein Kinases (MAPKs) play a key role in this action. MAPK pathways are activated in response to a large array of stimuli from growth factors to environmental stress [1-2]. Furthermore, they are evolutionarily conserved from yeast to human [3]. These pathways are organized around a core signaling module composed of three kinases that are sequentially activated (Fig. 1). The first kinase is a serine/threonine kinase called MAP kinase kinase kinase or MKKK. When activated, it phosphorylates a dual specificity kinase (MAP kinase kinase or MKK) on two serine or threonine residues. Phosphorylated MKK activates in turn the serine/threonine kinase MAP kinase or MAPK on the threonine and tyrosine residues of the consensus motif TXY (where T is threonine, X is glutamic acid, glycine or proline, and Y is tyrosine) [4]. Finally, activated MAPK phosphorylates diverse substrates including nuclear transcription factors, membrane proteins and cytoplasmic

* pouysseg@unice.fr

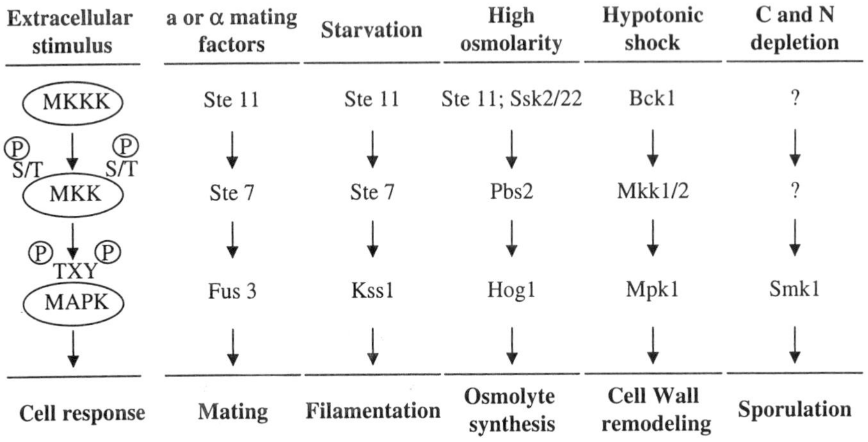

Figure 1. The MAPK modules in the budding yeast Saccharomyces cerevisiae.

substrates which are all modified on the threonine or serine of the PX(T/S)P consensus motif (where P is proline, X an indifferent amino acid and S is serine). Considering that MAPK pathways have been duplicated within the same cell to deliver specific biological responses, the cell has evolved various means that prevent inappropriate crosstalk between the different MAPK modules that possess a strong degree of homology. Two mechanisms ensure the specificity of activation: i) insulation of the protein kinases of a same module by scaffold proteins, and ii) presence of specific docking sites on the kinases and their substrates. Furthermore, anchor proteins and specific MAPK phosphatases have been characterized that regulate the spatio-temporal activation of the MAPK pathways. In the first part of this chapter we will rapidly review the currently known mammalian MAPK pathways. In the second part we will discuss the factors controlling the specificity in the MAPK modules and how these mechanisms have evolved from yeast to human. Finally, the key elements controlling the spatio-temporal activity of the growth factor response will be highlighted with a special emphasis at the level of the p42/p44 MAPK pathway.

THE MAMMALIAN MAPK PATHWAYS

Five MAPK modules have been identified in mammalian cells (Fig. 2). The three most described are the p42/p44 MAPK pathway which regulates cell growth and differentiation and the JNK and p38 MAPK pathways which play an important role in the response to environmental stress, inflammation and in apoptosis.

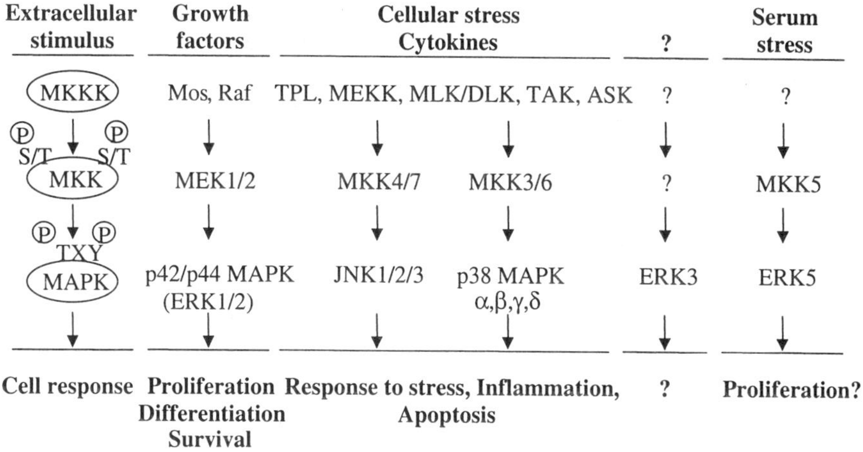

Figure 2. The mammalian MAPK modules

The p42/p44 MAPK (ERK) Module

This module is activated through receptor tyrosine kinases (RTKs; see Chapters 4 and 5), cytokine receptors (see Chapters 6-8), G-protein coupled receptors (see Chapters 9-11) and integrins [3]. Activation of these membrane receptors lead to activation of various exchange factors that in turn lead to the activation of the small G protein Ras which directly activates the cascade: Raf > MEK1/2 > p42/p44 MAPK. Activated p42/p44 MAPK phosphorylates numerous substrates including the membranous EGF receptor, the cytoplasmic kinase p90^{rsk} or the nuclear transcription factor Elk-1. The p42/p44 MAPK pathway plays a role in the control of proliferation and in the response to growth factors. Its activation has been demonstrated to be critical for the re-entry of quiescent cells into the cell cycle. Moreover, activation of p42/p44 MAPK is also involved in the differentiation of cells such as the PC12 cells which form neurite outgrowths upon p42/p44 MAPK pathway activation by NGF.

The JNK Module

The JNKs are activated by inflammatory cytokines such as TNF-α and IL-1β and a wide array of cellular stress including UV light, γ-irradiation, hydrogen peroxide (H_2O_2), heat and osmotic shock [5]. The exact sensors of cell stress have not yet been clearly identified, therefore the pathways that lead from stress to activation of the MAPK module are still unclear. The module can be activated by numerous MKKK proteins (MEKK1-4, ASK1, TAK1, MLK, TPL1) [6] which converge on two MKKs: MKK4 and MKK7 which both phosphorylate the JNKs: JNK 1, 2 and 3. So far, only nuclear transcription factors have been identified as substrates for the JNK module, these include c-Jun, ATF-2, Elk-1 and p53. The JNK pathway is implicated in inflammation, in tissue morphogenesis and in the apoptotic response [7].

The p38 MAPK Module

The p38 MAPKs are activated by cellular stress (UV light, osmotic shock, heat shock, lipopolysaccharides, protein synthesis inhibitors) and some cytokines, such as IL-1 or TNF-α [3]. The cascade of activation is not fully understood but it appears to involve the low molecular weight GTP-binding proteins of the Rho family which activate the cascade TAK1/ASK1 > MKK3/6 > p38 MAPK $\alpha/\beta/\gamma/\delta$. In addition to transcription factors such as ATF-2 or Elk-1, p38 MAPK also phosphorylates MAPK-activated protein (MAPKAP) kinases 2 and 3 which in turn phosphorylate heat-shock proteins. p38 MAPK is implicated in cytokine-stimulated cellular proliferation and in apoptosis induced by cellular stress or death receptor activation [8].

HOW TO ENSURE THE SPECIFICITY OF ACTIVATION

MAPK modules are very conserved in their general organization but also at the protein level with a high percentage of similarity in the primary sequence of the different MAPKs (60% between p42/p44 MAPK and either JNK or p38 MAPK). Furthermore, modules can be activated by overlapping sets of extracellular stimuli, and some components of these modules are involved in distinct signaling pathways. To ensure substrate specificity and prevent unwanted crosstalk between unrelated MAPK modules, cells have developed two mechanisms. First, scaffold proteins create multienzyme complexes specifically composed of the kinases of a given module. These complexes insulate the module from activation by irrelevant stimuli and favor the rapid passage of the signal through the cascade. Second, specific docking sites on MAPKs which serve for the binding of substrates, activators and regulators increase the fidelity and the efficiency of the enzymatic reactions.

MAPK Scaffold Proteins in Yeast

Scaffold proteins for the MAPK pathways were first identified in yeast. Amongst them, Ste5, a protein that is essential for the mating pathway is the best characterized to date [9].

Binding of pheromone to its G-protein coupled receptor induces the dissociation of the Gβγ subunit from the GTP bound Gα subunit (see Chapter 9). The Gβγ subunit composed of Ste4 and Ste18 then activates the MKKKK Ste20 and interacts with Ste5 resulting in specific activation of the module Ste11 > Ste7 > Fus3. Ste5 simultaneously binds Ste11, Ste7 and Fus3 through distinct regions but not Kss1, involved in the filamentation module which shares the same activators as Fus3 (see Fig. 1)[10].

Another well described scaffold protein in yeast is Pbs2 involved in the response to high osmolarity. Two independent osmosensors Sho1 and Sln1 relay the signal to the MAPK Hog1. Contrary to Ste5, Pbs2 is a protein kinase (MKK) part of the module for which it serves as a scaffold. Pbs2 ensures the specificity of activation of the cascade downstream of Sho1 by binding Sho1, Ste11 and Hog1 [11]. On the other hand, the cascade downstream of the Sln1 osmosensor proceeds through Ssk2 and Ssk22, Pbs2 and Hog1 but does not require a scaffold protein for activation.

These two examples illustrate the crucial role of scaffold proteins in preventing inappropriate crosstalk among distinct MAPK pathways. Indeed, Ste11 and Ste7 are shared by the pathways leading to mating and filamentation. Ste11 is also part of the pathway activated by high osmolarity. Hence, Ste11 is part of three different MAPK modules. The specificity of activation is nevertheless maintained by scaffold proteins. In response to pheromone, Ste5 restricts Ste11 to phosphorylate Ste7 which leads to Fus3 activation. In response to high osmolarity, Pbs2 forms a complex which restricts Ste11 to the Ste11 > Pbs2 > Hog1 pathway. Finally, in response to starvation, Ste11 activates Ste7 which leads to Kss1 activation without the involvement of any scaffold protein.

MAPK Scaffold Proteins in Mammals

Many attempts to identify scaffold proteins in mammals on the basis of their homology with the yeast proteins have failed and it now seems clear that these proteins have not been conserved throughout evolution. Moreover, the characterization of a protein as a scaffold is complicated by experimental artefacts. Indeed, when overexpressed, a scaffold protein usually disrupts the stoechiometry between the different partners and prevents rather than increases the interaction between the partners of a given signaling module. Nevertheless, scaffold proteins have also been identified for the JNK and the p42/p44 MAPK pathways in mammalian cells using two hybrid screens with the different MAPKs as baits [9-12].

The JNK Pathway

The first scaffold protein for the JNK pathway was identified as a JNK inhibitory protein. It has now be named JIP1 (JNK interacting protein 1) and is the first scaffold protein of a family of three JIPs: JIP1, JIP2 and JIP3 [13]. JIP1 and 2 are exclusively located in the cytoplasm where JNK is activated before translocation to the nucleus. JIPs bind HPK1 (a MKKKK homologue to Ste20), MLK3 and DLK which are members of the MLK group of MKKKs, MKK7 (but not MKK4 which is common to the JNK and p38 MAPK pathways) and JNK. Therefore JIP scaffold complexes represent only one mechanism of JNK activation [14].

Other scaffold proteins have been identified for the JNK module and more particularly for the pathways leading to JNK3 activation. JSAP scaffold protein increases JNK3 activation through the MEKK1 > MKK4 > JNK3 pathway. It interacts with the different partners through distinct regions and leads to a significant enhancement of the JNK3 activation without affecting JNK1 or JNK2 activity [15]. Very recently an unexpected scaffold protein for the activation of JNK3 as been identified. β-arrestin 2, originally identified as a protein involved in the desensitization and internalization of the G-protein coupled receptors was characterized as a receptor-regulated scaffold protein for JNK3 activation. Indeed β-arrestin 2 forms a complex with ASK1 (MKKK), MKK4 and JNK3 which results in an enhanced phosphorylation of JNK3. Interestingly, unlike the complexes formed with the JIPs, the β-arrestin 2 mediated complex is regulated by agonist stimulation of the G-protein coupled receptor and is very specific for the activation of JNK3 without affecting JNK1 or JNK2 activity [16]. Hence, formation of such complexes prevents inappropriate crosstalk between the various MAPK pathways and allows individual JNK isoforms to exhibit distinct patterns of regulation in response to specific stimuli.

The p42/p44 MAPK Pathway

A two hybrid screen using MEK1 as a bait identified MP1 (MEK Partner 1) as a scaffold protein that specifically binds MEK1 and p44 MAPK to the exclusion of MEK2 and p42 MAPK, thereby enhancing the activation of p44 MAPK [17]. Again this shows that in mammalian cells small adapter proteins can provide flexibility in regulating the efficiency and specificity of activation through the MAPK modules including between two closely related isoforms such as p42 and p44 MAPK.

Another apparent scaffold protein for this pathway is the Ksr protein. Ksr (Kinase suppressor of Ras) was first identified by genetic screens in

Drosophila melanogaster and *Caenorhabditis elegans* as an activator of the Ras pathway since mutations in Ksr resulted in attenuation of Ras-mediated signaling. A mammalian homologue has been isolated that interacts with MEK1 and p42 MAPK thereby acting as a scaffold protein to maintain specificity and increase signaling through this cascade [18-19].

Although scaffold proteins are usually described as "simple" adapter proteins that increase the efficiency of activation through a given MAPK module, they are also subject to strict regulation. They can form dimers (as demonstrated for JIPs and Ste5) [14-20], they can be phosphorylated [10], they can also have a potential activity as shown for Ksr which has a kinase domain with high homology with the kinase domain of Raf. Finally they are also subject to a tight spatial regulation as demonstrated for Ste5 for which nuclear shuttling is required for its recruitment to the plasma membrane and the activation of the mating MAPK cascade [21].

MAPKs Interact with Their Substrates, Activators and Regulators Through Specific and Conserved Docking Sites

The three major MAPKs phosphorylate their substrates on the consensus T/SP sequence and many potential substrates contain this motif. Cells have therefore developed specific docking sites that direct a particular MAPK to its relevant substrate. Docking sites also increase the local concentration of the kinase, hence favoring substrate phosphorylation. Many docking sites have been identified independently on MAPK interacting proteins including substrates, scaffold proteins and phosphatases [22-23]. All these docking sites interact with a cluster of negatively charged amino acids on the C-terminus of the MAPK and contain a cluster of positively charged amino acids conserved on MAPK activators (MKK), substrates and regulators such as the MAPK phosphatases (MKPs) [24]. This implies that the interaction of these proteins with the MAPKs are mutually exclusive, thereby providing a molecular mechanism for the sequential and specific activation and inactivation of MAPKs.

However, the specificity of interaction may not be determined solely by this docking site since substituting by mutation the docking site on p42/p44 MAPK into the docking site present on p38 MAPK still allows the binding of MEK to p42/p44 MAPK while no binding of MKK6 (the upstream activator of p38 MAPK) could be detected [24]. Therefore, there must be other components that direct the specificity of activation by one specific MAPK and even by one specific isoform of MAPK [25]. For example, the efficiency of binding to c-Jun varies between the different JNK isoforms [26]. On the contrary, Elk-1 can be phosphorylated by both p42/p44 MAPK and JNK [27]. Alignment of docking sites has revealed a LXL motif and hydrophobic stretches near the positively charged cluster of amino acids. The spacing and organization of these different motifs on the different MAPK interacting proteins is a feature that could account for the differential phosphorylations observed [28].

Moreover, the activation of the different partners also modulates the affinity of the interaction. Indeed, the phosphatases of the PTP-SL family bind inactive p42/p44 MAPK and the complex dissociates after p42/p44 MAPK activation [29]. On the other hand, the association of p42/p44 MAPK with Elk-1 is enhanced upon p42/p44 MAPK activation [30].

The conservation of these specific docking sites throughout evolution from yeast to human is undoubtedly an evidence that such docking are crucial for efficient and specific activation of the MAPK modules. However, the exact components that direct the specificity and the possible involvement of other structural motifs need to be analyzed carefully to provide a clear picture of the mechanisms of activation.

SPATIO-TEMPORAL REGULATION OF THE MAPK PATHWAYS

In addition to the mechanisms described previously which ensure the specificity and the efficiency of activation, spatio-temporal parameters are crucial to elicit the proper response to a given stimulus. In this section we will analyze the role of proteins that bind to different components of the module and anchor them in a specific compartment where activation either proceeds or, to the contrary, is prevented. In addition to this spatial regulation, we will also develop the temporal regulation which involves specific phosphatases for controlling the intensity of the MAPK signal and its duration. In particular, we will outline the key elements of the p42/p44 MAPK pathway.

Spatio-Temporal Regulation of the p42/p44 MAPK Pathway

Temporal Regulation

The duration of p42/p44 MAPK activation is crucial for the determination of the biological response following receptor activation. Indeed, in response to mitogenic stimuli such as serum or growth factors, p42/p44 MAPK of resting cells (G0-arrested) is activated following a biphasic time course with a first sharp phase of activity that lasts 20 to 30 minutes, followed by a second phase of lower but sustained activity that persists several hours (up to 6 hours) [31]. This long term activation which accompanies G1 progression is necessary for the entry of quiescent cells into the S-phase of the cell cycle. On the other hand, non mitogenic stimuli such as PMA, angiotensin II or serotonin also activate the p42/p44 MAPK pathway but with a different time course since only the first peak of activity is observed and the activity returns very rapidly to basal levels [32]. There is therefore a direct correlation between the length of p42/p44 MAPK activation and the ability of a stimuli to lead to DNA replication and proliferation. Shortening or specific ablation of this p42/p44 MAPK activation phase by overexpression of a MAPK phosphatase [33] or MAPK antisense [34] leads to inhibition of cell cycle entry. Conversely, forced activation of the p42/p44 MAPK cascade using Raf inducible constructs does not prevent DNA replication and mitosis, however cells are blocked in G1 phase of the second cell cycle (unpublished observations). These results illustrate that a tight control of the duration of the p42/p44 MAPK signal is crucial for the control of proliferation.

In other cells however, a prolonged p42/p44 MAPK activation is required for differentiation. This is particularly well illustrated with the PC12 cell line. NGF but not EGF induce neurite outgrowths in these cells. NGF induce a persistent activation of p42/p44 MAPK whereas EGF is a very transient activator of this pathway in these cells. Overexpression of EGF receptors restore in these cells a more persistent p42/p44 MAPK activation and neurite outgrowths formation in response to EGF [35].

All of these examples show that the temporal regulation of p42/p44 MAPK activity is crucial to elicit the appropriate response at the gene level. However, all of the determinants of this regulation have not been fully identified so far. Phosphatases that inactivate p42/p44 MAPK have been identified but their exact contribution to the different phases of inactivation is not yet clearly established.

Tyrosine and serine/threonine specific phosphatases can inactivate p42/p44 MAPK by dephosphorylating the tyrosine and threonine residues sequentially. Alternatively, dual specificity phosphatases can dephosphorylate both residues concomitantly. In yeast, it has been shown that Fus3 inactivation can be operated either by a dual specificity phosphatase (Msg5) or by tyrosine specific phosphatases (Ptp3 and Ptp2) [36]. In mammalian cells, PP2A (a serine/ threonine specific phosphatase) has been involved in the rapid inactivation of p42/p44 MAPK in association with a tyrosine specific phosphatase [37] that could be a member of the PTP-SL family [38-39]. However, dual specificity phosphatases of the MAPK phosphatases family and in particular those very specific for p42/p44 MAPK are good candidates for setting the intensity and duration of the p42/p44 MAPK activation [40]. To date, several MKPs have been identified. They show diverse subcellular localization and specificity for the MAPKs and their precise role in the spatio-temporal regulation of MAPKs remains to be clearly established [39-41].

Spatial Regulation

p42/p44 MAPK is located in the cytoplasm of quiescent cells. Following activation of the Raf > MEK pathway, activated p42 and p44 MAPK are the only two kinases of the module that can translocate to the nucleus [42]. Although the mechanisms of translocation are not fully understood [43-45], activation of the p42/p44 MAPK module is sufficient to trigger their translocation. We have established that this nuclear translocation is required for the transduction of mitogenic signals. Preventing p42/p44 MAPK nuclear translocation without impairing its activation nor its ability to phosphorylate cytoplasmic substrates inhibits growth factor induced gene expression and cell cycle entry [46]. Hence, the nucleus appears to be a preferential site for p42/p44 MAPK localization for the propagation of the mitogenic stimuli.

p42/p44 MAPK is anchored in the cytoplasm of quiescent cells through interaction with a specific docking site on inactive MEK [47]. Following activation, phosphorylated MEK and active p42/p44 MAPK lose their affinity and p42/p44 MAPK is free to move and phosphorylate its substrates in the cytoplasm and in the nucleus. Other proteins such as the tyrosine specific phosphatases of the PTP-SL family have also been proposed to play a role in cytoplasmic anchoring of inactive p42/p44 MAPK [29]. Following p42/p44 MAPK activation, PTP-SL is phosphorylated and p42/p44 MAPK is released.

Following mitogenic stimulation we can observe a progressive accumulation of p42/p44 MAPK in the nucleus which is maximal after 3 hours of serum stimulation in CCL39 fibroblasts. This nuclear accumulation is dependent on protein synthesis and is reinforced in presence of the proteasome inhibitor LLnL. Therefore, it has been proposed that this nuclear anchoring is dependent on the neosynthesis of nuclear anchors which have a short half life and are degraded by the proteasome [48]. These nuclear anchors have not yet been identified. In addition, nuclear p42/p44 MAPK could also be anchored in the nucleus through specific interaction with its multiple nuclear substrates such as Elk-1 which binds active p42/p44 MAPK [30].

Proposed Model for the Spatio-Temporal Regulation of p42/p44 MAPK Activity

In order to follow simultaneously the subcellular localization and the activity of p42/p44 MAPK during mitogenic stimulation, we performed immunolabeling with a phospho-specific antibody that recognizes exclusively the dually phosphorylated, active forms of p42/p44 MAPK [49]. Growth factors induce a rapid (after 5 min of serum stimulation) nuclear translocation and accumulation of active p42/p44 MAPK. After 15 min of serum stimulation, the activated p42/p44 MAPK signal can be detected homogeneously throughout the cell. Thereafter, contrary to predictions, the nuclear signal progressively disappears while a progressive nuclear accumulation of p42/p44 MAPK is detected with a global p42/p44 MAPK antibody. This is particularly clear after 3 hours of stimulation, as shown in Fig. 3, with p42/p44 MAPK massively anchored in the nucleus but in an inactive form as revealed by the absence of labeling with the anti activated p42/p44 MAPK antibody.

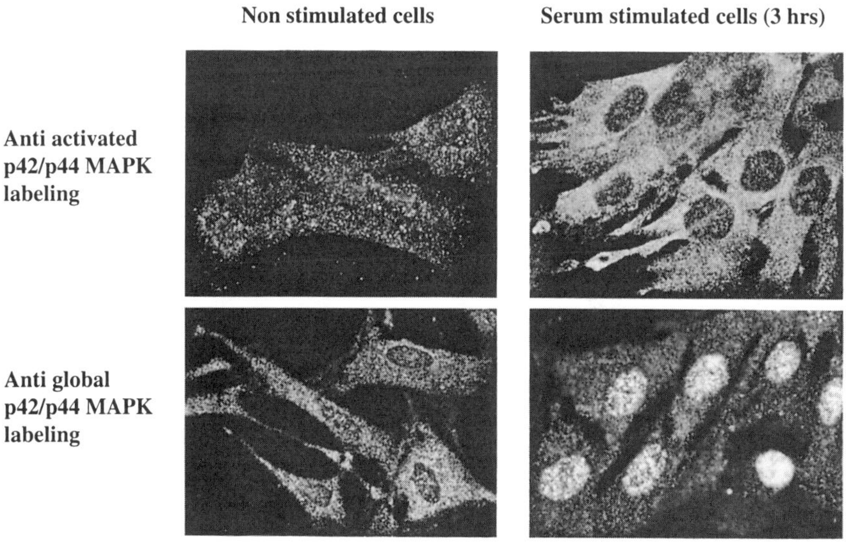

Figure 3. Nuclear sequestration of inactive p42/p44 MAPK after long term mitogenic stimulation.

This observation clearly reveals two mechanisms of regulation of p42/p44 MAPK during long term growth stimulation: 1) nuclear sequestration of p42/p44 MAPK away from its cytoplasmic activator MEK and 2) dephosphorylation of nuclear p42/p44 MAPK by specific nuclear phosphatases. Therefore we propose that these two mechanisms contribute to mitogenic signal termination [50].

In summary, in our proposed model, inactive p42/p44 MAPK is anchored in the cytoplasm of resting cells by MEK. Upon stimulation, the p42/p44 MAPK-MEK complex dissociates and active p42/p44 MAPK translocates to the nucleus where it induces the neosynthesis of specific nuclear anchors and phosphatases. These nuclear anchors and phosphatases progressively accumulate and are then responsible for the progressive anchoring and inactivation of p42/p44 MAPK in the nucleus. We have shown that p42/p44 MAPK constantly shuttles between the cytoplasm and the nucleus throughout the stimulation even in two extreme situations: in resting cells where p42/p44 MAPK is nuclear and after three hours of serum stimulation when the module is fully activated and p42/p44 MAPK is anchored in the nucleus (Fig. 4) [50]. The shuttling mechanism that allows rapid p42/p44 MAPK nuclear entry and exit is certainly NLS and NES mediated. MEK has been proposed as a possible candidate for this shuttling [51], however p42/p44 MAPK substrates ($p90^{RSK}$, transcription factors) could also actively participate in this constant shuttling.

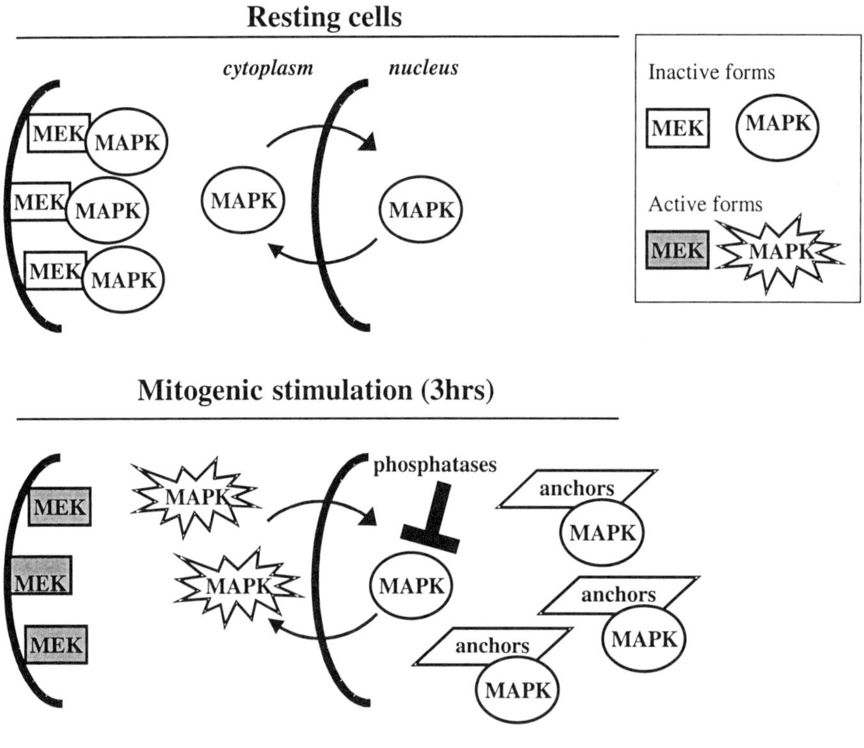

Figure 4. Proposed model for the nucleo-cytoplasmic shuttling of p42/p44 MAPK

Spatio-Temporal Regulation in Other MAPK Pathways

The p42/p44 MAPK pathway is the best characterized in terms of spatio-temporal regulation. However, we have clear indications that similar mechanisms also exist in other MAPK pathways.

It has been shown that JNK is localized in the cytoplasm of non-stimulated cells and translocates to the nucleus upon stimulation by UV irradiation [52]. Moreover, in non-stimulated cells JNK3 and β arrestin 2 are detected in the cytoplasm while after stimulation by angiotensin II, a significant portion of β-arrestin 2 is detected in endosomal vesicles probably associated with the internalized receptors. Active JNK3 was found to be colocalized with β-arrestin 2 in the endosomal vesicles while most of the JNK3 pool accumulated in the nucleus in an inactive state [16]. Although these experiments were performed with cells overexpressing the different partners of the pathway, we can predict that endogenous proteins are regulated similarly and that similar to the p42/p44 MAPK, the nucleus is a specific site for JNK3 signal termination.

p38 MAPK is located in the nucleus of non-stimulated cells and is exported in the cytoplasm after stress stimulation. Interestingly, this export is dependent on the phosphorylation of MAPKAP kinase 2 which induces a conformational change responsible for the unmasking of a nuclear export sequence. p38 MAPK nuclear export upon stimulation is then directly driven by its substrate MAPKAP kinase 2 [53].

Spatio-temporal parameters are also important in lower eucaryotes. For example, in yeast, the subcellular localization of Hog1, the yeast homologue of p38 MAPK is determined by two protein phosphatases: nuclear Ptp2 and cytoplasmic Ptp3 [54]. In this case the proteins responsible for the anchoring and the inactivation (i.e. the spatio-temporal regulation) are the same entity.

CONCLUSIONS

MAPK signaling pathways are conserved throughout evolution from yeast to human and they respond to very diverse extracellular stimuli to induce specific biological responses. Therefore they are key components of the cell communication network. To ensure the specificity and efficiency of activation, cells have developed scaffolding proteins and specific docking sites to avoid inappropriate crosstalk between the different pathways.

Moreover, we have shown that spatio-temporal parameters are crucial for generating the appropriate response to a specific signal. These parameters are determined by anchoring proteins that modulate the subcellular localization of MAPKs during stimulation and by phosphatases which inactivate MAPKs by specific dual dephosphorylation of the critical threonine and serine residues. In this chapter we have exemplified the spatio-temporal control of p42/p44 MAPK activation following growth factor stimulation, however these mechanisms are probably important for all MAPK mediated responses.

What is striking is that p42/p44 MAPK itself controls, like an oscillator, its temporal activity. p42/p44 MAPK controls its nuclear translocation, the synthesis of the MAPK phosphatases [40], the half life of those MAPK phosphatases [55] and the synthesis of nuclear anchor(s) [48]. All these steps contribute to a MAPK "auto-control" mechanism, setting the intensity and the duration of the MAPK signal. It is very likely that every hormone signaling process has a similar "build-in" auto-control mechanism.

Acknowledgments

The authors thank Dr Ellen Van Obberghen-Schilling for carefully reading the manuscript

REFERENCES

1. **Schaeffer HJ, Weber MJ**. 1999 Mitogen-activated protein kinases: specific messages from ubiquitous messengers. Mol Cell Biol 19: 2435-44.
2. **Robinson MJ, Cobb MH**. 1997 Mitogen-activated protein kinase pathways. Curr Opin Cell Biol 9: 180-6.
3. **Widmann C, Gibson S, Jarpe MB, Johnson GL**. 1999 Mitogen-activated protein kinase: conservation of a three-kinase module from yeast to human. Physiol Rev 79: 143-80.
4. **Dhanasekaran N, Premkumar Reddy E**. 1998 Signaling by dual specificity kinases. Oncogene 17: 1447-55.
5. **Ichijo H**. 1999 From receptors to stress-activated MAP kinases. Oncogene 18: 6087-93.
6. **Fanger GR, Gerwins P, Widmann C, Jarpe MB, Johnson GL**. 1997 MEKKs, GCKs, MLKs, PAKs, TAKs, and tpls: upstream regulators of the c-Jun amino-terminal kinases? Curr Opin Genet Dev 7: 67-74.
7. **Ip YT, Davis RJ**. 1998 Signal transduction by the c-Jun N-terminal kinase (JNK)--from inflammation to development. Curr Opin Cell Biol 10: 205-19.
8. **Tibbles LA, Woodgett JR**. 1999 The stress-activated protein kinase pathways. Cell Mol Life Sci 55: 1230-54.
9. **Whitmarsh AJ, Davis RJ**. 1998 Structural organization of MAP-kinase signaling modules by scaffold proteins in yeast and mammals. Trends Biochem Sci 23: 481-5.
10. **Elion EA**. 1995 Ste5: a meeting place for MAP kinases and their associates. Trends Cell Biol 5: 322-7.
11. **Posas F, Saito H**. 1997 Osmotic activation of the HOG MAPK pathway via Ste11p MAPKKK: scaffold role of Pbs2p MAPKK. Science 276: 1702-5.

12. **Garrington TP, Johnson GL**. 1999 Organization and regulation of mitogen-activated protein kinase signaling pathways. Curr Opin Cell Biol 11: 211-8.
13. **Whitmarsh AJ, Cavanagh J, Tournier C, Yasuda J, Davis RJ**. 1998 A mammalian scaffold complex that selectively mediates MAP kinase activation. Science 281: 1671-4.
14. **Yasuda J, Whitmarsh AJ, Cavanagh J, Sharma M, Davis RJ**. 1999 The JIP group of mitogen-activated protein kinase scaffold proteins. Mol Cell Biol 19: 7245-54.
15. **Ito M, Yoshioka K, Akechi M, Yamashita S, Takamatsu N, Sugiyama K, Hibi M, Nakabeppu Y, Shiba T, Yamamoto KI**. 1999 JSAP1, a novel jun N-terminal protein kinase (JNK)-binding protein that functions as a Scaffold factor in the JNK signaling pathway. Mol Cell Biol 19: 7539-48.
16. **McDonald PH, Chow CW, Miller WE, Laporte SA, Field ME, Lin FT, Davis RJ, Lefkowitz RJ**. 2000 beta-arrestin 2: A receptor-regulated MAPK scaffold for the activation of JNK3. Science 290: 1574-7.
17. **Schaeffer HJ, Catling AD, Eblen ST, Collier LS, Krauss A, Weber MJ**. 1998 MP1: a MEK binding partner that enhances enzymatic activation of the MAP kinase cascade. Science 281: 1668-71.
18. **Yu W, Fantl WJ, Harrowe G, Williams LT**. 1998 Regulation of the MAP kinase pathway by mammalian Ksr through direct interaction with MEK and ERK. Curr Biol 8: 56-64.
19. **Stewart S, Sundaram M, Zhang Y, Lee J, Han M, Guan KL**. 1999 Kinase suppressor of Ras forms a multiprotein signaling complex and modulates MEK localization. Mol Cell Biol 19: 5523-34.
20. **Yablonski D, Marbach I, Levitzki A**. 1996 Dimerization of Ste5, a mitogen-activated protein kinase cascade scaffold protein, is required for signal transduction. Proc Natl Acad Sci U S A 93: 13864-9.
21. **Mahanty SK, Wang Y, Farley FW, Elion EA**. 1999 Nuclear shuttling of yeast scaffold Ste5 is required for its recruitment to the plasma membrane and activation of the mating MAPK cascade. Cell 98: 501-12.
22. **Camps M, Nichols A, Gillieron C, Antonsson B, Muda M, Chabert C, Boschert U, Arkinstall S**. 1998 Catalytic activation of the phosphatase MKP-3 by ERK2 mitogen-activated protein kinase. Science 280: 1262-5.
23. **Smith JA, Poteet-Smith CE, Malarkey K, Sturgill TW**. 1999 Identification of an extracellular signal-regulated kinase (ERK) docking site in ribosomal S6 kinase, a sequence critical for activation by ERK in vivo. J Biol Chem 274: 2893-8.
24. **Tanoue T, Adachi M, Moriguchi T, Nishida E**. 2000 A conserved docking motif in MAP kinases common to substrates, activators and regulators. Nat Cell Biol 2: 110-6.
25. **Brunet A, Pouyssegur J**. 1996 Identification of MAP kinase domains by redirecting stress signals into growth factor responses. Science 272: 1652-5.
26. **Gupta S, Barrett T, Whitmarsh AJ, Cavanagh J, Sluss HK, Derijard B, Davis RJ**. 1996 Selective interaction of JNK protein kinase isoforms with transcription factors. EMBO J 15: 2760-70.
27. **Yang SH, Whitmarsh AJ, Davis RJ, Sharrocks AD**. 1998 Differential targeting of MAP kinases to the ETS-domain transcription factor Elk-1. EMBO J 17: 1740-9.
28. **Sharrocks AD, Yang SH, Galanis A**. 2000 Docking domains and substrate-specificity determination for MAP kinases. Trends Biochem Sci 25: 448-53.
29. **Zuniga A, Torres J, Ubeda J, Pulido R**. 1999 Interaction of mitogen-activated protein kinases with the kinase interaction motif of the tyrosine phosphatase PTP-SL provides substrate specificity and retains ERK2 in the cytoplasm. J Biol Chem 274: 21900-7.
30. **Yang SH, Yates PR, Whitmarsh AJ, Davis RJ, Sharrocks AD**. 1998 The Elk-1 ETS-domain transcription factor contains a mitogen-activated protein kinase targeting motif. Mol Cell Biol 18: 710-20.

31. **Meloche S, Pages G, Pouyssegur J**. 1992 Functional expression and growth factor activation of an epitope-tagged p44 mitogen-activated protein kinase, p44mapk. Mol Biol Cell 3: 63-71.
32. **Vouret-Craviari V, Van Obberghen-Schilling E, Scimeca JC, Van Obberghen E, Pouyssegur J**. 1993 Differential activation of p44mapk (ERK1) by alpha-thrombin and thrombin-receptor peptide agonist. Biochem J 289: 209-14.
33. **Brondello JM, McKenzie FR, Sun H, Tonks NK, Pouyssegur J**. 1995 Constitutive MAP kinase phosphatase (MKP-1) expression blocks G1 specific gene transcription and -phase entry in fibroblasts. Oncogene 10: 1895-904.
34. **Pages G, Lenormand P, L'Allemain G, Chambard JC, Meloche S, Pouyssegur J**. 1993 Mitogen-activated protein kinases p42mapk and p44mapk are required for fibroblast proliferation. Proc Natl Acad Sci U S A 90: 8319-23.
35. **Traverse S, Seedorf K, Paterson H, Marshall CJ, Cohen P, Ullrich A**. 1994 EGF triggers neuronal differentiation of PC12 cells that overexpress the EGF receptor. Curr Biol 4: 694-701.
36. **Keyse SM**. 1998 Protein phosphatases and the regulation of MAP kinase activity. Semin Cell Dev Biol 9: 143-52.
37. **Alessi DR, Gomez N, Moorhead G, Lewis T, Keyse SM, Cohen P**. 1995 Inactivation of p42 MAP kinase by protein phosphatase 2A and a protein tyrosine phosphatase, but not CL100, in various cell lines. Curr Biol 5: 283-95.
38. **Pulido R, Zuniga A, Ullrich A**. 1998 PTP-SL and STEP protein tyrosine phosphatases regulate the activation of the extracellular signal-regulated kinases ERK1 and ERK2 by association through a kinase interaction motif. EMBO J 17: 7337-50.
39. **Keyse SM**. 2000 Protein phosphatases and the regulation of mitogen-activated protein kinase signalling. Curr Opin Cell Biol 12: 186-92.
40. **Brondello JM, Brunet A, Pouyssegur J, McKenzie FR**. 1997 The dual specificity mitogen-activated protein kinase phosphatase-1 and -2 are induced by the p42/p44MAPK cascade. J Biol Chem 272: 1368-76.
41. **Camps M, Nichols A, Arkinstall S**. 2000 Dual specificity phosphatases: a gene family for control of MAP kinase function. Faseb J 14: 6-16.
42. **Lenormand P, Sardet C, Pages G, L'Allemain G, Brunet A, Pouyssegur J**. 1993 Growth factors induce nuclear translocation of MAP kinases (p42mapk and p44mapk) but not of their activator MAP kinase kinase (p45mapkk) in fibroblasts. J Cell Biol 122: 1079-88.
43. **Adachi M, Fukuda M, Nishida E**. 1999 Two co-existing mechanisms for nuclear import of MAP kinase: passive diffusion of a monomer and active transport of a dimer. EMBO J 18: 5347-58.
44. **Fukuda M, Gotoh I, Adachi M, Gotoh Y, Nishida E**. 1997 A novel regulatory mechanism in the mitogen-activated protein (MAP) kinase cascade. Role of nuclear export signal of MAP kinase kinase. J Biol Chem 272: 32642-8.
45. **Khokhlatchev AV, Canagarajah B, Wilsbacher J, Robinson M, Atkinson M, Goldsmith E, Cobb MH**. 1998 Phosphorylation of the MAP kinase ERK2 promotes its homodimerization and nuclear translocation. Cell 93: 605-15.
46. **Brunet A, Roux D, Lenormand P, Dowd S, Keyse S, Pouyssegur J**. 1999 Nuclear translocation of p42/p44 mitogen-activated protein kinase is required for growth factor-induced gene expression and cell cycle entry. EMBO J 18: 664-74.
47. **Fukuda M, Gotoh Y, Nishida E**. 1997 Interaction of MAP kinase with MAP kinase kinase: its possible role in the control of nucleocytoplasmic transport of MAP kinase. EMBO J 16: 1901-8.

48. **Lenormand P, Brondello JM, Brunet A, Pouyssegur J**. 1998 Growth factor-induced p42/p44 MAPK nuclear translocation and retention requires both MAPK activation and neosynthesis of nuclear anchoring proteins. J Cell Biol 142: 625-33.
49. **Yung Y, Dolginov Y, Yao Z, Rubinfeld H, Michael D, Hanoch T, Roubini E, Lando Z, Zharhary D, Seger R**. 1997 Detection of ERK activation by a novel monoclonal antibody. FEBS Lett 408: 292-6.
50. **Volmat V, Camps M, Arkinstall S, Pouysségur J, Lenormand P**. 2001 The nucleus, a site for signal termination by sequestration and inactivation of p42/p44 MAP Kinases. J Cell Sci, in press.
51. **Adachi M, Fukuda M, Nishida E**. 2000 Nuclear export of MAP kinase (ERK) involves a MAP kinase kinase (MEK)-dependent active transport mechanism. J Cell Biol 148: 849-56.
52. **Cavigelli M, Dolfi F, Claret FX, Karin M**. 1995 Induction of c-fos expression through JNK-mediated TCF/Elk-1 phosphorylation. EMBO J 14: 5957-64.
53. **Engel K, Kotlyarov A, Gaestel M**. 1998 Leptomycin B-sensitive nuclear export of MAPKAP kinase 2 is regulated by phosphorylation. EMBO J 17: 3363-71.
54. **Mattison CP, Ota IM**. 2000 Two protein tyrosine phosphatases, Ptp2 and Ptp3, modulate the subcellular localization of the Hog1 MAP kinase in yeast. Genes Dev 14: 1229-35.
55. **Brondello JM, Pouyssegur J, McKenzie FR**. 1999 Reduced MAP kinase phosphatase-1 degradation after p42/p44MAPK-dependent phosphorylation. Science 286: 2514-7.

Section II

Receptor Tyrosine Kinases

4

The EGF Receptor Signaling System

A Model for Growth Factor Receptor Signaling

Alan Wells*

Department of Pathology, University of Pittsburgh, and Pathology and Laboratory Services, Pittsburgh VAMC, Pittsburgh, PA 15261, USA

INTRODUCTION

The receptor for epidermal growth factor (EGF) and related ligands (EGFR; also referred to as HER and c-erbB1) is expressed on cells of many lineages. Receptor binding by at least five genetically distinct ligands activates the intrinsic kinase resulting in EGFR tyrosyl-phosphorylating itself and numerous intermediary effector molecules, including closely-related c-erbB receptor family members. This action initiates a multitude of signaling pathways, some of which attenuate receptor signaling. The integrated biological responses to EGFR signaling include mitogenesis or apoptosis, enhanced cell motility, protein secretion, and differentiation or dedifferentiation; the outcome depends both on the proteomic complement of cell and other operative extracellular signals. Implicated in organ morphogenesis, maintenance and repair, upregulated EGFR signaling also correlates with a wide variety of tumors, especially at progression to invasion and metastasis. Thus, EGFR and its resultant signals are therapeutic targets in cancer and wound healing treatments.

Sensing clues in the external milieu is critical for individual cells in organ development and functioning as the cells must respond appropriately to their localization and in concert with the needs of the animal. One major family of cell surface molecules that communicate such information is comprised of transmembrane receptors with intrinsic tyrosine kinase activity (RPTK). Cell reponses initiated by this class of receptors span the spectrum from cell growth

* wellsa@msx.upmc.edu

and death to migration and differentiation or dedifferentiation. These receptors are triggered by polypeptide hormones activating their kinases that are directed against tyrosine residues located both within the receptor itself (autophosphorylation) and on downstream target molecules. This kinase activity, with a possible few minor exceptions, is required for all cellular responses. The pleiotropic cell responses are actuated *via* still ill-defined pathways and modulated not only by the differentiation status of the cell but by the many other external stimuli acting upon that cell. This review will briefly describe the prototypal member of this large superfamily of cell receptors and its activating ligands, and then discuss the main cell responses and their biochemical controls elicited during physiological and pathological signaling of this receptor.

EGFR FAMILY MEMBERS

The EGF receptor (EGFR) is considered the prototypal member of this large superfamily of RPTK. It was the first cellular transmembrane protein described to possess tyrosine kinase activity and the first member of this superfamily to be sequenced. Co-incidently, the gene structure of EGFR appears to represent an archetypal pattern for this superfamily of extracellular sensors that control basic cell functions.

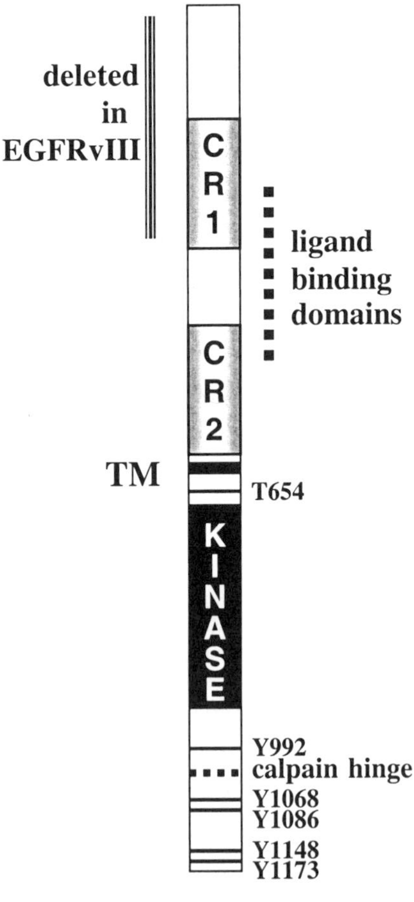

Figure 1. EGFR structure

EGFR and Related Receptors

EGFR (c-erbB1) is produced as a single major isoform, from a single 26 exon gene located across 110 kb on chromosome 7p11-13. This is the sole or overwhelmingly predominant receptor for multiple distinct ligands (below). The protein product of this gene is most often an 1186 amino acid mature transmembrane glycoprotein (Fig. 1). After cleavage of the amino-terminal signal sequence, the 622 amino acid extracellular domain containing two cysteine-rich domains comprises the ligand binding domain, that binds the various ligands seemingly at overlapping but slightly differing sites. There is a single α-helical 22 amino acid transmembrane pass. The intracellular 542 amino acids can be grouped into three domains. The juxtamembrane domain (~50 amino acids) serves primarily as a site for feedback attenuation by PKC and ERK MAP kinases, though there is evidence that a motif within this region may link to heterotrimeric G proteins [1]. Next comes a contiguous ~250 amino acid tyrosine kinase (SH1) domain which also presents targets for transmodulation by other kinases [2]. A unique 229 amino acid long carboxy-terminal tail contains 5 autophosphorylation motifs which link to proteins containing SH2 or PTB domains, at least 3 internalization motifs comprised of a tight turn each, and sites for transphosphorylation and proteolytic activation and degradation. This tail also functions as an autoinhibitory substrate; in the absence of either autophosphorylation or removal, ligand-activated EGFR is unable to phosphorylate substrates.

The EGFR autophosphorylation motifs are structurally similar [3] and functionally redundant [4], in distinction to many other RPTK. If one removes or negates individual autophosphorylated tyrosines others subsume that specific interactive function and there is little qualitative change in the downstream pathways triggered. This simple architecture and flexible interchange of redundant motifs bespeaks an archetypal gene.

There are two established and one predicted variants. A splice variant proximal to the transmembrane domain generates a secreted EGFR which can act as a dominant-negative in experimental situations [5]. The *in vivo* role of this remains to be determined but has been suggested as a negative titrater of signaling, a soluble binding protein extending the life-span of EGFR ligands, or an extracellular sink for predeposited ligands. The presence of seemingly full-length EGFR in the nucleoplasm has lead to suggestions of a transmembrane-negative splice variant similar to one reported for the related *k-sam* gene; however, this species has yet to be positively identified [6]. An EGFR variant (EGFRvIII) gaining attention as a therapeutic target, lacks amino acids 6-273 (exons 2-7) and the first cysteine-rich domain [7]. This was first found as a tumor-specific gene rearrangement but subsequent work suggested that this rearranged gene replicates a physiological variant present during development [8]. EGFRvIII fails to bind ligand but displays constitutive low level kinase activity. This variant is present in a significant fraction of glioblastoma multiforms and other cancers, but always in conjunction with wild-type EGFR.

The c-erbB subfamily of RPTK comprises four receptors sharing a similar architecture with distinct ligand-binding and signaling properties. ErbB3 and erbB4 serve as receptors for heregulin and neuregulin. Interestingly, no ligand has been defined for erbB2. This led to the discovery that a major function of these receptors appears to be as downstream effectors of each other. These receptors hetero-aggregate, cross-phosphorylate, and modulate signaling from each other in specific pairings [9]. For instance, EGFR (erbB1) will interact with erbB2 and erbB3 but not erbB4, but erbB4 will pair with erbB2. Of particular note, erbB3 lacks kinase activity, and thus serves as a docking protein to recruit a broader spectrum of downstream effectors after phosphorylation by EGFR or erbB2; this may be of physiological and pathological relevance as the erbB3 tyrosyl-phosphorylation sites likely present a docking specificity different from other erbB family members [3].

Expression and Degradation of EGFR

EGFR, and to a lesser extent the other erbB receptors, are present on all epithelial cells and stromal fibroblasts and select glial and smooth/cardiac muscle cells. The transcriptional controls of these receptors are less well characterized but divergent; EGFR is transcribed from a TATA-less promoter [10] whereas ErbB2 is driven from a TATA-box [11]. EGFR transcription appears to decline with cellular aging which may be responsible for the low level of responsiveness of near senescent fibroblasts [12]. Interestingly, EGFR transcription may be promoted by mutant p53 [13] suggesting a second mechanism by which this widespread cancer-associated genetic alteration might promote mitogenesis.

These receptors are all glycosylated and otherwise post-translationally modified; these are required for full functioning. In polar epithelial cells, EGFR is restricted to the basolateral aspects, allowing for paracrine signaling from fibroblast-derived TGFα and other matrix-associated EGFR ligands. This compartmentalized presentation of EGFR limits autocrine signaling, as many of these epithelial organs, particularly throughout the genito-urinary system, secrete copious amounts of EGF into the lumens [14]. Thus, a breakdown of cell polarity or cell-cell type junctions, as occurs during tumorigenesis, would lead to autocrine signaling. ErbB2 appears to be retained in intracellular vesicles unless overexpressed [15]. This movement to the surface would promote interactions with other members of the EGFR family, the presumed mode of activation of this orphan receptor [9].

Signal attenuation is critical to appropriate control of these plieotropic signaling receptors. A number of such mechanisms exist. Within seconds EGFR is dephosphorylated with subsequent shutdown of signaling [16]; this is reversed by ligand binding. Longer term and irreversible, activated EGFR undergoes internalization *via* a saturable endocytic system utilizing both specific adaptins and sorting nexins complexing with redundant tight-turn motifs in the carboxy-terminal tail. Occupancy and kinase activity dictate EGFR fate; EGF, remaining bound in the acidic late endosomal compartment, directs EGFR to degradation whereas TGFα, which displays a pH-sensitive dissociation, results in receptor recycling. In sum, ligand-induced internalization leads to net removal of either EGFR (for nondissociative ligands like EGF) or ligand (for dissociative ligands like TGFα) [17]. Thus, different ligands alter the strength and lifesapn of EGFR signals, which provides a rationale for the existence of multiple genetically distinct ligands.

EGFR Ligands

Numerous cell- and virally-encoded ligands bind to and activate EGFR with seemingly indistiguishable affinities and binding sites. This is misleading as differences in pH dependencies and co-receptors alters temporal and spatial signaling from EGFR. Furthermore, as all the ligands are produced as membrane-bound precursors, there is some question as to whether the ligands themselves serve as signaling receptors, especially in the case of pro-EGF. The reason for expression as pro-ligands is unknown. One theory holds that the post-cleavage transmembrane portion of the ligand serves as a 'counting' marker for the cell to determine signaling capacity whereas a second, and not exclusive, model suggests that membrane-bound pro-ligands cannot function as cell autocrine ligands and thus prevents premature activation of EGFR [19].

Further confounding the situation is that the ligands are produced essentially by the same fibroblasts and epithelial cells that express EGFR. Spatial segregation, ligand expressed on the apical surface of polarized epithelial cells, and expression in a pro-ligand form presumably prevent cell autocrine signaling. Paracrine signaling would be promoted by such spatially-restricted signaling (Fig. 2), with autocrine signaling occuring only when the cell-cell junctions are breached as during wound repair. However, this begs a larger question. Pro-EGF is expressed at highest levels on the lumenal surface of organs that lead to the exterior, such as renal tubules and prostate and mammary glands [14]. The expression of EGF in these locales might be required to repair minor traumatic tears in the epithelial lining downstream of these structures, particularly for renal EGF. It has been suggested that EGF might act as an exocrine growth factor that promotes the healing and/or development of the recipient epithelial lining, especially for neonatal oral and upper gastro-intestinal tracts.

All of the proven EGFR ligands are mature proteins of >50 amino acids with six conserved cysteines that form three disulfide bonds and three loops [20]. The first loop encodes high affinity binding. This structure makes these ligands relatively protease insensitive and acid soluble. Epidermal growth factor (EGF) was one of the first peptide growth factors isolated, even before the elucidation of the receptor [21]. This 1218 amino acid transmembrane protein has 7 other EGF-like repeats with the mature EGF being the most membrane-proximal. EGF was first isolated as urogastrone, a peptide of variable size (up to 40 kDa) from urine; this results from shedding of the extracellular domain of the pro-form from the renal tubules. EGF binds tightly to its receptor both at the cell surface and during endosomal acidification, resulting in shunting EGFR towards degradation [22]. EGF is produced and secreted into the lumen by epithelial cells.

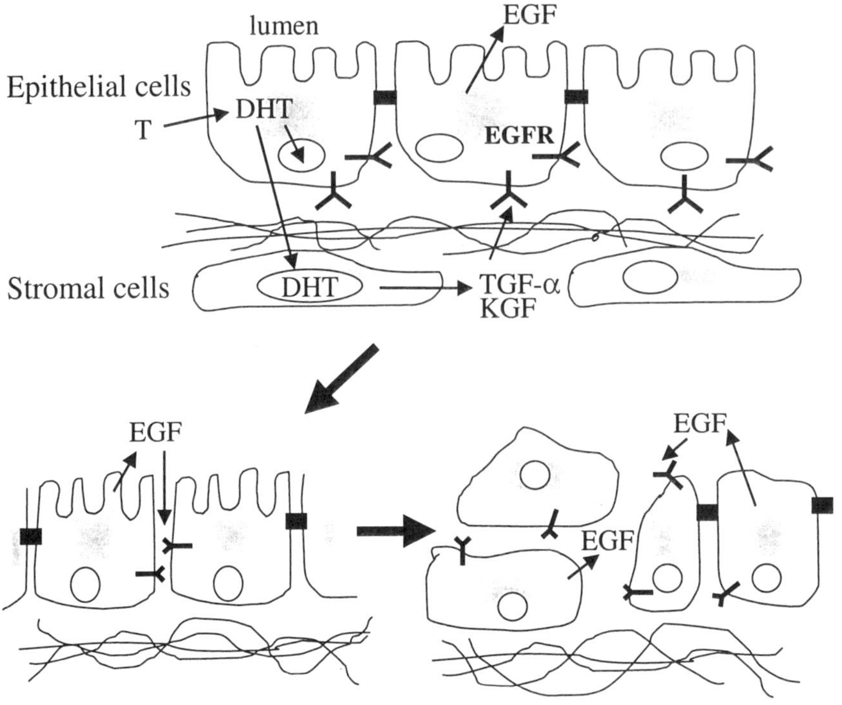

Figure 2. EGFR signaling is altered from stromal-epithelial paracrine to autocrine as prostate epithelial cells lose their cell-cell cohesiveness. The membrane-anchored 'forks' are EGFR.

Adapted from [117]

TGFα was isolated as half of the active fraction of sarcoma growth factor from ras-transformed cells (TGFβ was the other component). The 160 amino acid proform is clipped to release the mature TGFα from the membrane upon PKC phosphorylation of its cytoplasmic tail [23]. TGFα demonstrates a sharp pH-dependency in affinity for EGFR with release upon endocytic acidification with the result that internalized EGFR are recycled [22]. Thus, TGFα is noted as upregulated in advanced tumors to produce autocrine signaling loops [24, 25], as it spares the receptor [17]. In addition, TGFα appears to be the predominant EGFR ligand produced by fibroblasts and responsible for stromal-epithelial paracrine signaling.

More recently, two EGFR ligands have been described that require heparin co-binding for stable interactions with the receptor, Amphiregulin [26] and HB-EGF [27]. These two ligands also originate as transmembrane proforms with an amino-terminal heparin-binding domain adjacent to the EGF-like repeat. Whether the heparin-binding is solely directed at cell surface heparan sulfate proteoglycan, or contributes to a stable pre-deposition in matrix is under investigation. These factors are produced by epithelial cells and appear responsible for physiological epithelial cell autocrine and paracrine signaling.

An emerging hypothesis has select EGF-like repeats in matrix constituents acting as low affinity ligands for EGFR. These ubiquitous repeats are mainly of unknown function. However, anti-adhesive activity has been mapped to the repeats in tenascin [28]. The repeats are approximately 40 amino acids long, lacking the first loop that is responsible for high affinity EGFR binding [20]. Our data suggest that at least some of these repeats can, at high concentrations, trigger signaling through EGFR.

BIOCHEMICAL SIGNALS TO CELL BIOLOGICAL PROCESSES

EGFR, like all ErbB family members and other RPTK, triggers a multitude of biochemical pathways. The biological responses encompass mitogenesis and apoptosis, migration and differentiation, and dedifferentiation. These disparate and mutually exclusive responses can occur in the same cell type, presumably due to the interplay with other signals and cell context. Known parameters include cell density, position within a colony, and matrix and cytokine signals. Furthermore, evidence is accumulating that both the timing and localization of ligand binding affects the cellular outcome. While the molecular bases of these responses is only now being defined [29], the current status will be reviewed below.

Ligand binding triggers the receptor's intrinsic tyrosine kinase activity leading to auto- and exo-phosphorylation. Early work demonstrated that one-to-one stoichiometry of ligand and receptor was sufficient for kinase activation [30] but more recent data suggest that ligand induces homo- and hetero-aggregation among members of the ErbB family [9]. Kinase active EGFR cross phosphorylate kinase inactive mutants of EGFR and other ErbB members. In fact, at high levels, ErbB2 may become the preferred aggregation partner for EGFR [31]. Thus, biochemical responses secondary to EGFR activation may be due to signaling from other ErbB members; cross phosphorylation of ErbB3 may explain the recruitment of PI3-kinase signaling noted only in certain circumstances [24, 32].

A few guiding principles about connecting biochemical signals to biological responses have emerged from this forest of inter-related cascades. First, pathways can be shown to be required but not sufficient for a particular response; one example is ERK-mediated calpain activity in cell motility [33]. Second, specific effectors often contribute to multiple, conflicting responses; ERK activation is required for both proliferation and migration induced by EGF [34]. Third, temporal and spatial control of EGFR signaling dictates the outcome, likely due to altering the balance of signaling between competing pathways. Last, other biological and biophysical signals modulate the response to EGFR; proliferative responses require a tensed substratum [35].

Downstream Biochemical Pathways

Downstream from the receptors are a plethora of familiar pathways, including those that involve PLCγ and subsequent calcium- and PKC-mediated cascades, Ras activation leading to various MAP kinases, other small GTPases such as Rho and Rac, multiple STAT isoforms, and heterotrimer G proteins, as well as others to a lesser extent including PI3 kinase, PLD, and Src. Below we will highlight three major, and seemingly distinct, signaling pathways triggered by EGFR activation. This does not imply that the multitude of other pathways are unimportant or their contributions are minor. On the contrary, during tumor progression EGFR activation of PI3-kinase may play a prominent role in metastatic egress [32, 36] and Rho and Rac functioning are critical to actin cytoskeletal reorganization in response to RPTK including EGFR [37]. However, this discussion is limited so as to enumerate the pathways most vdirectly linked to EGFR and its resultant biological and pathological cell responses.

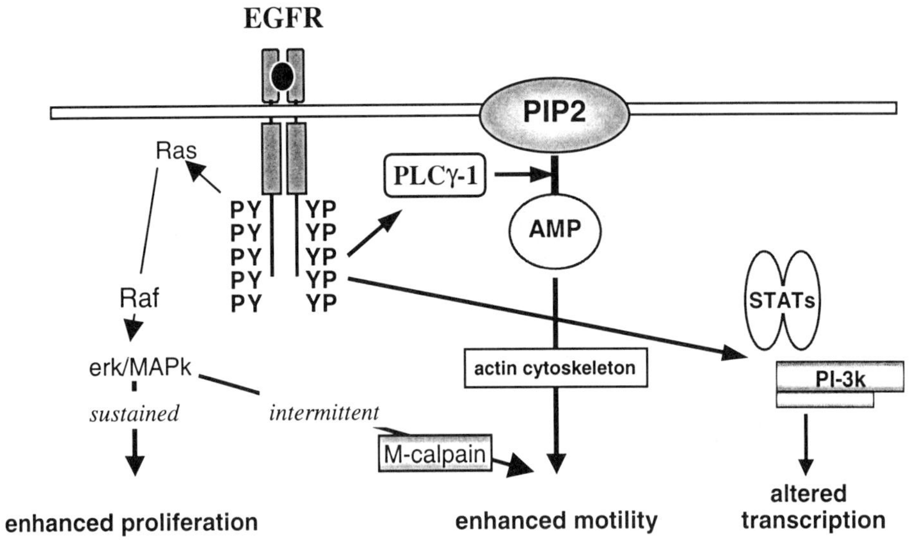

Figure 3. Schematic representation of EGFR signaling pathways to cell responses.

PLCγ Signaling Pathway

EGFR stongly activates phospholipase C-γ (PLCγ), with downstream linkages to cell motility and tumor invasion (Fig 3). We have shown that EGF-induced cell motility is inhibited by decreasing PLCγ activity [4]. Fibroblasts devoid of PLCγ-1 fail to demonstrate the full motility response from EGFR [38]. PLCγ activation also plays a required role in ErbB2, PDGF-, IGF-1- and HGF-induced motility [39-44]. The FGFR family of RPTK appears not to require PLCγ for motility; in endothelial cells which respond to bFGF with chemotaxis, PLCγ activation seems not to be required but phospholipase A2 activity is [45, 46], suggesting that some modes of induced motility may bypass the function subsumed by PLCγ activation. Of interest, integrins appear to actuate haptokinesis independent of PLCγ signaling [4, 47]. Thus, PLCγ likely represents a target that distinguishes between RPTK- and integrin-mediated motility.

Other external signals likely impact EGFR activation of this pathway by altering activities of phosphatidylinositol transfer protein (PITP) and phosphatidylinositol 4' kinase [48, 49]. PI4-kinase is recruited and activated by integrin-containing complexes during haptokinesis [50] suggesting a molecular mechanism by which adhesion-related signals may 'prime' the cell to respond motogenically to growth factors.

Activated PLCγ hydrolyzes PIP2 and alters the submembrane milieu by removing specific docking sites for numerous actin cytoskeleton-modifying proteins. Gelsolin, cofilin, and profilin bind PIP2 and appear to be in inactive states [51-54]. Upon PIP2 hydrolysis these would then interact with the nearby submembrane cytoskeleton to disrupt existing fibers and nucleate new cytoskeletal growth [55-57]. PIP2 is also required for the functioning of other molecules that affect cell shape and motility including α-actinin, vinculin and ezrin [58-61].

Deconstruction of the biophysical processes of motility determined that this pathway enables sessile fibroblasts to attain a motile morphology [62]. A similar structural reorganization likely occurs in epithelial cells as these cells also require PLCγ for EGFR-mediated motility [42, 44].

The roles, if any, for other pathways triggered by PLCγ signaling are still undetermined. PIP2 hydrolysis generates IP3 and DAG with subsequent calcium fluxes and PKC activation. While these have pleiotr^pic effects in cells, their functions in EGFR signaling have not be proven with the exception of PKC-mediated inhibition of EGFR signaling [63-65]. Recently, we have found that the expression of a small number of genes are regulated by and contribute to tumor invasion enhanced by this EGFR-PLCγ pathway [66]. This provides for yet-to-be-described signaling pathways.

ERK MAP Kinase Signaling

EGFR ranks as one of the most robust activators of the ERK MAP kinases. This likely occurs through the well-described grb2/shc > ras > raf > MEK > ERK cascade [67] (Fig 3; see also Chapter 3). The activation of this pathway occurs through direct binding of the SH2 domains of grb2 or shc to EGFR or hetero-phosphorylated ErbB2 [68]. ERK activation leads to both proliferation and migration. The proliferation response occurs through the translocation of ERK to the nucleus and subsequent transcriptional activations [69, 70]. ERK1/2 MAP kinases have been shown to be necessary but not sufficient for inducing proliferation [71, 72]. This suggested that additional signaling pathways also contributed to the proliferation response. Other ERK isoforms (ERK5/Bmk1) may also be required for cell proliferation in response to EGFR activation [73], but it appears that these also are activated downstream of ras signaling [74].

The biochemical basis of the ERK-associated motility response is only now being deciphered. ERK signaling is responsible for the EGFR-mediated loss of defined focal adhesions and transient de-adhesion in sessile fibroblasts [34]. Activation of M-calpain (calpain II) accomplishes this process and enables retraction of the trailing edge of the cell [33]. This may be in distinction to the role of ERK in integrin-mediated haptokinesis, in which the downstream operative response occurs through myosin light chain kinase (MLCK) [75], though there is no data at present to determine whether EGFR signaling also increases MLCK activity [76].

Lastly, the importance of signal attenuation is highlighted by the fact that ERK feedback attenuates EGFR signaling by phosphorylating threonine 669 [77, 78].

STAT Signaling

Activation of STATs (see Chapter 6) represents an independent signaling pathway that diverges immediately post-EGFR signaling. EGFR directly activates STATs 1, 3 and 5, independent of intermediary tyrosine kinases such as JAK involved in cytokine receptor signal transduction [79, 80](see Chapter 6). How the STATs interact with EGFR is undefined but does not require SH2 binding to autophosphoylated tyrosines, as an EGFR mutant lacking all these sites still activates STATs [80]. This STAT signaling has been linked to cell proliferation [81], and has been reported as required for EGFR-mediated proliferation of carcinoma cells [82, 83]. Our recent unpublished data suggest that STAT signaling, STAT3 in particular, may also be critical for induced motility. This transcriptional activation likely contributes to the *de novo* transcription required for sustained cell migration [84-86].

Modulatory Signals

Signaling from EGFR and other RPTKs does not occur in a vacuum; the cellular response results from an orchestration of events initiated by adhesion receptors, hormones, cytokines, channels and physical stimuli. In many ways, RPTKs may be simplistically considered as superimposed controlling elements for which other signaling mechanisms provide opportunities. For example, PDGF and EGF elicit mitogenic responses only when cells are adherent on stressable substrata [35, 87-90]. Specific matrix components also determine which genes are transcribed [91-95], and thus alter the proteome keyboard upon which growth factor signals play. All these provide a mechanism by which adhesion receptors may 'prime' the cell for responses to growth factors.

Much of this signaling occurs from adhesion receptors, particularly integrins. One biochemical pathway thought to be important for such transmodulation involves integrin low-level activation of the ERK MAP kinases [75], which is necessary for EGFR-mediated responses [34, 96].

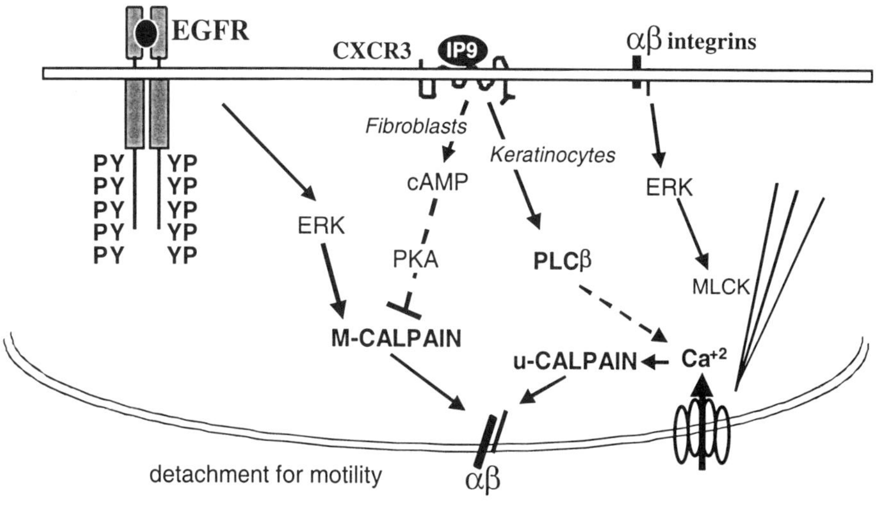

Figure 4. Central role of calpain in integrating signals from diverse receptors.

Another convergent biochemical event appears to be calpain activation [33, 97, 98]; though it appears that integrins and chemokines activate the μ-calpain isoform, and thus the convergence is likely at the target molecules (Fig 4). This crosstalk is bidirectional as RPTK signaling alters the adhesion receptor profiles [99-102].

Cross-signalings from other stimuli also modulate EGFR-mediated responses. While the literature is rife with documented examples, few pathways have been elucidated that also correspond to specific cell responses. One of the earliest examples described concerns negative transmodulation of EGFR activation by PDGF *via* PKC phosphorylation [65]. More recently, we have described ELR-negative CXC chemokine abrogation of EGFR-mediated deadhesion occuring by PKA-dependent inhibition of M-calpain [103] (Fig 4).

Ligand-Independent Signaling

A fascinating recent development demonstrates that EGFR relays signaling from a host of disparate cell surface receptors [104]. Originally, it was observed that mitogenic signaling, involving ERK MAP kinase (Chapter 3), from the G-protein-coupled receptors for endothelin-1 (Chapter 11), LPA, and thrombin was prevented by EGFR inhibition [105]. Further investigations determined that these receptors function by releasing HB-EGF and activating an autocrine loop [106]. Others have shown that motility of B82 fibroblasts signaled by integrins and the related PDGF receptor also requires EGFR signaling [107, 108], though the molecular mechanisms underlying these transactivations remain to be deciphered. These developments suggest that EGFR might subsume a role as a central touchpoint for cell surface signaling, not just from its own ligands but from a variety of receptors.

PHYSIOLOGY AND PATHOBIOLOGY OF EGFR SIGNALING

Despite the copious information about the biochemical controls and pathways and cellular responses invoked by EGFR signaling, the precise physiological roles effected by EGFR have remained ill-defined. This question has been approached by extrapolation from *in vitro* experiments, *in vivo* perturbations such as disrupting EGFR regulation by adding or blocking ligands, and, most recently, by genetic engineering.

EGFR was first linked to epithelial tissue maturation, as injections of urogastrone caused precocious eye opening and tooth eruptions in mouse pups [109]. This developmental role for EGFR signaling was supported almost 30 years later with the use of EGFR knockout mice [110-112]. The animals die in the neonatal period due to severe immaturity of several epithelial organs. Two aspects of this proved surprising. First, that the animals survived to birth with all organs being present, at least as ill-developed but recognizable tissues, was unexpected. Second, the pattern and severity of the developmental retardation was dictated by the genetic background of the mouse strain. These two points suggest that in the face of germline homozygous EGFR deletion, other RPTK may be recruited to subsume, at least partially, the roles of EGFR.

In adult animals, EGFR signaling has been postulated as intimately involved in wound repair, a process that might be viewed as neo-organogenesis. Reduction of EGF levels during hepatic regeneration impairs the response. Topical application of EGF speeds corneal wound repair and slightly improves dermal wound healing. The secretion of EGF in saliva and urine is proposed to be present to stimulate repair of epithelial breaks by gaining access to basolateral EGFR at the sites of discontinuous epithelium. Thus, a role for EGFR in organogenesis is gaining acceptance.

In this context, the operative cellular response has been assumed to be proliferation. However, during organ development cells must also migrate and differentiate. Parsing the rate-limiting response has been complicated by the pleiotropic nature of EGFR and many of its downstream effectors. PLCγ has presented a rare opportunity in that it is required for motility but not mitogenesis. Of interest, the phenotype of mice expressing an inducible dominant-negative PLCγ fragment in the breast and prostate is indistinguishable from mice presenting dominant-negative EGFR in these same tissues [113]; in both animals the branching morphogenesis is severely retarded. This suggests that the operational response might be that of migration.

EGFR signaling also serves metabolic functions. Postpartum milk production and actual milk letdown is at least in part controlled by EGFR signaling [114]. EGFR signaling also increases glucose utilization and secretion of many proteins, but the organismal relevance of these findings has yet to be established.

The first broadly accepted function of EGFR signaling, and that which catapulted this molecule into a position of prominence, is in cancer. Due to the initial identification of EGFR as the proto-oncogene of the transforming *v-erbB* oncogene [115], numerous studies have investigated signaling from this system in a wide variety of human tumors. EGFR is the receptor most often found upregulated in a wide variety of human tumors [116]. Upregulated signaling through EGFR is seen at all levels - amplified EGFR gene, increased mRNA, increased tyrosyl-phosphorylated EGFR protein, and increase production of ligand (promoting an autocrine positive signaling loop).

The upregulation of signaling results in establishment of either an autocrine signaling loop [117] or enhanced sensitivity to a low level of paracrine ligand [118]. This could promote various stages of cancer (Fig 2, above). Increased signaling could lead to reduced cell-cell cohesion, and loss of polarity, further promoting an autocrine signaling loop leading to dysplasia in secretory epithelial organs; parenthetically, prevention of this autocrine loop may be one reason why cadherin/catenin functioning can act as a tumor suppressor [119]. Mitogenic responses would promote both tumor proliferation *in situ* or lessen context dependency to enable metastatic growth. Lastly, EGFR-mediated motility may promote tumor invasiveness.

The role of EGFR signaling in promoting tumor proliferation is supported by experimental systems. The early studies expressed ligand in an autocrine fashion leading to cellular transformation and tumor formation [120]. Overexpression of EGFR also produced transformed cells that grew in athymic mice [118]. Subsequently, transgenic mice producing TGFα develop carcinomas of the target organs [121-123]. These studies formed the intellectual basis for currently emerging therapies aimed at EGFR signaling [25]. EGFR has been the target of numerous therapies, ranging from therapeutic and imaging antibodies to toxin-linked ligands to enhancement of targeting for gene therapy vectors [124]. Approaches that have shown promise in clinical trials include an anti-receptor ligand-blocking antibody and small molecule inhibitors of the tyrosine kinase activity. Phase I and II trials have shown promise in the form of tumor regression or nonprogression.

Intriguingly, when human tumors were examined closely for EGFR expression, the correlations were not with tumor growth but invasion and progression. For instance, EGFR signaling is upregulated in the invasive glioblastoma multiforms but not gliomas [125], invasive but not superficial bladder carcinomas [126], and advanced (invasive) gastric carcinomas [127]. This has suggested that a key role for EGFR signaling is in tumor invasion, and that the operative response is migration not proliferation [128]. EGFR-mediated migration can be approached by abrogating EGFR activation of PLCγ. *In vitro,* abrogation of this pathway prevents invasion of extracellular matrices [44, 129]. In mouse xenograft models of human prostate carcinoma, inhibition of this signaling pathway prevents tumor invasion, albeit tumor growth remains unaffected [130, 131]. This likely represents the pathological aspect of the physiological role of branching morphogenesis.

The EGFR variant EGFRvIII may represent a tumor-specific antigen. In adults it is present only on tumor cells, but always in concert with non-deleted EGFR. The EGFR mutant has only a fraction of the kinase activity of wild-type EGFR but signals constitutively; a situation akin to *v-erbB* signaling [8, 132]. Thus, transformation is driven likely more by disregulation of EGFR signaling rather than supraphysiological levels of signaling recruiting new pathways. Antibodies directed at the splice junction, and thereby unique to this variant, have been proposed as therapeutic agents.

ErbB2 warrants special mention in context of cancer. ErbB2 (or HER2) was originally cloned as the *neu* oncogene generated in ethyl-nitrosourea-induced glioblastomas/astrocytoma (incorrectly identified as neuroblastomas) [133]. The transforming oncogene *neu* contained a mutation in the the transmembrane domain that conferred dimerization and activation. This receptor appears to be overexpressed, but not mutated, in a wide variety of carcinoma, and breast cancer in particular. The overexpression, and appearance on the cell surface correlates strongly with progression to the metastatic and invasive states. This has been used as a biomarker to predict breast cancer spread and recurrence [134]. Overexpression of ErbB2 and its originally-defined mutated variant *neu*, leads to cell transformation that can be reversed by antibody-mediated downregulation of the receptor [135]. Based on these basic findings an antibody was developed against the human ErbB2 extracellular domain. This antibody, manufactured by Genentech under the name Herceptin, has proved successful at reversing or limiting tumor growth of a significant subset of breast carcinomas. Current standard approach to breast cancer now includes staining for ErbB2 expression to determine the appropriateness of Herceptin treatment. Additional trials are underway to determine the use of Herceptin in other tumors.

EGFR ligands are ideally suited for industrial use. Mature soluble recombinant forms of EGF and TGFα are easily generated in abundance in bacteria. Being acid-soluble, isolation in bulk as pure factors is routine. Furthermore, these ligands are promiscuous across species. These properties make them ideal as a defined media supplement in large-scale cell culturing required for the bio-pharmaceutical industry. Furthermore, specific alterations and modifications have been shown to alter the signaling properties to extend the bioactivity or spare cell receptors [22]. Since EGFR is present on all epithelial and most mesenchymal cells, these ligands represent a near-universal cell growth and survival factor.

Acknowledgements

Supported by grants from the National Institute of General Medical Sciences (NIH), National Cancer Institute (NIH), Veterans Administration, and the Bioengineering Division of the National Science Foundation. Insightful discussions and ideas were provided by members of my laboratory and colleagues, including DA Lauffenburger, HS Wiley, J Grandis and T Turner.

REFERENCES

1. **Sun H, Chen Z, Poppleton H, Scholich K, Mullenix J, Weipz GJ, Fulgham DL, Bertics PJ, Patel TB** 1997 The juxtamembrane, cytosolic region of the epidermal growth factor receptot is involved in association with α-subunit of Gs. J Biol Chem 272:5413-5420
2. **Stover DR, Becker M, Liebetanz J, Lydon NB** 1995 Src phosphorylation of the epidermal growth factor receptor at novel sites mediates receptor interaction with src and p85α. J Biol Chem 270:15591-15597
3. **Songyang Z, Shoelson SE, Chaudhuri M, Gish G, Pawson T, Haser WG, King F, Roberts T, Ratnofsky S, Lechleider RJ, Neel BG, Birge RB, Fajardo JE, Chou MM, Hanafusa H, Schaffhausen B, Cantley LC** 1993 SH2 domains recognize specific phosphopeptide sequences. Cell 72:767-778

4. **Chen P, Xie H, Sekar MC, Gupta KB, Wells A** 1994 Epidermal growth factor receptor-mediated cell motility: phospholipase C activity is required, but MAP kinase activity is not sufficient for induced cell movement. J Cell Biol 127:847-857
5. **Flickinger TW, Maihle NJ, Kung H-J** 1992 An alternatively processed mRNA from the avian c-*erbB* gene encodes a soluble, truncated form of the receptor that can block ligand-dependent transformation. Mol Cell Biol 12:883-893
6. **Marti U, Wells A** 2000 The nuclear accumulation of a variant epidermal growth factor receptor (EGFR) lacking the transmembrane domain requires co-expression of a full length EGFR. Mol Cell Biol Res Com 3:8-14
7. **Sampson JH, Crotty LE, Lee S, Archer GE, Ashley DM, Wkstrand CJ, Hale LP, Small C, Dranoff G, Friedman AH, Friedman HS, Bigner DD** 2000 Unarmed, tumor-specific monoclonal antibody effectively treats brain tumors. Proc Natl Acad Sci (USA) 97:7053-7058
8. **Nishikawa R, Ji X-D, Harmon R, Lazar CS, Gill GN, Cavenee WK, Huang H-JS** 1994 A mutant epidermal growth factor receptor common in human glioma confers enhanced tumorigenicity. Proc Natl Acad Sci (USA) 91:7727-7730
9. **Carraway KL, Cantley LC** 1994 A neu acquaintance for erbB3 and erbB4: a role for receptor heterodimerization in growth signaling. Cell 78:5-8
10. **Haley J, Whittle M, Bennett P, Kinchington K, Ullrich A, Waterfield MD** 1987 The human EGF receptor gene: structure of the 110kb locus and identification of sequences regulating its transcription. Oncogene Res 1:375-396
11. **Ishii S, Imamoto F, Yamanashi Y, Toyoshima K, Yamamoto T** 1987 Characterization of the promoter region of the human *c-erbB-2* protooncogene. Proc Natl Acad Sci (USA) 84:4374-4378
12. **Shiraha H, Gupta K, Drabik KA, Wells A** 2000 Aging fibroblasts present reduced epidermal growth factor (EGF) responsiveness due to preferential loss of EGF receptors. J Biol Chem. 275:19343-19351.
13. **Ludes-Meyers JH, Subler MA, Shivakumar CV, Munoz RM, Jiang P, Bigger JE, Brown DR, Deb SP, Deb S** 1996 Transcriptional activation of the human epidermal growth factor receptor promoter by human p53. Mol Cell Biol 16:6009-6019
14. **Marti U, Burwen SJ, Jones AL** 1989 Biological effects of epidermal growth factor, with emphasis on the gastrointestinal tract and liver: an update. Hepatology 9:126-138
15. **DePotter CR, Quatacker J** 1993 The $p185^{erbB2}$ protein is localized on cell organelles involved in cell motility. Clin Exp Metastas 11:453-461
16. **Hernandez-Sotomayor SMT, Artega CL, Soler C, Carpenter G** 1993 Epidermal growth factor stimulates substrate-selective protein-tyrosine phosphatase activity. Proc Natl Acad Sci (USA) 90:7691-7695
17. **Reddy CC, Wells A, Lauffenburger DA** 1996 Receptor-mediated effects on ligand availability influence relative mitogenic potencies of epidermal growth factor and transforming growth factor α. J Cell Physiol 166:512-522
19 .**Dong J, Opresko LK, Dempsey PJ, Lauffenburger DA, Coffey RJ, Wiley HS** 1999 Metalloprotease-mediated ligand release regulates autocrine signaling through the epidermal growth factor receptor. Proc Natl Acad Sci (USA) 96:6235-6240
20. **Tadaki DK, Niyogi SK** 1993 The functional importance of hydrophobicity of the tyrosine at position 13 of human epidermal growth factor in receptor binding. J Biol Chem 268:10114-10119
21. **Carpenter G, Cohen S** 1979 Epidermal growth factor. Ann Rev Biochem 48:193-216
22. **Reddy CC, Niyogi SK, Wells A, Wiley HS, Lauffenburger DA** 1996 Re-engineering epidermal growth factor for enhanced potency. Nature Biotechnol 14:1696-1699

23. **Bosenberg MW, Pandiella A, Massague J** 1992 The cytoplasmic carboxy terminal amino acid specifies cleavage of membrane TGFα into soluble growth factor. Cell 71:1157-1165
24. **Wells A** 2000 Tumor invasion: role of growth factor-induced cell motility. Advances in Cancer Res 78:31-101
25. **Wells A** 2000 The epidermal growth factor receptor – a new target in cancer therapy. Signal 1:4-10
26. **Shoyab M, Plowman GD, McDonald VI, Bradley JG, Todaro GJ** 1989 Structure and function of human amphiregulin: a member of the epidermal growth factor family. Science 243:1074-1076
27. **Higashiyama S, Abraham JA, Klagsbrun M** 1993 Heparin-binding EGF-like growth factor stimulation of smooth muscle cell migration: dependence on interactions with cell surface heparan sulfate. J Cell Biol 122:933-940
28. **Murphy-Ullrich J** 1995 Anti-adhesive proteins of the extracellular matrix: thrombospondin, tenascin, and SPARC. Trends Glycosci Glyc 7:89-100
29. **Wells A, Gupta K, Chang P, Swindle S, Glading A, Shiraha H** 1998 Epidermal growth factor receptor-mediated motility in fibroblasts. Microscopy Res Techniq 43:395-411
30. **Weber W, Bertics PJ, Gill GN** 1984 Immunoaffinity purification of the epidermal growth factor receptor: stoichiometry of binding and kinetics of self-phosphorylation. J Biol Chem 259:14631-14636
31. **Qian X, LeVea CM, Freeman JK, Dougall WC, Greene MI** 1994 Heterodimerization of epidermal growth factor receptor and wild-type or kinase-deficient Neu: a mechanism of interreceptor kinase activation and transphosphorylation. Proc Natl Acad Sci (USA) 91:1500-1504
32. **Hill K, Welti S, Yu J, Murray JT, Yip S-C, Condeelis JS, Segall JE, Backers JM** 2000 Specific requirement for the p85-p110α phosphatidylinositol 3-kinase during epidermal growth factor-stimulated actin nucleation in breast cancer cells. J Biol Chem 275:3741-3744
33. **Glading A, Chang P, Lauffenburger DA, Wells A** 2000 Epidermal growth factor receptor activation of calpain is required for fibroblast motility and occurs via an ERK/MAP kinase signaling pathway. J Biol Chem 275:2390-2398
34. **Xie H, Pallero MA, Gupta D, Chang P, Ware MF, Witke W, Kwiatkowski DJ, Lauffenburger DA, Murphy-Ullrich JE, Wells A** 1998 EGF receptor regulation of cell motility: EGF induces disassembly of focal adhesions independently of the motility-associated PLCγ signaling pathway. J Cell Sci 111:615-624
35. **Assoian R** 1997 Anchorage-dependent cell cycle progression. J Cell Biol 136:1-4
36. **Segall JE, Tyerech S, Boselli L, Masseling S, Helft J, Chan A, Jones J, Condeelis J** 1996 EGF stimulates lamellipod extension in metastatic mammary adenocarcinoma cells by an actin-dependent mechanism. Clin Exp Metastas 14:61-72
37. **Hall A** 1998 Rho GTPases and the actin cytoskeleton. Science 279:509-514
38. **Ji Q-S, Ermini S, Baulida J, Sun F-L, Carpenter G** 1998 Epidermal growth factor signaling and mitogenesis in Plcg1 null mouse embryonic fibroblasts. Mol Biol Cell 9:749-757
39. **Kundra V, Escobedo JA, Kazlauskas A, Kim HK, Rhee SG, Williams LT, Zetter BR** 1994 Regulation of chemotaxis by the platelet-derived growth factor receptor-β. Nature 367:474-476
40. **Derman MP, Chen JY, Spokes KC, Songyang Z, Cantley LG** 1996 An 11-amino acid sequence from c-met initiates epithelial chemotaxis via phosphatidylinositol 3-kinase and phospholipase C. J Biol Chem 271:4251-4255
41. **Bornfeldt KE, Raines EW, Nakano T, Graves LM, Krebs EG, Ross R** 1994 Insulin-like growth factor-1 and platelet-derived growth factor-BB induce directed migration of

human arterial smooth muscle cells via signalling pathways that are distinct from those of proliferation. J Clin Invest 93:1266-1274
42. **Polk DB** 1998 Epidermal growth factor receptor-stimulated intestinal epithelial cell migration requires phospholipase C activity. Gastroenterology 114:493-502
43. **Xie H, Turner T, Wang M-H, Singh RK, Siegal GP, Wells A** 1995 In vitro invasiveness of DU-145 human prostate carcinoma cells is modulated by EGF receptor-mediated signals. Clin Exp Metastas 13:407-419
44. **Kassis J, Moellinger J, Lo H, Greenberg N, Kim H-G, Wells A** 1999 A role for phospholipase C-γ-mediated signaling in tumor cell invasion. Clin Cancer Res 5:2251-2260
45. **Sa G, Fox PL** 1994 Basic fibroblast growth factor-stimulated endothelial cell movement is mediated by a pertussis-sensitive pathway regulating phospholipase A2 activity. J Biol Chem 269:3219-3225
46. **Clyman RI, Peters KG, Chen YQ, Escobedo J, Williams LT, Ives HE, Wilson E** 1994 Phospholipase C-γ activation, phosphotidylinositol hydrolysis, and calcium mobilization are not required for FGF receptor-mediated chemotaxis. Cell Adhes Commun 1:333-342
47. **Carloni V, Romanelli RG, Pinzani M, Laffi G, Gentilini P** 1997 Focal adhesion kinase and phospholipase Cγ involvement in adhesion and migration of human hepatic stellate cells. Gastroenterology 112:522-531
48. **Hsuan JJ, Tan SH** 1997 Growth factor-dependent phosphoinositide signalling. Int J Biochem Cell Biol 29:415-435
49. **Kauffmann-Zeh A, Thomas GM, Ball A, Prosser S, Cunningham E, Cockcroft S, Hsuan JJ** 1995 Requirement for phosphatidylinositol transfer protein in epidermal growth factor signaling. Science 268:1188-1190
50. **Yauch RL, Berdichevski F, Harler MB, Reichner J, Hemler ME** 1998 Highly stoichiometric, stable, and specific association of integrin α3β1 with CD151 provides a major link to phosphatidylinositol 4-kinase, and may regulate cell migration. Mol Biol Cell 9:2751-2765
51. **Janmey PA, Lamb J, Allen PG, Matsudaira PT** 1992 Phosphoinositide-binding peptides derived from the sequences of gelsolin and villin. J Biol Chem 267:11818-11823
52. **Goldschmidt-Clermont PJ, Mechesky LM, Baldassare JJ, Pollard TD** 1990 The actin-binding protein profilin binds to PIP2 and inhibits its hydrolysis by PLC. Science 247:1575-1578
53. **Onoda K, Yu FX, Yin HL** 1993 gCap39 is a nuclear and cyotplasmic protein. Cell Motil Cytoskel 26:227-238
54. **Lamb JA, Allen PG, Tuan BY, Janmey PA** 1993 Modulation of gelsolin function: activation at low pH overrides Ca^{2+} requirement. J Biol Chem 268:8999-9004
55. **Weeds A, Maciver S** 1993 F-actin capping proteins. Curr Opin Cell Biol 5:63-9
56. **Chen P, Murphy-Ullrich J, Wells A** 1996 A role for gelsolin in actuating EGF receptor-mediated cell motility. J Cell Biol 134:689-698
57. **Theriot JA, Mitchison TJ** 1993 The three faces of profilin. Cell 75:835-838
58. **Fukami K, Furuhashi K, Inagaki M, Endo T, Hatano S, Takenawa T** 1992 Requirement of phosphatidylinositol 4,5-bisphosphate for α-actinin function. Nature 359:150-152
59. **Gilmore AP, Burridge K** 1996 Regulation of vinculin binding to talin and actin by phosphatidyl-inositol-4-5-bisphoshate. Nature 381:531-535
60. **Niggli V, Andreoli C, Roy C, Mangeat P** 1995 Identification of a phosphatidylinositol-4,5-bisphosphate-binding domain in the N-terminal region of ezrin. FEBS Lett 376:172-176

61. **Honda K, Yamada T, Endo R, Ino Y, Gotoh M, Tsuda H, Yamada Y, Chiba H, Hirohashi S** 1998 Actinin-4, a novel actin-bundling protein associated with cell motility and cancer invasion. J Cell Biol 140:1383-1393
62. **Wells A, Ware MF, Allen FD, Lauffenburger DA** 1999 Shaping up for shipping out: PLCγ signaling of morphology changes in EGF-stimulated fibroblast migration. Cell Motil Cytoskel 44:227-233
63. **Welsh JB, Gill GN, Rosenfeld MG, Wells A** 1991 A negative feedback loop attenuates EGF-induced morphological changes. J Cell Biol 114:533-543
64. **Chen P, Xie H, Wells A** 1996 Mitogenic signaling from the EGF receptor is attenuated by a motility-associated phospholipase C-γ/protein kinase C feedback mechanism. Mol Biol Cell 7:871-881
65. **Davis RJ, Czech MP** 1987 Stimulation of epidermal growth factor receptor threonine 654 phosphorylation by platelet-derived growth factor in protein kinase C-deficient human fibroblasts. J Biol Chem 262:6832-6841
66. **Manos EJ, Kim M, Kassis J, Chang B, Wells A, Jones DA** 2001 Prostin-1, a novel phospholipase C-γ regulated gene negatively associated with prostate tumor invasion. Oncogene 20:2781-2790
67. **Davis RJ** 1993 The mitogen-activated protein kinase signal transduction pathway. J Biol Chem 268:14553-14556
68. **Sasaoka T, Langlois WJ, Bai F, Rose DW, Leitner JW, Decker SJ, Saltiel AR, Gill GN, Kobayashi M, Draznin B, Olefsky JM** 1996 Involvement of ErbB2 in the signaling pathway leading to cell cycle progression from a truncated epidermal growth factor receptor lacking the C-terminal autophosphorylation sites. J Biol Chem 271:8338-8344
69. **Hochholdinger F, Baier G, Nogalo A, Bauer B, Grunicke HH, Uberall F** 1999 Novel membrane-targeted erk1 and erk2 chimeras which act as dominant negative, isotype-specific mitogen-activated protein kinase inhibitors of ras-raf-mediated transcriptional activation of c-*fos* in NIH 3T3 cells. Mol Cell Biol 19:8052-8065
70. **Groom LA, Sneddon AA, Alessi DR, Dowd S, Keyse SM** 1996 Differential regulation of the MAP, SAP and RK/p38 kinases by Pyst1, a novel cytosolic dual-specificity phosphatase. EMBO J 15:3621-3632
71. **Pages G, Lenormand P, L'Allemain G, Chambard J-C, Meloche S, Pouyssegur J** 1993 Mitogen-activated protein kinases $p42^{mapk}$ and $p44^{mapk}$ are required for fibroblast proliferation. Proc Natl Acad Sci (USA) 90:8319-8323
72. **Walker F, Kato A, Gonez LJ, Hibbs ML, Pouliot N, Levitzki A, Burgess AW** 1998 Activation of the ras/mitogen-activated protein kinase pathway by kinase-defective epidermal growth factor receptors results in cell survival but not proliferation. Mol Cell Biol 18:7192-7204
73. **Kato Y, Tapping RI, Huang S, Watson MH, Ulevitch RJ, Lee J-D** 1998 Bmk1/erk5 is required for cell proliferation induced by epidermal growth factor. Nature 395:713-716
74. **English JM, Pearson G, Hockenberry T, Shivakumar L, White MA, Cobb MB** 1999 Contribution of the ERK5/MEK5 pathway to Ras/Raf signaling and growth control. J Biol Chem 274:26563-26571
75. **Klemke RL, Cai S, Giannini AL, Gallagher PJ, deLanerolle P, Cheresh DA** 1997 Regulation of cell motility by mitogen-activated protein kinase. J Cell Biol 137:481-492
76. **Murthy KS, Grider JR, Kuemmerle JF, Makhlouf GM** 2000 Sustained muscle contraction induced by agonists, growth factors, and Ca(2+) mediated by distinct PKC isozymes. Am J Physiol 279:G201-210
77. **Morrison P, Takishima K, Rosner MR** 1993 Role of threonine residues in regulation of the epidermal growth factor receptor by protein kinase C and mitogen-activated protein kinase. J Biol Chem 268:15536-15543

78. **Haugh JM, Huang AC, Wiley HS, Wells A, Lauffenburger DA** 1999 Participation of internalized epidermal growth factor receptors in the activation of $p21^{ras}$. J Biol Chem 274:34350-34360
79. **Zhong Z, Wen Z, Darnell JE** 1994 Stat3: a STAT family member activated by tyrosine phosphorylation in response to epidermal growth factor and interleukin-6. Science 264:95-99
80. **David M, Wong L, Flavell R, Thompson SA, Wells A, Larner AC, Johnson GR** 1996 STAT activation by epidermal growth factor (EGF) and amphiregulin: requirement for the EGF receptor kinase but not for tyrosine phosphorylation sites or JAK1. J Biol Chem 271:9185-9188
81. **Williams JG** 2000 STAT signalling in cell proliferation and development. Curr Opin Gen Dev 0:503-507
82. **Grandis JR, Drenning SD, Chakraborty A, Zhou M-Y, Zeng Q, Pitt AS, Tweardy DJ** 1998 Requirement of Stat3 but not Statl activation for epidermal growth factor receptor-mediated cell growth in vitro. J Clin Invest 102:1385-1392
83. **Song JI, Grandis JR** 2000 STAT signaling in head and neck cancer. Oncogene 19:2489-2495
84. **Chen P, Gupta K, Wells A** 1994 Cell movement elicited by epidermal growth factor receptor requires kinase and autophosphorylation but is separable from mitogenesis. J Cell Biol 124:547-555
85. **Malliri A, Symons M, Hennigan RF, Hurlstone AFL, Lamb RF, Wheeler T, Ozanne BW** 1998 The transcriptional factor AP-1 is required for EGF-induced activation of rho-like GTPases, cytoskeletal rearrangements, motility, and in vitro invasion of A431 cells. J Cell Biol 143:1087-1099
86. **Kustikova O, Kramerov D, Grigorian M, Berezin V, Bock E, Lukanidin E, Tulchinsky E** 1998 Fra-1 induces morphological transformation and increases in vitro invasiveness and motility of epithelioid adenocarcinoma cells. Mol Cell Biol 18:7095-7105
87. **Cybulsky AV, McTavish AJ, Cyr MD** 1994 Extracellular matrix modulates epidermal growth factor receptor activation in rat glomerular epithelial cells. J Clin Invest 94:68-78
88. **Bornfeldt KE, Raines EW, Graves LM, Skinner MP, Krebs EG, Ross R** 1995 Platelet-derived growth factor. Distinct signal transduction pathways associated with migration versus proliferation. Ann NY Acad Sci 766:416-430
89. **Alford D, Baeckstrom D, Geyp M, Pitha P, Taylor-Papadimitriou P** 1998 Integrin-matrix interactions affect the form of the structures developing from human mammary epithelial cells in collagen or fibrin gels. J Cell Sci 111:521-532
90. **Lin Y-C, Grinnell F** 1993 Decreased level of PDGF-stimulated receptor autophosphorylation by fibroblasts in mechanically relaxed collagen matrices. J Cell Biol 122:663-672
91. **Xu J, Clark RAF** 1996 Extracellular matrix alters PDGF regulation of fibroblast integrins. J Cell Biol 132:239-249
92. **Xu J, Zutter MM, Santoro SA, Clark RAF** 1998 A three-dimensional collagen lattice activates NF-kB in human fibroblasts: role in integrin α_2 gene expression and tissue remodeling. J Cell Biol 140:709-719
93. **Zhu X, Ohtsubo M, Bohmer RM, Roberts JM, Assoian RK** 1996 Adhesion-dependent cell cycle progression linked to the expression of cyclin D1, activation of cyclin E-cdk2, and phosphorylation of the retinoblastoma protein. J Cell Biol 133:391-403
94. **Bohmer R-M, Scharf E, Assoian RK** 1996 Cytoskeletal integrity is required throughout the mitogen stimulation phase of the cell cycle and mediates the anchorage-dependent expression of cyclin D1. Mol Biol Cell 7:101-111

95. **Roovers K, Davey G, Zhu X, Bottazzi ME, Assoian RK** 1999 $\alpha 5 \beta 1$ integrin controls cyclin D1 expression by sustaining mitogen-activated protein kinase activity in growth factor-treated cells. Mol Biol Cell 10:3197-3204
96. **Miyamoto S, Teramoto H, Gutkind JS, Yamada KM** 1996 Integrins can collaborate with growth factors for phosphorylation of receptor tyrosine kinases and MAP kinase activation: roles of integrin aggregation and occupancy of receptors. J Cell Biol 135:1633-1642
97. **Huttenlocher A, Palecek SP, Lu Q, Zhang W, Mellgren RL, Lauffenburger DA, Ginsburg MH, Horwitz AF** 1997 Regulation of cell migration by the calcium-dependent protease calpain. J Biol Chem 272:32719-32722
98. **Lee J, Ishihara A, Oxford G, Johnson B, Jacobson K** 1999 Regulation of cell movement is mediated by stretch-activated calcium channels. Nature 400:382-386
99. **Chen JD, Kim JP, Zhang K, Sarret Y, Wynn KC, Kramer RH, Woodley DT** 1993 Epidermal growth factor (EGF) promotes human keratinocyte locomotion on collagen by increasing the alpha 2 integrin subunit. Exp Cell Res 209:216-223
100. **Ye J, Xu RH, Taylor-Papadimitriou J, Pitha PM** 1996 Sp1 binding plays a critical role in erb-B2- and v-ras-mediated downregulation of α_2-integrin expression in human mammary epithelial cells. Mol Cell Biol 16:6178-6189
101. **Zhang M, Singh RK, Wang M-H, Wells A, Siegal GP** 1996 Epidermal growth factor modulates cell attachment to hyaluronic acid by the cell surface glycoprotein CD44. Clin Exp Metastas 14:268-276
102. **Zhang M, Wang M-H, Singh RK, Wells A, Siegal GP** 1997 Epidermal growth factor induces CD44 expression through a novel regulatory element in mouse fibroblasts. J Biol Chem 272:14139-14146
103. **Shiraha H, Gupta K, Glading A, Wells A** 1999 Chemokine transmodulation of EGF receptor signaling: IP-10 inhibits motility by decreasing EGF-induced calpain activity. J Cell Biol 146:243-253
104. **Carpenter G** 1999 Employment of the epidermal growth factor receptor in growth factor-independent signaling pathways. J Cell Biol 145:697-702
105. **Daub H, Weiss TU, Wallasch C, Ullrich A** 1996 Role of transactivation of the EGF receptor in signalling by G-protein-coupled receptors. Nature 379:557-560
106. **Prenzel N, Zwick E, Daub H, Leserer M, Abraham R, Wallasch C, Ullrich A** 1999 EGF receptor transactivation by G-protein-coupled receptors requires metalloproteinase cleavage of proHB-EGF. Nature 402:884-888
107. **Li J, Lin ML, Wiepz GJ, Guadarrama AG, Bertics PJ** 1999 Integrin-mediated migration of murine B82L fibroblasts is dependent on the expression of an intact epidermal growth factor receptor. J Biol Chem 274:11209-11219
108. **Li J, Kim Y-N, Bertics PJ** 2000 Platelet-derived growth factor-stimulated migration of murine fibroblasts is associated with epidermal growth factor receptor expression and tyrosine phosphorylation. J Biol Chem 275:2951-2958
109. **Adamson ED** 1990 EGF receptor activities in mammalian development. Mol Reprod Dev 27:16-22
110. **Miettinen PJ, Berger JE, Meneses J, Phung Y, Pedersen RA, Werb Z, Derynck R** 1995 Epithelial immaturity and multiorgan failure in mice lacking epidermal growth factor receptor. Nature 376:337-341
111. **Threadgill DW, Dlugosz AA, Hansen LA, Tennenbaum T, Lichti U, Yee D, LaMantia C, Mourton T, Herrup K, Harris RC, Barnard JA, Yuspa SH, Coffey RJ, Magnuson T** 1995 Targeted disruption of mouse EGF receptor: effects of genetic background on mutant phenotype. Science 269:230-234
112. **Sibilia M, Wagner EF** 1995 Strain dependent epithelial defects in mice lacking the EGF receptor. Science 269:234-238

113. **Xie W, Paterson AJ, Chin E, Nabell LM, Kudlow JE** 1998 Targeted expression of a dominant negative epidermal growth factor receptor in the mammary gland of transgenic mice inhibits pubertal mammary duct development. Mol Endocrinol 11:1766-1781
114. **Okamoto S, Oka T** 1984 Evidence for physiological function of epidermal growth factor: pregestational sialoadenectomy of mice decreases milk production and increases offspring mortality during lactation period. Proc Natl Acad Sci (USA) 81:6059-6063
115. **Downward J, Yarden Y, Mayes ELV, Scrace G, Totty N, Stockwell P, Ullrich A, Schlessinger J, Waterfield MD** 1984 Close similarity of the epidermal growth factor receptor and *v-erb-B* oncogene protein sequences. Nature 307:521-527
116. **Aaronson SA** 1991 Growth factors and cancer. Science 254:1146-1153
117. **Kim H, Turner T, Kassis J, Souto J, Wells A** 1999 EGF receptor signaling in prostate development. Histol Histopathol 14:1175-1182
118. **Masui H, Wells A, Lazar CS, Rosenfeld MG, Gill GN** 1991 Enhanced tumorigenesis of NR6 cells which express non-downregulating epidermal growth factor receptors. Cancer Res 51:6170-6175
119. **Simcha I, Geiger B, Yehuda-Levenberg S, Salomon D, Ben-Ze'ev A** 1996 Suppression of tumorigenicity by plakoglobin: an augmenting effect of N-cadherin. J Cell Biol 133:199-209
120. **DiFiore PP, Pierce JH, Fleming TP, Hazan R, Ullrich A, King CR, Schlessinger J, Aaronson SA** 1987 Overexpression of the human EGF receptor confers an EGF-dependent tranformed phenotype to NIH 3T3 cells. Cell 51:1063-1070
121. **Matsui Y, Halter SA, Holt JT, Hogan BL, Coffey RJ** 1990 Development of mammary hyperplasia and neoplasia in MMTV-TGF alpha transgenic mice. Cell 61:1147-1155
122. **Sandgren EP, Luetteke NC, Palmiter RD, Brinster RL, Lee DC** 1990 Overexpression of TGFα in transgenic mice: induction of epithelial hyperplasia, pancreatic metaplasia, and carcinoma of the breast. Cell 61:1121-1135
123. **Jhappan C, Stahle C, Harkins RN, Fausto N, Smith GH, Merlino GT** 1990 TGFα overexpression in transgenic mice induces liver neoplasia and abnormal development of the mammary gland and pancreas. Cell 61:1137-1146
124. **Voldborg BR, Damstrup L, Spang-Thomsen M, Poulsen HS** 1997 Epidermal growth factor receptor (EGFR) and EGFR mutations, functions and possible role in clinical trials. Ann Oncol 8:1197-1206
125. **Libermann TA, Nusbaum HR, Razon N, Kris R, Lax I, Soreq H, Whittle N, Waterfield MD, Ullrich A, Schlessinger J** 1985 Amplification, enhanced expression and possible rearrangement of EGF receptor gene in primary human brain tumours of glial origin. Nature 313:144-147
126. **Neal DE, Marsh C, Bennett MK, Abel PD, Hall RR, Sainsbury JRC, Harris AL** 1985 Epidermal-growth-factor receptors in human bladder cancer: comparison of invasive and superficial tumours. Lancet 1(8425):366-368
127. **Yasui W, Sumiyoshi H, Hata J, Kameda T, Ochiai A, Ito H, Tahara E** 1988 Expression of epidermal growth factor receptor in human gastric and colon carcinomas. Cancer Res 48:137-141
128. **Kassis J, Lauffenburger DA, Turner T, Wells A** 2001 Tumor invasion as dysregulated cell motility. Seminars in Cancer Biology 11:105-117
129. **Khoshyomn S, Penar PL, Rossi J, Wells A, Abramson DL, Bhushan A** 1999 Inhibition of PLC gamma-1 activation blocks glioma cell motility and invasion of fetal rat brain aggregates. Neurosurgery 44:568-577
130. **Turner T, Chen P, Goodly LJ, Wells A** 1996 EGF receptor signaling enhances *in vivo* invasiveness of DU-145 human prostate carcinoma cells. Clin Exp Metastas 14:409-418

131. **Turner T, VanEpps-Fung M, Kassis J, Wells A** 1997 Molecular inhibition of PLCγ signaling abrogates DU-145 prostate tumor cell invasion. Clin Cancer Res 3:2275-2282
132. **Gilmore T, DeClue JE, Martin GS** 1985 Protein phosphorylation at tyrosine is induced by the *v-erbB* gene product *in vivo* and *in vitro*. Cell 40:609-618
133. **Bargmann CI, Hung MC, Weinberg RA** 1986 Multiple independent activations of the *neu* oncogene by a point mutation altering the transmembrane domain of p185. Cell 45:649-657
134. **Slamon DJ, Clark GM, Wong SG, Levin WJ, Ullrich A, McGuire WL** 1987 Human breast cancer: correlation of relapse and survival with amplification of the HER-2/*neu* oncogene. Science 235:177-182
135. **Drebin JA, Link VC, Weinberg RA, Greene MI** 1986 Inhibition of tumor growth by a monoclonal antibody reactive with an oncogene-encoded tumor antigen. Proc Natl Acad Sci (USA) 83:9129-9133

5

Insulin and Insulin-Like Growth Factor-1 Receptors and Signaling Pathways: Similarities and Differences

Derek LeRoith[1*], Michael J. Quon[2], Yehiel Zick[3]

[1]Clinical Endocrinology Branch, NIDDK, [2]Cardiology Branch, NHLBI, NIH, Bethesda MD, 20892-1758 and [3]The Weizmann Institute of Science, Rehovot, Israel.

INTRODUCTION

Insulin and the insulin-like growth factors (IGFs) play critical roles in normal physiology. Pathological perturbations in these hormone systems result in serious medical consequences. Insulin is considered to be primarily a "metabolic" hormone with important effects on the normal homeostasis of glucose, fat and protein. If insulin production or insulin action at target tissues are impaired, diabetes and numerous concomitant conditions of insulin resistance including hypertension and hypertriglyceridemia will ensue. The IGFs, on the other hand, primarily control growth and development of the organism. IGFs control growth by facilitating cellular proliferation in terminally differentiated cells and by inhibiting apoptosis. In adult tissues, these growth factors also have various tissue-specific functions. While there are clear differences in the physiological roles of insulin and the IGFs, these hormones/growth factors also share many similarities. The hormones have similar tertiary structures, and they share ~50% amino acid identity. Both insulin and IGFs interact with specific cell surface receptors that represent one sub-class of the tyrosine kinase receptor family. The signaling cascades

* derek@helix.nih.gov

induced by activation of the insulin receptor (IR) and IGF-I receptor (IGF-IR) are highly overlapping, with some subtle differences, as will be discussed in more detail below.

This review will summarize the current state of knowledge of the signaling pathways involved in both insulin and IGF action. We will also compare and contrast the structures of the IR and IGF-IR. Finally, we will address the differences in biological action between these factors by analyzing certain signaling pathways that are apparently specific to each hormone.

STRUCTURE OF INSULIN AND IGF-I RECEPTORS

The IR, the IGF-IR, and the insulin-receptor-related receptor (IRR) belong to a homologous family of receptor tyrosine kinases (RTKs). Each of these receptor family members are produced by proteolytic cleavage of a high molecular-mass polypeptide chain precursor, and then processed to produce two chains, termed α and β, linked by disulfide bonds to form biologically active $\alpha_2\beta_2$ heterotetrameric receptor complexes. Insulin and IGF-I each bind with a high affinity to their cognate receptors, and with a much lower affinity to each other's receptors. However, the IRR is an "orphan receptor" with no known ligand as yet identified. Overall, the IGF-IR shares extensive similarity with the IR at the level of precursor and subunit size, as well as at the level of structural topology and primary sequence. Nevertheless, these two receptors differ from each other in several fundamental features. These include their extracellular ligand-binding domains as well as specific intracellular regions that confer upon these receptors their unique physiological functions. Some manifestations of these differences include certain distinct substrate specificities of the two receptors.

The molecular organization of the IR and IGF-IR was largely elucidated upon cloning of their cDNA and genes [1,2]. More recently, further structural detail has become available as a result of X-ray crystallography of various receptor domains [3-5]. Model structures, based on these observations were consistent with previous biochemical and immunological studies [6]. The IR is a product of a single gene containing 22 exons, and encodes a protein of 1355 amino acids, with a predicted Mr of 153,917 [2]. Exons 1-11 encode the entire α subunit and exons 12-22 encode the β subunit. Alternative splicing of exon 11 occurs in hematopoietic tissues, whereas the longer transcript predominates in liver, muscle and adipose tissues. Similarly, the IGF-IR gene codes for a precursor protein of 1337 residues which generates 135 kDa (α) and 90 kDa (β) fully glycosylated receptor subunits [1]. The IGF-IR gene also contains a similar number of exons.

Functional Autonomy of the Receptor Subunits

Although the IR and IGF-IR are covalently-linked heterotetramers, the extracellular and intracellular domains of these proteins function independently. This was demonstrated by the finding that truncated forms of the IR and IGF-IR that express only their extracellular domains effectively bind ligands [7,8], whereas overexpression of isolated intracellular regions of the receptor kinase yields active enzymes. Furthermore, chimeric receptors comprising the extracellular regions of the IR, linked to the transmembrane and intracellular domains of the epidermal growth factor (EGF) receptor enable insulin to stimulate tyrosine kinase activity of the EGF receptor. Taken together, these findings indicate that intact structural domains of RTKs preserve their functional autonomy irrespective of their new molecular context. The data further suggest that RTKs have a common evolutionary origin, and presumably evolved from a gene coding for a ligand-binding protein that fused to a gene coding for a tyrosine kinase.

The IR and IGF-I receptors as glycoproteins

The IR and IGF-IR are glycoproteins whose extracellular domains contain N-linked and O-linked sugar moieties, attached to the protein through asparagine or serine residues. Both the α and β subunits are heavily glycosylated. This is reflected in the apparent molecular weights of the IR and IGF-IR on SDS-PAGE, which are >50% higher than those predicted from the amino acid sequences. The function of these sugar moieties has not been fully elucidated, although several studies indicate that mutations of amino acids that serve as potential sugar-binding sites alters IR kinase activity [9,10].

Topology and Domain Structure of the IR and IGF-IR

The extracellular domain of the IR and IGF-IR

The extracellular portions of the IR and IGF-IR consist of the entire α-subunit and about one third of the β-subunit. These regions of the receptor form the ligand binding domain. The ligands, IGF-I and insulin, share a common architecture and competitively cross-react with both receptors. It is therefore likely that the ligands bind to the receptors in a structurally equivalent fashion.

The α-subunit has been shown to be the primary ligand-binding domain. Studies of naturally occurring mutations in the IR together with studies employing domain-swapping and alanine scanning mutagenesis of the IR and IGF-IR have delineated regions of the α-subunits which contribute to ligand binding. Chimeric receptors derived from the IGF-IR and IR have also been used to identify specific regions of the receptors that account for ligand specificity. These chimeras have also provided an efficient means of identifying some parts of the hormone-binding site. For example, replacing residues 1–62 of IGF-IR with 1–68 of IR confers insulin-binding ability on IGF-IR, whereas substitution of IR 198–300 with IGF-IR 191–290 allows this receptor to bind IGF-I [11].

Characterization of the crystal structure of the first three domains of the IGF-IR (residues 1-459) [3] have further improved our understanding of the ligand-binding site. These domains, which are termed L1, cysteine-rich (Cys-rich), and L2, share structural similarities with other growth factor receptors. While, these regions cannot bind ligand on their own, high affinity ligand binding to the IR (1-468) or to the IGF-IR (1-459) can be restored by the addition of residues 704-719 of the IR or residues 691-706 of the IGF-IR. Each of these regions encodes a predicted α-helical domain from the carboxyl-terminus of the α subunits [7,8]. These results suggest that the carboxyl-terminal region of the α subunits of the IR and IGF-IR are interchangeable, implying that this domain is part of the common ligand binding site.

The 1-459 fragment of IGF-IR adopts an extended bilobed structure with the L domains at either end, connected by a rod-shaped Cys-rich domain. The L domains (residues 1–150 and 300–460) consist of a single-stranded right-handed β-helix, capped on the ends by short helices and disulfide bonds. Facing the central space formed by the L1- Cys-rich- L2 domains is a surface of the L1 domain containing a patch of residues that form a putative hormone binding "footprint" based on alanine scanning mutagenesis [3]. The Cys-rich domain consists of 16.5% cysteine residues and 72% hydrophilic residues, which suggests that this region is embedded in an hydrophilic environment. This region is probably rigid, as some of these Cys residues form intramolecular or intermolecular disulfide bridges. There is only 48% homology between the Cys-rich domains of the IR and IGF-1R, but there is a complete conservation of the Cys residue spacing between these two receptors.

Little is known about the structure of the rest of the extracellular region, but three repeats of a fibronectin type-III domain (FnIII) have been identified. The first FnIII domain can be deleted without seriously compromising insulin binding. However, deletion of residues 450-601 results in a blunted tyrosine autophosphorylation response, suggesting that this domain may be involved in the transmission of the insulin binding signal to the tyrosine kinase.

The crystal structure of the N-terminal part of IGF-IR has shown that there is considerable divergence in the sequences in the L1 and L2 domains of the IR and IGF-IR in the ligand-binding domains of these receptors. This suggests that these regions may define ligand specificity within a common, compact, structural unit formed by a network of disulfide bridges. Ligand binding involves interaction with both α subunits within the native receptor heterotetramer, adding further complexity to the ligand-binding site. At least two separate regions in the α-chain of the IR, residues 1-120 in domain L1 and 704-716 in the insert region of the FnIII1 domain, are essential for insulin binding [12,13]. In contrast, the Cys rich domain of IGF-IR is critical for IGF binding [11]. These findings suggest a molecular basis for the different ligand-binding specificities of the IR and IGF-IR.

The transmembrane domain

The transmembrane-domain of the β-subunits is an α-helix with seven turns. This region of the receptor is thought to function primarily as passive lipid anchor. The IR β-subunit contains 23 hydrophobic amino acids (with 22 in the IGF-IR), flanked at the carboxyl-terminal end by a short sequence of basic residues (Arg-Lys-Arg). This feature of the primary sequence of the cytoplasmic region is shared by many membrane proteins, including the EGF (Arg-Arg-Arg) and low density lipoprotein (LDL) (Lys-Asn-Trp-Arg-Leu-Lys) receptors. In addition to their potential interactions with the head groups of negatively charged phospholipids, these basic residues may represent a stop-transfer signal anchoring the insulin and IGF-IR in the membrane during biosynthesis.

The intracellular cytoplasmic domain

The intracellular regions of the IR and IGF-IR contain a protein tyrosine kinase domain, the hallmark feature of the RTK family. The cytoplasmic domain of the IR and IGF-IR includes the carboxyl-terminal 403 and 407 amino acids, respectively, of their β-subunits. These domains share a high degree of homology not only between the IR and IGF-IR, but also with other growth factor receptors, like the EGF receptor. The highest homology (84%) is within the tyrosine kinase catalytic domain that includes the ATP and substrate binding sites. The intracellular region can be divided into several sub-domains, as detailed below.

i. The juxtamembrane region. These ~50 amino acid regions share 61% homology between the IR and IGF-IR. These domains also include at least one autophosphorylation site (Tyr^{972} in the IR and Tyr^{950} in the IGF-IR) which resides in an L-X_4-NPXYXSXSD motif. Replacement of this Tyr with Phe or Ala impairs receptor signal transmission, and abolishes both the metabolic and growth-promoting effects of the IR and IGF-IR. This occurs in the presence of normal autophosphorylation in other regions of the receptor despite the fact that the kinase is fully active *in vitro* [14,15]. The functional effects of these mutations appears to be due to an inability of the mutant receptors to interact with receptor substrates such as insulin receptor substrate-1 (IRS-1) and Crk-II [16,17]. Indeed, overexpression of IRS-1 can rescue certain biological functions in cells overexpressing this mutant IR ($Tyr^{972} \rightarrow Phe^{972}$). These findings suggest that the juxtamembrane regions of the IR and IGF-IR are involved in the interactions of these receptors with substrate proteins. Further studies revealed that the phosphotyrosine binding (PTB) domain within IRS and Shc proteins serves as the binding domain that associates these proteins with the juxtamembrane region of IR [18,19]. The juxtamembrane domain also contains several motifs implicated in regulating IR and IGF-IR internalization. These include the NPXY, GPLY and the di-Leu motifs [20].

ii. The tyrosine kinase domain. The kinase regions of IR and IGF-IR share 84% homology. Crystal structure of the kinase region of IR, both in its basal [5] and activated [4] states revealed that it is composed of two lobes with a single connection between them, similar to the kinase core of Ser/Thr kinases. The N-terminal lobe constitutes the ATP-binding site. The C-terminal lobe contains the active site (catalytic loop), the three autophosphorylation sites (activation loop) and the kinase-insert region.

iii. The carboxyl-terminus domain. The carboxyl-terminal (CT) regions of IR and IGF-IR share only limited (44%) homology [1]. They contain two (IR) and three (IGF-IR) autophosphorylation sites, whose role in receptor signaling is still unresolved [21-23]. Mutations of these Tyr residues augments insulin-dependent activation of mitogen activated protein kinase (MAP kinase) and phosphoinositide 3'-kinase (PI3K), suggesting that these tyrosine residues negatively regulate the growth-promoting effects of insulin [24]. Similarly, cells expressing a receptor mutant in which the carboxyl-terminal 43 amino acids (including these Tyr residues) were deleted ($IR_{\Delta}43$) exhibit impaired metabolic effects [25,26] and impaired induction of *c-fos*, but have augmented mitogenic signaling [25,26]. Deletion of 108 amino acids from the CT region of IGF-IR inhibits its tumorigenic activity [27], while replacement of the IR CT domain with that of the IGF-1R, severely affects insulin-stimulated responses [28]. These, and additional findings, indicate that the equivalent CT domains of IR and the IGF-IR are not interchangeable, and that within the carboxyl-terminus, specific regions have important influences on various functions of IR and IGF-IR [22,29]. These differences could account, at least in part, for the different biological effects of the two receptors.

SIGNALING PATHWAYS

Insulin and IGF-1 initiate their biological actions upon binding to specific cell surface receptors, as outlined above. Signaling pathways from these receptors regulate a diverse array of functions, including glucose transport, mitogenesis, and various enzymatic pathways. Below, we give an overview of how insulin and IGF-1 receptor signaling follow a general paradigm for RTK signal transduction.

Ligand Binding

RTKs are activated when ligand binds to the extracellular portion of the cell surface receptor. In the case of monomeric receptors (e.g., EGF and PDGF receptors – see chapter 4), ligand binding results in receptor dimerization, a necessary first step in RTK activation [30,31]. Certain RTKs, including the IR and IGF-IR exist in a pre-dimerized form even in the absence of ligand. A rapid conformational change in the receptor occurs when insulin binds to specific regions of the α-subunit that results in activation of the tyrosine kinase domain [32].

Autophosphorylation and Activation of Tyrosine Kinase

The kinase regions of various RTKs share homology within both the catalytic domain and the ATP binding site [31]. When the receptor is activated, the kinase domain in one half of the receptor dimer phosphorylates cytoplasmic tyrosine residues in the activation loop of the other half of the receptor dimer. This trans-autophosphorylation results in a large increase in kinase activity of the receptor. In the basal state, the activation loop of the insulin receptor occludes the catalytic site so that access to ATP and substrates are blocked. Upon autophosphorylation, this region swings out of the catalytic site to give unrestricted access to ATP and substrates [4].

Proximal Signaling Events

The autophosphorylated receptor can engage other signaling molecules *via* interactions between phosphotyrosine motifs on the receptor and src homology-2 (SH2) domains located on downstream signaling molecules, as shown in Fig 1. SH2 domains are protein regions that share homology with a noncatalytic region of the src proto-oncogene product (see chapter 1). Many molecules that mediate RTK signaling contain SH2 domains, including PI3K, growth factor receptor bound protein 2 (GRB-2), SH2-containing phosphatase-2 (SHP-2),

GTPase activating protein (GAP), and phospholipase C-γ (PLC-γ), among others. Specific SH2 binding motifs are defined by the three amino acid residues on the C-terminal side of the phosphotyrosine residue. These motifs provide specificity for interaction with particular SH2 domains [33]. The phosphotyrosine sites that engage specific SH2 domains of various signaling molecules have been well mapped for the EGF and PDGF receptors.

Although the autophosphorylated IR can also directly interact with SH2 domains in molecules such as PI3K, SHP-2 and GAP [34], these interactions are not critical for insulin and IGF-1 signaling. Instead, a number of insulin receptor substrate proteins, such as IRS-1, IRS-2, IRS-3, IRS-4, Shc, and GAB-1 are essential to functionally couple IR and IGF-IR signaling with downstream SH2-domain containing signaling molecules [35,36].

The various members of the IRS family share several common features, including amino-terminal pleckstrin homology (PH) domains and PTB domains that are important for mediating interactions with the insulin receptor [36]. In particular, the PTB domain binds to the phosphorylated Tyr^{972} in the NPEY motif of the juxtamembrane region of the insulin receptor β-subunit [36]. The carboxyl-terminal portion of IRS proteins contain multiple tyrosine- containing motifs that undergo phosphorylation by the IR and IGF-IR and serve as docking sites for SH2-domain containing proteins. These phosphotyrosine motifs are highly conserved between the various IRS proteins. For example, multiple YXXM motifs that bind to the SH2 domains in the p85 regulatory subunit of PI3K are present in all IRS family members. Thus, phosphorylated IRS proteins can form signaling complexes consisting of several SH2-domain containing molecules.

Recently, important differences between the various IRS family members have begun to be delineated, suggesting that these molecules do have some functional specificity. For example, IRS-3 is approximately 50% shorter than IRS-1 and IRS-2 and, unlike IRS-1 and IRS-2, IRS-3 does not contain a phosphotyrosine motif predicted to bind to the SH2 domain of Grb-2 [33]. Furthermore, the phenotype of transgenic mice homozygous for null alleles of IRS-1 is distinct from the phenotype of IRS-2 knockout mice [37,38]. IRS-1 knockout mice are mildly insulin resistant but do not develop diabetes. In contrast, IRS-2 knockout mice display both insulin resistance and severe pancreatic β-cell defects leading to abnormal insulin secretion and the development of diabetes. Thus, the various IRS family members most likely have some overlapping functions but are clearly not completely interchangeable.

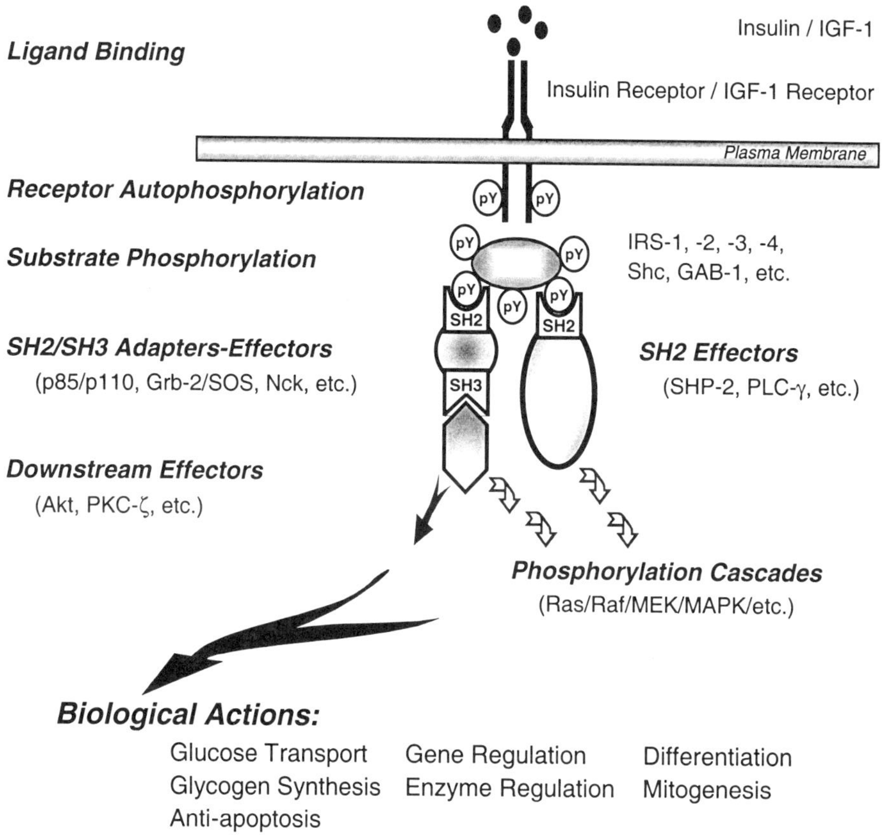

Figure 1. Ligand binding to the IR and IGF-IR results in autophosphorylation on tyrosine residues of both receptors. SH2 domain-containing proteins such as the IRS family members, Shc, and the regulatory subunit (p85) of PI3K bind to these phosphotyrosine residues. This leads to regulation of various downstream effectors. The downstream effectors can include SH3-domain containing proteins and various Ser/Thr protein kinase signaling systems. These signaling cascades lead to various physiological outcomes, depending on the cell and tissue type.

Signaling Proteins with SH2 and SH3 Domains

Many signaling molecules downstream of RTKs and their substrates contain SH2 and/or SH3 domains that mediate protein-protein interactions. As mentioned above, SH2 domains interact specifically with phosphorylated tyrosine motifs. SH3 domains bind with high affinity to particular proline rich sequences. Some SH2 domain containing proteins (e.g., SHP-2, PLC-γ) are effector molecules that possess intrinsic catalytic activity that is regulated by interactions of the SH2 domain with phosphotyrosine motifs on other proteins (e.g., IRS-1). These SH2 domains facilitate localization of signaling molecules to particular compartments within the cell. Other SH2/SH3 domain containing proteins (e.g., Grb-2, Nck, and the p85 regulatory subunit of PI3K) function strictly as adapter proteins, and have no intrinsic catalytic activity. These adapters mediate the formation of specific signaling complexes *via* simultaneous interactions of multiple SH2/SH3 domains on the adapter protein with both upstream and downstream signaling molecules. Two major effectors of pathways that are activated by a number of RTKs, including the IR and IGF-IR, are Ras and PI3K. For example, Grb-2 is pre-bound to the guanine nucleotide exchange factor SOS (two SH3 domains of Grb-2 bind proline rich regions of SOS). When phosphotyrosine motifs on IRS-1 or Shc bind to the SH2 domain of Grb-2, the pre-bound SOS catalyzes the exchange of GTP for GDP on Ras leading to its activation. Similarly, the p85 regulatory subunit of PI3K is pre-associated with the p110 catalytic subunit. Upon insulin stimulation, phosphorylated YXXM motifs on IRS proteins engage the SH2 domains of p85 leading to activation of p110 (for reviews, see [23,39]).

Distal Phosphorylation Cascades

Distal RTK signaling pathways are difficult to dissect because numerous branching pathways emerge from single effectors. In addition, multiple upstream inputs often converge upon single branch points. Furthermore, the presence of negative feedback mechanisms adds to the complexity of signal transduction. However, some distal signaling mechanisms such as downstream phosphorylation cascades are common to many growth factor receptors, including the IR and IGF-IR. For example, upon stimulation of virtually all growth factor signaling pathways, the Ras GTPase directly activates Raf, a serine/threonine kinase that phosphorylates and activates MEK or MAP kinase kinase. MEK, in turn, phosphorylates MAP kinase, which can then phosphorylate transcription factors such as Elk-1, thereby leading to induction of various immediate early genes such as the protooncogenes c-jun and c-fos. Insulin signaling pathways downstream from PI3K also include serine/threonine phosphorylation cascades. Phospholipid products generated by PI3K activate phosphoinositide dependent kinase-1 (PDK1) by binding to its PH domain. Upon activation of PDK-1, Thr^{308} and Ser^{473} of Akt (another serine/threonine kinase) become phosphorylated, contributing to activation of Akt. Akt, in turn, phosphorylates and inactivates glycogen synthase kinase-3 (GSK-3), thereby stimulating glycogen synthase [40]. Interestingly, some downstream kinases in insulin signaling pathways (e.g., Akt, GSK-3, and PKCs) can also phosphorylate upstream signaling components such as IRS-1 on serine residues [41]. Although serine phosphorylation of IRS-1 is generally associated with impairment of its function [42], phosphorylation of IRS-1 by Akt may enhance signaling by IRS-1 [41]. Thus, complex positive and negative feedback systems can regulate insulin and IGF-1 signaling.

Protein Tyrosine Phosphatases

Another aspect of regulation common to RTK signaling is the dephosphorylation of RTKs and their substrates by protein tyrosine phosphatases (PTPases – see chapter 2). The number and diversity of PTPases rivals that of the RTKs [43]. The PTPases are classified into two broad categories: cytoplasmic proteins containing a single catalytic PTPase domain and transmembrane "receptor-like" PTPases that typically contain tandem PTPase domains. Cytoplasmic PTPases such as SHP-2 contain SH2 domains that confer specificity while the receptor-like PTPases have extracellular domains that presumably interact with specific ligands.

The mechanisms by which the various PTPases selectively dephosphorylate specific substrates is not completely understood. However, it is known that certain PTPases are selective for specific RTKs [44]. The transmembrane PTPases, PTP-α, PTP-ε and LAR (see chapter 2) have all been implicated as modulators of insulin action [45]. In particular, LAR interacts with and dephosphorylates the insulin receptor in intact cells. Among the nontransmembrane PTPases, PTP1B and SHP-2 both modulate insulin signaling. PTP1B dephosphorylates the insulin receptor both *in vitro* and in intact cells and regulates both mitogenic and metabolic actions of insulin [45,46]. PTP1B knockout mice are more sensitive to insulin and resistant to obesity, as compared to wild-type littermates [47]. Binding of the SH2 domains of SHP-2 to phosphotyrosine motifs on either the IR or on IRS-1 results in activation of SHP-2 phosphatase activity. A number of studies have shown that SHP-2 participates in Ras and MAP kinase dependent pathways as a positive mediator of mitogenic actions of insulin and other growth factors [48]. Studies showing that expression of dominant negative SHP-2 in transgenic mice causes insulin resistance are also consistent with a role for SHP-2 as a positive mediator of insulin action [49].

SPECIFICITY

One of the more perplexing aspects of insulin and IGF-I signaling is understanding how these apparently similar pathways can lead to divergent biological responses. In this section we attempt to deal with this by bringing examples involving specific effects of insulin or IGF-I.

The plasma membrane glycoprotein pp120, also referred to as C-CAM or Caecam 1, is an ecto-ATPase protein [50] that has multiple functions. Of particular interest is the role played by pp120 in the actions of insulin. pp120 is a substrate of the IR that becomes tyrosine phosphorylated upon activation of the receptor. Furthermore, pp120 plays a role in the endocytosis of the insulin/insulin receptor complex [51]. Apparently, pp120-induced internalization of the insulin receptor complex reduces the mitogenic response to insulin in certain cells, and conversely, in the absence of pp120, mitogenic responses to insulin may be artificially enhanced [52]. The tyrosine phosphorylation of pp120 is essential for this function as demonstrated by mutant forms of pp120 that failed to phosphorylate and function on insulin receptor endocytosis. Interestingly, pp120 is a substrate for the activated IR tyrosine kinase, but not of the IGF-IR. On the other hand, substitution of the carboxyl-terminus of the IGF-IR by the equivalent region of the IR results in a chimeric receptor that can both phosphorylate pp120 and be internalized, two functions which are absent in the wild-type IGF-IR [51]. At the most distal region of the IR are two tyrosine residues, Tyr^{1316} and Tyr^{1322} that are autophosphorylated by the activated IR. Only Tyr^{1322} is conserved in the IGF-IR. Interestingly, substitution of Tyr^{1316} in the IR abrogates the tyrosine phosphorylation of pp120 and its ability to internalize the IR [50]. Thus, pp120 appears to be an important endogenous substrate that is specific for the IR.

The carboxyl-terminus is the region that is the least conserved between the two receptors. Tyr^{1250} and Tyr^{1251} in the IGF-IR are not present in the IR, and these residues are important for the mitogenic response of the IGF-IR [53]. Furthermore, deletion of the carboxyl-terminus of the IGF-IR markedly inhibits the ability of the receptor to transform cells [54].

To further investigate the specificity of the IR and IGF-IR, Siddle and co-workers performed a number of elegant studies where they utilized chimeric receptors [55]. These chimeras consisted of an extracellular portion of the neurotrophin receptor TrkC fused to the intracellular portion of either the IR or the IGF-IR and were stably transfected into 3T3-L1 adipocytes. Since these cells do not express endogenous TrkC receptors, stimulation of the cells with neurotrophin-3 specifically activates the recombinant RTK, thereby eliminating potential complications due to regulation of endogenous receptors. These studies revealed that activation of the TrkC-IR tyrosine kinase resulted in a higher level of IRS-1 phosphorylation, a significantly greater association of IRS-1 with PI3K, and an increase in translocation of GLUT4, compared with the cells expressing the TrkC-IGF-IR chimera. In contrast, cells expressing the TrkC-IGF-IR chimera displayed a higher level of Shc phosphorylation, Shc-Grb2 association, and activation of MAP kinase, as compared to cells expressing TrkC-IR. These findings are consistent with those from previous studies suggesting that the IR plays a stronger role in stimulating metabolic effects in these cells, and the IGF-IR has a greater effect on mitogenic pathways [55]. Similarly, Accili and coworkers demonstrated that glycogen synthesis is more effectively stimulated *via* the IR than the IGF-IR in hepatocytes and 3T3-L1 adipocytes [56]. The precise molecular basis of the different functional specificities between the IR and IGF-IR remains speculative, although a number of possibilities have been put forward. One possibility is that the divergent carboxyl-terminal domains of the two receptors may interact with distinct sets of substrate proteins [28,57,58]. Another possibility is that differences in endocytosis and subsequent cellular trafficking of the IR and IGF-IR may lead to specific cellular localizations of the two receptor subtypes [59,60].

Signaling *via* the Jak/Stat pathway and regulation by SOCS

The JAK (Janus kinase) and Stat (signal transduction activator of transcription) pathways have been primarily described as the major signaling pathways used by the cytokine family of receptors (see Chapter 6). However, the IR and IGF-IR can also activate these pathways. Interestingly, these two receptors utilize different subsets of substrates, suggesting that these systems confer some specificity to IR and IGF-IR signaling. Using the yeast two-hybrid system, Van Obberghen and coworkers showed that Stat5B binds to the IR at tyrosine 972 *via* an SH2 domain, and that the IR induces tyrosine phosphorylation of Stat5B and thereby stimulates gene transcription [61]. Furthermore, SOCS-3 (suppressor of cytokine signaling-3) inhibits this process by directly interacting with the IR and competing with Stat5B binding at Tyr^{972}. In contrast, the IGF-IR phosphorylates JAK1, JAK2, and Stat3 [62]. In the case of the IGF-IR, SOCS-1 and SOCS-2 bind the receptor [63] and, when activated, these proteins inhibit the activation of Stat3, *via* inhibition of JAK activity [62]. Thus, while there is a certain degree of overlap in the use of the JAK/Stat system by the IR and IGF-IR, there also appears to be considerable specificity of these two receptors for regulating the JAK/Stat system.

Differential regulation of other downstream targets by the IR and IGF-IR

Both insulin and IGF-I induce the expression of vascular endothelial growth factor (VEGF) [64,65]. Intriguingly, the signaling pathways involved in stimulating VEGF gene expression by insulin and IGF-I are different [66]. In

NIH-3T3 cells that overexpress the IR, VEGF gene expression is induced by exposure to 100 nM insulin. This effect is mediated *via* the PI3K pathway, as it is blocked by wortmannin, an inhibitor of PI3K. In contrast, wortmannin had no effect on the stimulation of VEGF gene expression induced by 10 nM IGF-I in NIH-3T3 cells that overexpress IGF-IR [66]. IGF-I receptor-induced expression of VEGF was inhibited by pretreatment with PD 98059, a specific inhibitor of the MEK/MAPK signaling cascade. However, PD 98059 had no effect on insulin-induced VEGF expression.

Another example of the differential roles of insulin and IGF-I signaling has been shown in murine skin keratinocytes [67]. These cells express both the IR and IGF-IR, which remain unchanged during differentiation. Insulin stimulation induces the expression of classic differentiation markers, such as keratin 1 and 10, whereas IGF-I stimulation inhibited expression of these markers [67]. These results suggest that the IR was causing differentiation of keratinocytes, whereas activation of the IGF-IR was more mitogenic in these cells.

Activation of the IR and IGF-IR can also stimulate c-Jun NH2-terminal kinase (JNK) [68,69]. In Rat-1 fibroblasts, which express high levels of the IR, insulin activates JNK *via* a Ras-mediated pathway that requires the protein-tyrosine phosphatase SHP-2. Expression of a dominant negative form of Ha-Ras (RasN17) blocked this effect, whereas LY294002, a specific inhibitor of PI3K pathway, failed to do so [69]. In contrast, in 293 human embryonic kidney cells, activation of JNK by anisomycin or TNF-α was inhibited by IGF-I. This effect was blocked in the presence of LY294002, but not by the specific MEK inhibitor PD98059. Expression of the dominant negative form of Akt/PKB, a protein kinase downstream of PI3K, also inhibited the IGF-I suppression of JNK [68].

Finally, activation of the IR causes nuclear export of the forkhead transcription factor, FKHR, resulting in a suppression of IGF-binding protein–1 gene expression. This effect is mediated by activation of Akt, which phosphorylates FKHR on Ser^{253}. This is followed by phosphorylation of FKHR at Thr^{24} and Ser^{316} by Akt. In contrast, IGF-I fails to phosphorylate Thr^{24} on FKHR. These differences may help to further explain the differential effects of insulin and IGF-I on gene expression [70].

SUMMARY

Since binding of insulin and IGF-I to their respective receptors triggers distinct cellular responses, the IR and IGF-IR must differ not only in the structure of their ligand-binding pockets, but also in intracellular processes that regulate signal transduction. The structural features of these two receptors, as outlined in this chapter, support this idea. Characterization of the crystal structure of the α and β subunits will help to define the common and unique structural characteristics of the IR and IGF-IR. Delineation of the intracellular signaling pathways targeted by these two receptors has begun to suggest a cellular basis for the selectivity of biological responses mediated by the IR and IGF-IR. Additional studies in this area should further improve our understanding of how insulin and IGF-I regulate their respective metabolic and mitogenic pathways.

REFERENCES

1. **Ullrich, A., Gray, A., Tam, A.W., Yang-Feng, T., Tsubokawa, M., Collins, C., Henzel, W., Le Bon, T., Kathuria, S., and Chen, E.** (1986) Insulin-like growth factor I receptor primary structure: comparison with insulin receptor suggests structural determinants that define functional specificity. Embo J 5, 2503-2512.
2. **Ullrich, A., Bell, J.R., Chen, E.Y., Herrera, R., Petruzzelli, L.M., Dull, T.J., Gray, A., Coussens, L., Liao, Y.C., and Tsubokawa, M.** (1985) Human insulin receptor and its relationship to the tyrosine kinase family of oncogenes. Nature 313, 756-761.
3. **Garrett, T.P., McKern, N.M., Lou, M., Frenkel, M.J., Bentley, J.D., Lovrecz, G.O., Elleman, T.C., Cosgrove, L.J. and Ward, C.W.** (1998) Crystal structure of the first three domains of the type-1 insulin-like growth factor receptor. Nature 394, 395-399.
4. **Hubbard, S.R.** (1997) Crystal structure of the activated insulin receptor tyrosine kinase in complex with peptide substrate and ATP analog. Embo J 16, 5572-5581.
5. **Hubbard, S.R., Wei, L., Ellis, L. and Hendrickson, W.A.** (1994) Crystal structure of the tyrosine kinase domain of the human insulin receptor. Nature 372, 746-754.
6. **Kahn, C.R., Baird, K.L., Flier, J.S., Grunfeld, C., Harmon, J.T., Harrison, L.C., Karlsson, F.A., Kasuga, M., King, G.L., Lang, U.C., Podskalny, J.M. and Van Obberghen, E.** (1981) Insulin receptors, receptor antibodies, and the mechanism of insulin action. Recent Prog Horm Res 37, 477-538.
7. **Kristensen, C., Wiberg, F.C., Schaffer, L. and Andersen, A.S.** (1998) Expression and characterization of a 70-kDa fragment of the insulin receptor that binds insulin. Minimizing ligand binding domain of the insulin receptor. J Biol Chem 273, 17780-17786.
8. **Kristensen, C., Wiberg, F.C. and Andersen, A.S.** (1999) Specificity of insulin and insulin-like growth factor I receptors investigated using chimeric mini-receptors. Role of C-terminal of receptor alpha subunit. J Biol Chem 274, 37351-37356.
9. **Leconte, I., Carpentier, J.L. and Clauser, E.** (1994) The functions of the human insulin receptor are affected in different ways by mutation of each of the four N-glycosylation sites in the beta subunit. J Biol Chem 269, 18062-18071.
10. **Elleman, T.C., Frenkel, M.J., Hoyne, P.A., McKern, N.M., Cosgrove, L., Hewish, D.R., Jachno, K.M., Bentley, J.D., Sankovich, S.E. and Ward, C.W.** (2000) Mutational analysis of the N-linked glycosylation sites of the human insulin receptor. Biochem J 347 Pt 3, 771-779.

11. **Schumacher, R., Mosthaf, L., Schlessinger, J., Brandenburg, D. and Ullrich, A.** (1991) Insulin and insulin-like growth factor-1 binding specificity is determined by distinct regions of their cognate receptors. J Biol Chem 266, 19288-19295.
12. **Kurose, T., Pashmforoush, M., Yoshimasa, Y., Caroll, R., Schwartz, G.P., Burke, G.T., Katsoyannis, P.G., Steiner, D.F.** Cross-linking of a B25 azidophenylananine insulin derivative to the carboxyl-terminal region of the alpha-subunit of the insulin receptor. (1994) Identification of a new insulin-binding domain in the insulin receptor. J Biol Chem 269, 29190-29197
13. **Williams, P.F., Mynarcik, D.C., Yu, G.Q., Whittaker, J.** (1995) Mapping of an NH2-terminal ligand binding site of the insulin receptor by alanine scanning mutagenesis. J Biol Chem 270, 3012-3016
14. **Kaburagi, Y., Momomura, K., Yamamoto-Honda, R., Tobe, K., Tamori, Y., Sakura, H., Akanuma, Y., Yazaki, Y. and Kadowaki, T.** (1993) Site-directed mutagenesis of the juxtamembrane domain of the human insulin receptor. J Biol Chem 268, 16610-16622.
15. **White, M.F., Livingston, J.N., Backer, J.M., Lauris, V., Dull, T.J., Ullrich, A. and Kahn, C.R.** (1988) Mutation of the insulin receptor at tyrosine 960 inhibits signal transmission but does not affect its tyrosine kinase activity. Cell 54, 641-649.
16. **Keegan, A.D., Nelms, K., White, M., Wang, L.M., Pierce, J.H. and Paul, W.E.** (1994) An IL-4 receptor region containing an insulin receptor motif is important for IL-4-mediated IRS-1 phosphorylation and cell growth. Cell 76, 811-820.
17. **Koval, A.P., Blakesley, V.A., Roberts Jr, C.T., Zick, Y. and Leroith, D.** (1998) Interaction in vitro of the product of the c-Crk-II proto-oncogene with the insulin-like growth factor I receptor. Biochem J 330, 923-932.
18. **Eck, M.J., Dhe-Paganon, S., Trub, T., Nolte, R.T. and Shoelson, S.E.** (1996) Structure of the IRS-1 PTB domain bound to the juxtamembrane region of the insulin receptor. Cell 85, 695-705.
19. **Paz, K., Voliovitch, H., Hadari, Y.R., Roberts, C.T., Jr., LeRoith, D. and Zick, Y.** (1996) Interaction between the insulin receptor and its downstream effectors. Use of individually expressed receptor domains for structure/function analysis. J Biol Chem 271, 6998-7003.
20. **Haft, C.R., Klausner, R.D. and Taylor, S.I.** (1994) Involvement of dileucine motifs in the internalization and degradation of the insulin receptor. J Biol Chem 269, 26286-26294.
21. **Tavare, J.M. and Siddle, K.** (1993) Mutational analysis of insulin receptor function: consensus and controversy. Biochim Biophys Acta 1178, 21-39.
22. **LeRoith, D., Werner, H., Beitner-Johnson, D. and Roberts, C.T., Jr.** (1995) Molecular and cellular aspects of the insulin-like growth factor I receptor. Endocr Rev 16, 143-163.
23. **Cheatham, B. and Kahn, C.R.** (1995) Insulin action and the insulin signaling network. Endocr Rev 16, 117-142.
24. **Pang, L., Milarski, K.L., Ohmichi, M., Takata, Y., Olefsky, J.M. and Saltiel, A.R.** (1994) Mutation of the two carboxyl-terminal tyrosines in the insulin receptor results in enhanced activation of mitogen-activated protein kinase. J Biol Chem 269, 10604-10608.
25. **Maegawa, H., McClain, D.A., Freidenberg, G., Olefsky, J.M., Napier, M., Lipari, T., Dull, T.J., Lee, J. and Ullrich, A.** (1988) Properties of a human insulin receptor with a COOH-terminal truncation. II. Truncated receptors have normal kinase activity but are defective in signaling metabolic effects. J Biol Chem 263, 8912-8917.
26. **Thies, R.S., Ullrich, A. and McClain, D.A.** (1989) Augmented mitogenesis and impaired metabolic signaling mediated by a truncated insulin receptor. J Biol Chem 264, 12820-12825.

27. **Surmacz, E., Sell, C., Swantek, J., Kato, H., Roberts, C.T., Jr., LeRoith, D. and Baserga, R.** (1995) Dissociation of mitogenesis and transforming activity by C-terminal truncation of the insulin-like growth factor-I receptor. Exp Cell Res 218, 370-380.
28. **Faria, T.N., Blakesley, V.A., Kato, H., Stannard, B., LeRoith, D. and Roberts, C.T., Jr.** (1994) Role of the carboxyl-terminal domains of the insulin and insulin-like growth factor I receptors in receptor function. J Biol Chem 269, 13922-13928.
29. **Kalloo-Hosein, H.E., Whitehead, J.P., Soos, M., Tavare, J.M., Siddle, K. and O'Rahilly, S.** (1997) Differential signaling to glycogen synthesis by the intracellular domain of the insulin versus the insulin-like growth factor-1 receptor. Evidence from studies of TrkC-chimeras. J Biol Chem 272, 24325-24332.
30. **Heldin, C.H.** (1995) Dimerization of cell surface receptors in signal transduction. Cell 80, 213-223.
31. **van der Geer, P., Hunter, T. and Lindberg, R.A.** (1994) Receptor protein-tyrosine kinases and their signal transduction pathways. Annu Rev Cell Biol 10, 251-337.
32. **Lee, J., Pilch, P.F., Shoelson, S.E. and Scarlata, S.F.** (1997) Conformational changes of the insulin receptor upon insulin binding and activation as monitored by fluorescence spectroscopy. Biochemistry 36, 2701-2708.
33. **Songyang, Z., Shoelson, S.E., Chaudhuri, M., Gish, G., Pawson, T., Haser, W.G., King, F., Roberts, T., Ratnofsky, S., and Lechleider, R.J..** (1993) SH2 domains recognize specific phosphopeptide sequences. Cell 72, 767-778.
34. **Staubs, P.A., Reichart, D.R., Saltiel, A.R., Milarski, K.L., Maegawa, H., Berhanu, P., Olefsky, J.M. and Seely, B.L.** (1994) Localization of the insulin receptor binding sites for the SH2 domain proteins p85, Syp, and GAP. J Biol Chem 269, 27186-27192.
35. **Lavan, B.E., Lane, W.S. and Lienhard, G.E.** (1997) The 60-kDa phosphotyrosine protein in insulin-treated adipocytes is a new member of the insulin receptor substrate family. J Biol Chem 272, 11439-11443.
36. **Myers, M.G., Jr. and White, M.F.** (1996) Insulin signal transduction and the IRS proteins. Annu Rev Pharmacol Toxicol 36, 615-658.
37. **Araki, E., Lipes, M.A., Patti, M.E., Bruning, J.C., Haag, B., 3rd, Johnson, R.S. and Kahn, C.R.** (1994) Alternative pathway of insulin signalling in mice with targeted disruption of the IRS-1 gene. Nature 372, 186-190.
38. **Withers, D.J., Gutierrez, J.S., Towery, H., Burks, D.J., Ren, J.M., Previs, S., Zhang, Y., Bernal, D., Pons, S., Shulman, G.I., Bonner-Weir, S. and White, M.F.** (1998) Disruption of IRS-2 causes type 2 diabetes in mice. Nature 391, 900-904.
39. **Nystrom, F.H. and Quon, M.J.** (1999) Insulin signalling: metabolic pathways and mechanisms for specificity. Cell Signal 11, 563-574.
40. **Alessi, D.R., James, S.R., Downes, C.P., Holmes, A.B., Gaffney, P.R., Reese, C.B. and Cohen, P.** (1997) Characterization of a 3-phosphoinositide-dependent protein kinase which phosphorylates and activates protein kinase Balpha. Curr Biol 7, 261-269.
41. **Paz, K., Liu, Y.F., Shorer, H., Hemi, R., LeRoith, D., Quon, M., Kanety, H., Seger, R. and Zick, Y.** (1999) Phosphorylation of insulin receptor substrate-1 (IRS-1) by protein kinase B positively regulates IRS-1 function. J Biol Chem 274, 28816-28822.
42. **Hotamisligil, G.S., Peraldi, P., Budavari, A., Ellis, R., White, M.F. and Spiegelman, B.M.** (1996) IRS-1-mediated inhibition of insulin receptor tyrosine kinase activity in TNF-alpha- and obesity-induced insulin resistance. Science 271, 665-668.
43. **Walton, K.M. and Dixon, J.E.** (1993) Protein tyrosine phosphatases. Annu Rev Biochem 62, 101-120.
44. **Lammers, R., Bossenmaier, B., Cool, D.E., Tonks, N.K., Schlessinger, J., Fischer, E.H. and Ullrich, A.** (1993) Differential activities of protein tyrosine phosphatases in intact cells. J Biol Chem 268, 22456-22462.

45. **Chen, H., Wertheimer, S.J., Lin, C.H., Katz, S.L., Amrein, K.E., Burn, P. and Quon, M.J.** (1997) Protein-tyrosine phosphatases PTP1B and syp are modulators of insulin-stimulated translocation of GLUT4 in transfected rat adipose cells. J Biol Chem 272, 8026-8031.
46. **Tonks, N.K., Cicirelli, M.F., Diltz, C.D., Krebs, E.G. and Fischer, E.H.** (1990) Effect of microinjection of a low-Mr human placenta protein tyrosine phosphatase on induction of meiotic cell division in Xenopus oocytes. Mol Cell Biol 10, 458-463.
47. **Elchebly, M., Payette, P., Michaliszyn, E., Cromlish, W., Collins, S., Loy, A.L., Normandin, D., Cheng, A., Himms-Hagen, J., Chan, C.C., Ramachandran, C., Gresser, M.J., Tremblay, M.L. and Kennedy, B.P.** (1999) Increased insulin sensitivity and obesity resistance in mice lacking the protein tyrosine phosphatase-1B gene. Science 283, 1544-1548.
48. **Milarski, K.L. and Saltiel, A.R.** (1994) Expression of catalytically inactive Syp phosphatase in 3T3 cells blocks stimulation of mitogen-activated protein kinase by insulin. J Biol Chem 269, 21239-21243.
49. **Maegawa, H., Hasegawa, M., Sugai, S., Obata, T., Ugi, S., Morino, K., Egawa, K., Fujita, T., Sakamoto, T., Nishio, Y., Kojima, H., Haneda, M., Yasuda, H., Kikkawa, R. and Kashiwagi, A.** (1999) Expression of a dominant negative SHP-2 in transgenic mice induces insulin resistance. J Biol Chem 274, 30236-30243.
50. **Soni, P., Lakkis, M., Poy, M.N., Fernstrom, M.A. and Najjar, S.M.** (2000) The differential effects of pp120 (Ceacam 1) on the mitogenic action of insulin and insulin-like growth factor 1 are regulated by the nonconserved tyrosine 1316 in the insulin receptor. Mol Cell Biol 20, 3896-3905.
51. **Najjar, S.M., Blakesley, V.A., Li Calzi, S., Kato, H., LeRoith, D. and Choice, C.V.** (1997) Differential phosphorylation of pp120 by insulin and insulin-like growth factor-1 receptors: role for the C-terminal domain of the beta-subunit. Biochemistry 36, 6827-6834.
52. **Formisano, P., Najjar, S.M., Gross, C.N., Philippe, N., Oriente, F., Kern-Buell, C.L., Accili, D. and Gorden, P.** (1995) Receptor-mediated internalization of insulin. Potential role of pp120/HA4, a substrate of the insulin receptor kinase. J Biol Chem 270, 24073-24077.
53. **Blakesley, V.A., Kalebic, T., Helman, L.J., Stannard, B., Faria, T.N., Roberts, C.T., Jr. and LeRoith, D.** (1996) Tumorigenic and mitogenic capacities are reduced in transfected fibroblasts expressing mutant insulin-like growth factor (IGF)-I receptors. The role of tyrosine residues 1250, 1251, and 1316 in the carboxy-terminus of the IGF-I receptor. Endocrinology 137, 410-417.
54. **Porcu, P., Ferber, A., Pietrzkowski, Z., Roberts, C.T., Adamo, M., LeRoith, D. and Baserga, R.** (1992) The growth-stimulatory effect of simian virus 40 T antigen requires the interaction of insulinlike growth factor 1 with its receptor. Mol Cell Biol 12, 5069-5077.
55. **Urso, B., Cope, D.L., Kalloo-Hosein, H.E., Hayward, A.C., Whitehead, J.P., O'Rahilly, S. and Siddle, K.** (1999) Differences in signaling properties of the cytoplasmic domains of the insulin receptor and insulin-like growth factor receptor in 3T3-L1 adipocytes. J Biol Chem 274, 30864-30873.
56. **Park, B.C., Kido, Y. and Accili, D.** (1999) Differential signaling of insulin and IGF-1 receptors to glycogen synthesis in murine hepatocytes. Biochemistry 38, 7517-7523.
57. **Tartare, S., Mothe, I., Kowalski-Chauvel, A., Breittmayer, J.P., Ballotti, R. and Van Obberghen, E.** (1994) Signal transduction by a chimeric insulin-like growth factor-1 (IGF-1) receptor having the carboxyl-terminal domain of the insulin receptor. J Biol Chem 269, 11449-11455.

58. **Esposito, D.L., Blakesley, V.A., Koval, A.P., Scrimgeour, A.G. and LeRoith, D.** (1997) Tyrosine residues in the C-terminal domain of the insulin-like growth factor-I receptor mediate mitogenic and tumorigenic signals. Endocrinology 138, 2979-2988.
59. **Chow, J.C., Condorelli, G. and Smith, R.J.** (1998) Insulin-like growth factor-I receptor internalization regulates signaling via the Shc/mitogen-activated protein kinase pathway, but not the insulin receptor substrate-1 pathway. J Biol Chem 273, 4672-4680.
60. **Ceresa, B.P., Kao, A.W., Santeler, S.R. and Pessin, J.E.** (1998) Inhibition of clathrin-mediated endocytosis selectively attenuates specific insulin receptor signal transduction pathways. Mol Cell Biol 18, 3862-3870.
61. **Emanuelli, B., Peraldi, P., Filloux, C., Sawka-Verhelle, D., Hilton, D. and Van Obberghen, E.** (2000) SOCS-3 is an insulin-induced negative regulator of insulin signaling. J Biol Chem 275, 15985-15991.
62. **Zong, C.S., Chan, J., Levy, D.E., Horvath, C., Sadowski, H.B. and Wang, L.H.** (2000) Mechanism of STAT3 activation by insulin-like growth factor I receptor. J Biol Chem 275, 15099-15105.
63. **Dey, B.R., Spence, S.L., Nissley, P. and Furlanetto, R.W.** (1998) Interaction of human suppressor of cytokine signaling (SOCS)-2 with the insulin-like growth factor-I receptor. J Biol Chem 273, 24095-24101.
64. **Warren, R.S., Yuan, H., Matli, M.R., Ferrara, N. and Donner, D.B.** (1996) Induction of vascular endothelial growth factor by insulin-like growth factor 1 in colorectal carcinoma. J Biol Chem 271, 29483-29488.
65. **Zelzer, E., Levy, Y., Kahana, C., Shilo, B.Z., Rubinstein, M. and Cohen, B.** (1998) Insulin induces transcription of target genes through the hypoxia-inducible factor HIF-1alpha/ARNT. Embo J 17, 5085-5094.
66. **Miele, C., Rochford, J.J., Filippa, N., Giorgetti-Peraldi, S. and Van Obberghen, E.** (2000) Insulin and insulin-like growth factor-I induce vascular endothelial growth factor mRNA expression via different signaling pathways [In Process Citation]. J Biol Chem 275, 21695-21702.
67. **Wertheimer, E., Trebicz, M., Eldar, T., Gartsbein, M., Nofeh-Moses, S. and Tennenbaum, T.** (2000) Differential roles of insulin receptor and insulin-like growth factor-1 receptor in differentiation of murine skin keratinocytes [In Process Citation]. J Invest Dermatol 115, 24-29.
68. **Okubo, Y., Blakesley, V.A., Stannard, B., Gutkind, S. and Le Roith, D.** (1998) Insulin-like growth factor-I inhibits the stress-activated protein kinase/c-Jun N-terminal kinase. J Biol Chem 273, 25961-25966.
69. **Fukunaga, K., Noguchi, T., Takeda, H., Matozaki, T., Hayashi, Y., Itoh, H. and Kasuga, M.** (2000) Requirement for protein-tyrosine phosphatase SHP-2 in insulin-induced activation of c-Jun NH(2)-terminal kinase. J Biol Chem 275, 5208-5213.
70. **Nakae, J., Barr, V. and Accili, D.** (2000) Differential regulation of gene expression by insulin and IGF-1 receptors correlates with phosphorylation of a single amino acid residue in the forkhead transcription factor FKHR. Embo J 19, 989-996.

Section III

Cytokine Receptors

6

The JAK-STAT Pathway

Warren J. Leonard*

Laboratory of Molecular Immunology, National Heart, Lung, and Blood Institute National Institutes of Health, Bethesda, Maryland 20892-1674, USA

OVERVIEW

The JAK-STAT pathway is one of the central signaling pathways used by interferons and type I cytokines. Janus family tyrosine kinases (JAK kinases) are cytoplasmic tyrosine kinases that associate with cytokine receptor chains. These kinases are activated by cytokine binding, and then can phosphorylate the receptors on tyrosine residues. This in turn allows the recruitment of signal transducer and activator of transcription (STAT) proteins *via* their SH2 domains to the phosphorylated receptor. The STAT proteins are then also phosphorylated on tyrosine residue(s) by the JAK kinases, which allows the dimerization and nuclear translocation of the STAT proteins. STAT proteins then bind to target DNA sequences, modulating gene expression of certain genes. Thus, STAT proteins are latent transcription factors: they initially exist in the cytosol, but translocate to the nucleus following cellular activation. Together with NF-κB, NF-AT, and SMADs, STATs represent one of four types of transcription factors that are rapidly translocated from cytosol to nucleus in a fashion that is dependent on phosphorylation/dephosphorylation events. This short chapter cannot comprehensively review all that is known of the JAK-STAT pathway. Therefore, the goal will be to overview the basic system and describe some of the features of JAK kinases and STAT proteins, focusing on *in vivo* data, when available, from various knockout models. Many recent reviews of the JAK/STAT pathway should be consulted for in-depth review of

* wjl@helix.nih.gov

other areas. We will begin with a brief discussion of "cytokines", as these are the molecules that initiate the activation of the JAK/STAT pathway.

TYPE I CYTOKINES, INTERFERONS AND THEIR RECEPTORS

The term "cytokines" refers to molecules that are made by cells and act on other cells [1]. The term is used to refer to immunologically related proteins, including those originally called lymphokines, interleukins, and colony stimulating factors. However, cytokines also include many polypeptide hormones and growth factors as well; thus, cytokines comprise a diverse group of molecules that together induce many important biological responses, including those related to cell growth, differentiation, and anti-apoptotic signals. Cytokines can act in either an autocrine or paracrine fashion, depending on whether the target cell is the same or different from the cytokine-producing cell. Two broad categories of cytokines have been described and are denoted as type I and type II cytokines [1]. Type II cytokines include type I interferons (IFNs) (IFN-α/β), type II IFNs (IFN-γ) as well as IL-10 and IL-22, whereas type I cytokines include the majority of the interleukins (e.g., IL-2, IL-3, IL-4, IL-5, IL-6, IL-7, IL-9, IL-11, IL-12, IL-13, IL-15, IL-21), as well as stem cell factor (SCF), thymic stromal lymphopoietin (TSLP), G-CSF, GM-CSF, leukemia inhibitory factor (LIF), oncostatin M (OSM), cardiotrophin-1 (CT-1), leptin, ciliary neurotrophic factor (CNTF), erythropoietin (EPO), thrombopoietin (TPO), growth hormone, and prolactin [generally reviewed in ref. 1]. The characterization of molecules as type I or type II cytokines is based on the three-dimensional structure of the cytokines and their receptors. For example, type I cytokines share a four α-helical bundle structure with an "up-up-down-down" topology; these can be subdivided into "short chain" or "long chain" type I cytokines, based both on the length of the four α helices, but also on other topological considerations as well [1]. One of the ways in which short-chain and long-chain cytokines differ is that the short-chain cytokines contain limited regions of β-sheet structures, whereas the long-chain cytokines do not. In addition to structural considerations shared by the cytokines, the receptors for type I cytokines also exhibit a number of shared structural features [1]. First, in general, the receptors for type I cytokines and also all type II cytokines lack intrinsic tyrosine kinase activity; however, exceptions do exist. For example, in the case of stem cell factor, the receptor is c-kit, a molecule with its own tyrosine kinase domain. A second generally shared feature is that in their extracellular domains, type I cytokine receptors typically contain four conserved cysteine residues, have a fibronectin type III module, and contain a Trp-Ser-X-Trp-Ser motif. Third, in their cytoplasmic domains, they contain tyrosine residues that can be phosphorylated; as noted above, some of these phosphotyrosines serve as docking sites for signaling molecules including STAT proteins. Fourth, they contain Box1/Box2 conserved regions, where the Box 1 region in particular is an interaction point for JAK kinases. Interferons and IL-10 also use the JAK-STAT pathway, which in fact was discovered in the context of interferons rather than type I cytokines [2-4]. Although the JAK-STAT pathway is the focus for this review, it is important to recognize that other multiple signaling pathways are also activated by cytokines. For example, cytokines can activate Src family kinases, phosphoinositol 3-kinase, Ras-MAP kinase-related pathways, and IRS protein-linked pathways, and can induce the expression of survival factors. The JAK-STAT pathway is one of the best-characterized pathways, but the focus on this pathway in this chapter is not to imply that it is necessarily a more important signaling mechanism for cytokines than other pathways. The relative importance of different pathways likely differs for different cytokines.

JAK KINASES AND STAT PROTEINS

As noted above, all type I and type II cytokines signal in part *via* the JAK-STAT pathway [2-4]. JAK kinases are large (approximately 120-140 kDa) cytoplasmic tyrosine kinases whose most distinctive feature is the presence of a pseudokinase domain in addition to a true catalytic domain. There are a total of four different mammalian JAK kinases, Jak1, Jak2, Jak3, and Tyk2. STAT proteins, or "signal transducers and activators of transcription" are latent transcription factors that initially exist in the cytosol. There are seven different STAT proteins, Stat1, Stat2, Stat3, Stat4, Stat5a, Stat5b, and Stat6, with Stat5a and Stat5b being particularly highly conserved. Schematics of JAK kinases and STAT proteins are shown in Fig. 1.

For historical reasons, the regions of JAK kinases are identified as JH or JAK homology regions. However, it is important to note that these regions have not all been correlated with functional domains. In addition to the importance of the C-terminal catalytic domain, the pseudokinase region, which resembles a kinase but lacks an ATP binding site, is also important in that JAK activity can be abrogated by mutations in this region [3]. The N-terminal region of JAK kinases has been shown to be required for their association with cytokine receptors, as was first demonstrated for the interactions of Jak2 with the erythropoietin and growth hormone receptors [5,6], but subsequently was extended to other JAK kinases and other cytokine receptors as well [3].

STAT proteins contain an N-terminal region that is important for tetramerization (discussed below), a coiled-coiled region, a DNA binding domain, an SH2 domain, and a C-terminal transactivation domain that contains

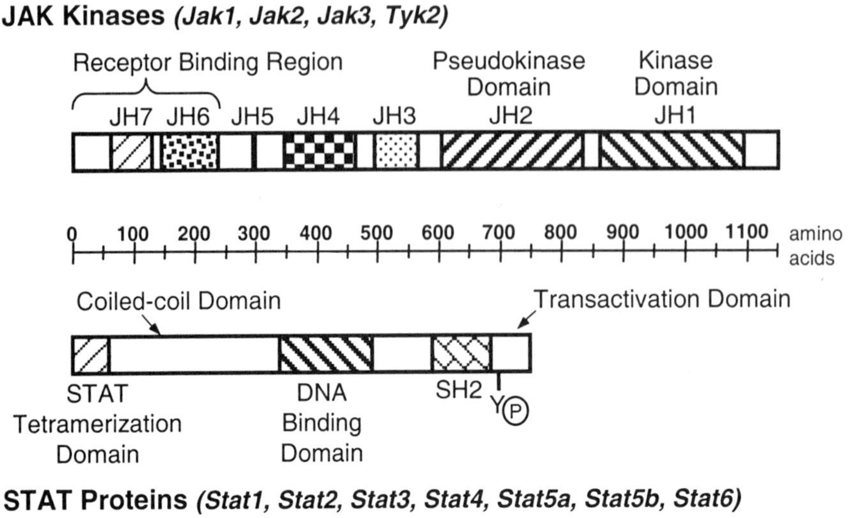

Figure 1. Schematic of JAK kinases and STAT proteins. The various domains of these proteins are described in the text. The coordinate system is approximate as the JAKs and STATs vary in size. Stat2 and Stat6 are larger than the other STAT proteins. Phosphorylation of the conserved tyrosine is required for dimerization of STAT proteins. The N-terminal region mediates oligomerization of STAT dimers.

a critical conserved tyrosine residue, approximately 700 amino acids from the N-terminus [2-4; see Fig. 1]. Consistent with this representing a gene family, the genes encoding STAT proteins are clustered on chromosomes: in the mouse, Stat1 and Stat4 are located on chromosome 1; Stat2 and Stat6 are located on chromosome 10, and Stat3, Stat5a, and Stat5b are located on chromosome 11 [7]. Human Stat5a and Stat5b have been mapped to chromosome 17 [8].

Table I summarizes the JAKs and STATs that are activated by various cytokines. Interestingly, Jak1, Jak2, and Tyk2 are constitutively expressed and ubiquitous, whereas Jak3 is relatively restricted to lympho-hematopoietic lineages and is inducible [3]. In general, different JAK kinases can associate with more than one IFN or cytokine receptor chain, but Jak3 is again distinctive in its selective interaction with the common cytokine receptor γ chain, γ_c, shared by the receptors for IL-2, IL-4, IL-7, IL-9, IL-15 and IL-21 [1, 3, 71]. The seven mammalian STAT proteins are ubiquitously expressed, although levels of different STAT proteins differ in their ratios in different tissues.

Table I: Activation of JAKs and STATs by Cytokines

Shown are the JAK kinases and the major STAT proteins activated by various cytokines.

Cytokines	Activated JAKs	Activated STATs
Type I cytokines		
Cytokines that share gp130		
IL-6, IL-11, OSM, CNTF, LIF, CT-1	Jak1, Jak2, Tyk2	Stat3
Cytokines that share β_c:		
IL-3, IL-5, GM-CSF	Jak2	Stat5a, Stat5b
Cytokines that share γ_c:		
IL-2, IL-7, IL-9, IL-15, IL-21*	Jak1, Jak3	Stat5a, Stat5b, Stat3, Stat1
IL-4	Jak1, Jak3	Stat6
IL-13	Jak1, Jak2, Tyk2	Stat6
TSLP	none	Stat5a, Stat5b
IL-12	Jak2, Tyk2	Stat4
Prolactin, EPO, TPO	Jak2	Stat5a, Stat5b
Growth Hormone	Jak2	Stat5a, Stat5b, Stat3
Type II cytokines		
IFN-α/β	Jak1, Tyk2	Stat1, Stat2
IFN-γ	Jak1, Jak2	Stat1
IL-10	Jak1, Tyk2	Stat3

Most cytokines are reviewed in detail in Ref. [1,3]; for signaling by IL-21, see Ref. [71,72].

THE JAK-STAT PATHWAY

The JAK-STAT pathway was first discovered in the context of interferons, and the development of our knowledge of the system was tremendously facilitated by a series of mutant cell lines that exhibited defects in signaling by interferons [9]. The basic scheme is summarized in Fig. 2. Following the binding of a cytokine or interferon to its receptor, receptor-associated JAK kinases are activated. In some cases, the JAK kinases are constitutively associated whereas in other cases, they are recruited or their association increases following ligand binding. Activation of the JAK kinases results in transphosphorylation of the JAK kinases; this transphosphorylation within the kinase activation loop in the JH1 catalytic domain region appears to be important for certain JAK kinases, such as Jak2, but may be less important for other JAK kinases, such as Jak3 [reviewed in 3]. In addition to transphosphorylation of JAK kinases, the cytokine receptor chains can be phosphorylated on tyrosine residues in their cytoplasmic domains. Some of these phosphotyrosines then can serve as docking sites for SH2-containing proteins, such as STAT proteins. The exact amino acid sequence in the vicinity of the tyrosine determines which STAT protein(s) are recruited [3]. For some receptors, more than one tyrosine can be a docking site for STAT proteins [3]. Each receptor-associated STAT protein can then be phosphorylated on a conserved tyrosine distal to the STAT protein SH2 domain. When the STAT proteins then dissociate from the receptor, they can form stable bivalent phosphotyrosine-SH2 domain dimers. These can either be homodimers or heterodimers, depending on the STAT proteins that are activated. The dimeric STAT complexes can then translocate into the nucleus, where they bind to target sequences. The binding of STAT dimers to DNA is typically direct, with the target sequences being denoted as $TTCN_mGAA$ (γ-interferon activated sequences, or GAS sites)[10]. However, in the case of Stat1/Stat2 heterodimers, the formation of which is induced by IFN-α/β, DNA binding requires the presence of a DNA binding protein, p48, which is a member of the IRF family of proteins. The Stat1/Stat2/p48 complex is known as ISGF3, and it binds to interferon stimulated response elements (ISRE; AGTTTNCNTTTCC) [11].

BINDING SITE PREFERENCES FOR STAT PROTEINS

As noted above, with the exception of ISGF3 (Stat1/Stat2/p48) complexes, other STAT homo- or hetero-dimers bind to GAS motifs. Although GAS motifs are typically $TTCN_3GAA$ motifs, the actual binding preferences are more diverse. In particular, the following preferences have been reported: Stat1, Stat3, and Stat4 have been reported to prefer TTCCSGGAA motifs (where S = G or C) [12, 13], indicating that two of the central three nucleotides affect optimal binding. Both Stat5a and Stat5b homodimers prefer binding to A/g T/A T T C C/T N G/a G A A A/T T/c motifs*, indicating that nucleotides flanking the central nine nucleotides are also important [14]. Stat6 optimally binds to TTCNTNGGAA motifs, preferring a spacer of 4 rather than 3 nucleotides in the central core [15]. Interestingly, while the binding preferences listed above are for STAT dimers, the binding preferences for tetramers (see below) appear to be less strict. In particular, a binding site selection with Stat5a

* upper case preferred over lower case

tetramers revealed that no sequences with two optimal Stat5a GAS motifs were selected. Instead, at least one of the "GAS" motifs was always imperfect, although either a "TTC" or "GAA" "half-GAS" motif was conserved. Such an observation is consistent with the idea that two suboptimal GAS motifs, which only can poorly bind Stat5a dimers, can cooperatively work to efficiently bind Stat5a tetramers.

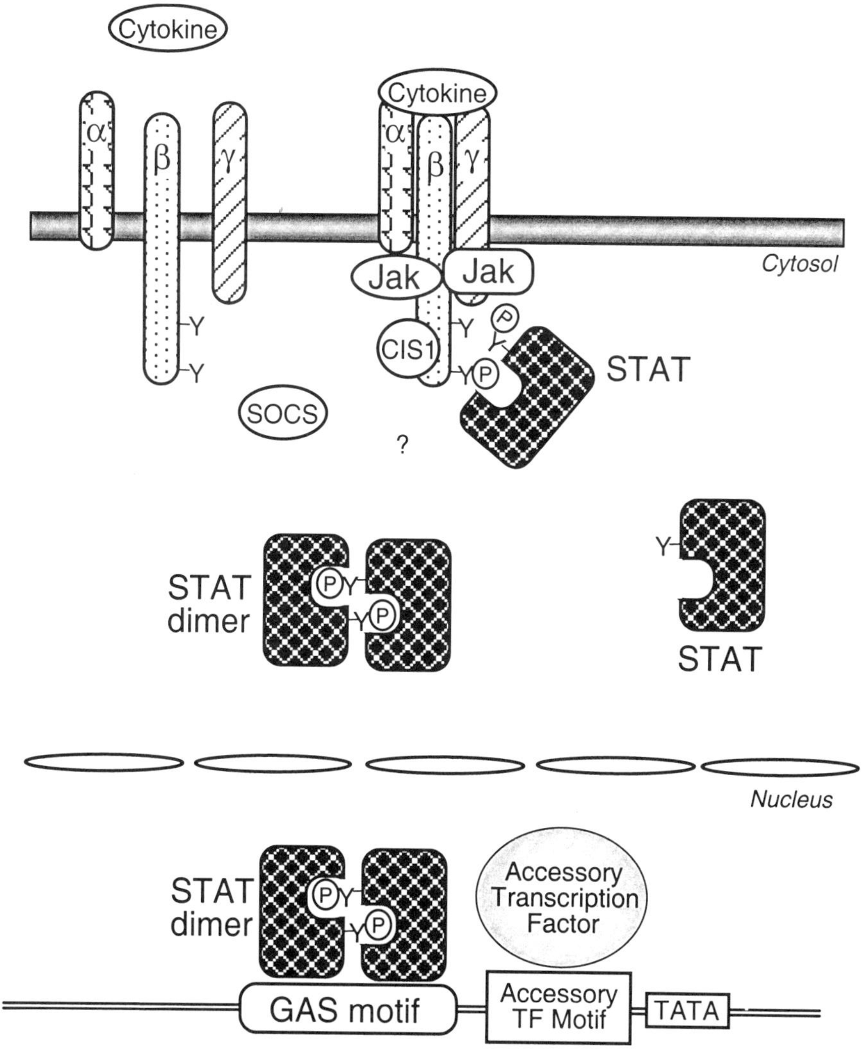

Figure 2: Schematic of the JAK-STAT pathway. Shown is a type I cytokine receptor system in which the cytokine receptor has three different chains. For some cytokines, the receptor is a homodimer or heterodimers. Shown is a SOCS protein and CIS1 as inhibitors.

SERINE PHOSPHORYLATION OF STAT PROTEINS

In addition to the basic activation and nuclear translocation of STAT proteins, two specific features that are intrinsic to STAT proteins have been demonstrated to affect their ability to augment transcription. First, a number of STAT proteins, including Stat1, Stat3, Stat4, Stat5a, and Stat5b have been demonstrated to be serine phosphorylated [16-19]. In the case of Stat1, Stat3, and Stat4, such serine phosphorylation has been shown to be required for potent activation and it occurs on a conserved serine residue in the C-terminal transactivation domain. MCM5 (minichromosome maintenance protein 5) has been shown to associate with Ser727 of Stat1 following the phosphorylation of this residue [20]. MAP kinase family proteins, including ERKs and p38, appear to be involved in some of the serine phosphorylation events. In contrast, the role of serine phosphorylation, if any, for Stat5a and Stat5b remains unclear. See ref. 21 for a general review on serine phosphorylation of STAT proteins.

TETRAMERIZATION AND TRANSCRIPTIONAL ACTIVATION OF STAT PROTEINS

Although a full understanding of how STAT proteins mediate transcriptional activation remains unclear, in addition to the importance of serine phosphorylation for certain STATs, the association with coactivator proteins is very important. In this regard, the CBP/p300 co-activators have been reported to be recruited to different STAT proteins [22-25]. Other transcription factors, such as Sp1 and c-Jun have been demonstrated to interact with STAT proteins [26, 27].

In addition to interacting proteins, STAT dimers can self-interact to form higher order oligomers [13, 28, 29]. Conserved residues in the N-terminal domains of STAT proteins mediate tetramerization or higher order oligomerization of STATs. The crystal structure of the N-domain has been solved and revealed a "hook-like" structure composed of eight helices. Central to the formation of STAT tetramers is a tryptophan residue that is conserved in all STAT proteins. The interaction mediated by the N-terminal regions allows oligomerization of STAT dimers. This unexpected finding is potentially very important. First, it can allow cooperative STAT protein recruitment to tandem lower affinity binding sites [30], thereby eliminating that need for higher affinity sites that exhibit lower specificity in general. Second, it may represent a mechanism for augmenting the ability to select among the different STAT proteins. Third, it could represent a mechanism for amplifying a signal, given that multiple STAT proteins can be recruited to tandem sites.

NEGATIVE REGULATION OF THE JAK-STAT PATHWAY

Biological signals cannot be allowed to continue unchecked, and the JAK-STAT pathway is similar. Negative regulation can be provided by turnover of receptor or JAK-STAT pathway components, of dephosphorylation by phosphatases, and by negative regulation by inhibition proteins. Stat1 was the first STAT to be suggested to be a target for degradation by a ubiquitin/proteasome-mediated mechanism [31]. Stat1, Stat3, and Stat5 have each been shown to have truncated forms lacking C-terminal regions [reviewed in ref. 32]. For Stat1 and Stat3, this occurs based on alternative splicing, whereas for Stat5 the mechanism is less clear. Shp1 can negatively regulate signaling by cytokines. Finally, there are two sets of proteins that are negative regulators of the JAK-STAT pathways. The first set of proteins includes the PIAS proteins, PIAS1 and PIAS3 [33], which negatively regulate Stat1 and Stat3 signaling, respectively. The second set of proteins includes the CIS/JAB/SOCS/SSI family of negative regulators. CIS1 (cytokine-induced SH2 containing protein) was originally identified as an IL-3 and erythropoietin-induced protein that was able to inhibit Stat5-dependent signaling [reviewed in 34]. Subsequently, a CIS-related protein was identified by three groups and variably denoted as JAB (JAK binding protein), SOCS1 (suppressor of cytokine signaling-1), and SSI-1 (STAT-induced STAT inhibitor-1) [34]. These proteins each contain a central SH2 domain as well as another conserved region generally noted as a SOCS box. In contrast to CIS1, which binds receptor chains, JAB/SOCS1/SSI-1 can bind to JAK kinases (see Fig. 2); thus, these related proteins may negatively regulate the JAK-STAT pathway by somewhat different mechanisms. There are now a total of eight members of this family of proteins. In addition, there are related proteins that contain a "SOCS" box, but not the SH2 domain. Studies are still underway to answer the following questions: (1) Which SOCS family proteins are responsible for inhibiting specific cytokine pathways? (2) Do all SOCS family proteins inhibit certain cytokine pathways? (3) Do SOCS family proteins have important roles unrelated to JAK kinases and STAT proteins? Interestingly, some important *in vivo* functions have been revealed by knockout studies [reviewed in 34, 35]. Mice lacking SOCS1 develop a complex fatal neonatal disease resulting from augmented responsiveness to IFNγ. Blocking IFNγ or making a double knockout mouse prevented this problem. Thus, SOCS1 is an essential modulator of IFNγ action. Mice lacking SOCS3 exhibit an embryonic lethality associated with a marked erythrocytosis, indicating the critical role played by SOCS3 in negatively regulating fetal liver hematopoiesis.

BIOLOGICAL ROLES OF JAKS AND STATS—LESSONS FROM KNOCKOUT MICE AND PATIENTS WITH IMMUNODEFICIENCY

In general, it has been assumed that if a cytokine activates a given JAK and STAT, then *in vivo* deletion of the JAK or STAT will result in the loss of signaling by that cytokine. This is sometimes the case, but because cytokines may activate more than one STAT protein and also typically activate more than one signaling pathway, loss of a STAT might not abrogate all signaling by a given cytokine. I will review the phenotypes of knockout mice, and where relevant, humans, with inactivating mutations in these proteins.

In vivo Roles of the Various JAK Kinases

Jak1 Deficiency

As noted in Table I, Jak1 is very broadly utilized and is known to play a role in signaling by interferons, gp130 family cytokines (IL-6, IL-11, ciliary neurotrophic factor, leukemia inhibitory factor, oncostatin M, and cardiotrophin-1), and γ_c-dependent cytokines (IL-2, IL-4, IL-7, IL-9, and IL-15). Consistent with this broad range of actions, the phenotype in Jak1-deficient mice is severe, resulting in perinatal lethality, although there are no obvious gross abnormalities [36]. Death apparently relates at least in part to neurological abnormalities. Examination of the immune system in these very young mice, however, additionally revealed defective signaling by γ_c-dependent cytokines, resulting in defective lymphopoiesis, whereas myelopoiesis is not affected.

Jak2 Deficiency

Jak2 is also broadly utilized, including IFN-γ, and by all type I cytokines whose receptors represent homodimers (e.g., growth hormone, prolactin, and erythropoietin), and hematopoietic cytokines such as IL-3, IL-5, and GM-CSF. The phenotype in Jak2-deficient mice is a severe fetal lethality, resulting from defective erythropoiesis [37, 38]. This confirms the essential non-redundant role played by Jak2 in signaling in response to erythropoietin.

Jak3 Deficiency

As noted above, Jak3 associates with γ_c and appears to be selectively activated by γ_c-dependent cytokines. Mutations in γ_c in humans result in X-linked severe combined immunodeficiency (XSCID) [39], a profound immunodeficiency also known as the "Bubble Boy Disease", in which T cells and natural killer (NK) cells either do not develop or are profoundly diminished in numbers, whereas B cells develop but are non-functional. Because Jak3 is "immediately downstream" from γ_c, it was hypothesized and then confirmed that mutations in Jak3 indeed also cause a form of SCID in humans in which T and NK cells are absent [40, 41]. The principal difference is that Jak3 is not on the X chromosome so that both boys and girls are found to have mutations in Jak3 as an autosomal recessive form of SCID. Analogous to the situation in humans, mice deficient in either γ_c [42, 43] or Jak3 [44, 45] exhibit a profound immunodeficiency with markedly attenuated thymic development and an absence of NK cells. The basis for the lack of T-cell development is due to defective IL-7 signaling, whereas the lack of NK-cells is due to defective IL-15 signaling.

Tyk2 Deficiency

Tyk2 was believed to be restricted to signaling in response to IFN-α/β and IL-12. Although signaling by these cytokines is diminished in Tyk2 knockout mice, it is not abrogated and it is primarily Stat3 activation that is affected. Moreover, signaling in response to IFN-γ was unexpectedly also affected [46, 47]. The partial phenotype may indicate that other JAK kinases can substitute for Tyk2, particularly in the "stressed" setting of a knockout mouse. The unexpected decrease in signaling by IFN-γ could potentially represent an indirect effect.

In vivo Roles of the Various STAT Proteins

Stat1 Deficiency

Stat1 is activated by interferons and a variety of other cytokines, including IL-6, IL-10, growth hormone, and thrombopoietin; however, Stat1 activation historically is most intimately tied to the interferons. Consistent with this special relationship to the interferons, the phenotype of the Stat1 knockout mouse indicates a rather selective defect in signaling by IFN-γ (which signals *via* Stat1 homodimers) and IFN-α/β (which signal *via* Stat1/Stat2 heterodimers). Stat1-deficient mice exhibit a severe defect in responses to viruses and microbial antigens, where host defense is normally dependent on interferons [48, 49]. Thus, *in vivo*, Stat1 is an essential non-redundant factor for interferons whereas other cytokines do not appear to absolutely require Stat1 for their actions - either Stat1 is not needed or other STAT proteins can substitute in the setting where Stat1 is absent.

Stat2 Deficiency

Stat2 is only known to be activated by IFN-α/β. Correspondingly, the major defect in Stat2 knockout mice is a defect in the response to IFN-α/β. Unexpectedly, however, the defect was more severe in Stat2-deficient fibroblasts than in macrophages, suggesting that there are tissue-specific differences in the responses to IFN-α/β [50].

Stat3 Deficiency

Stat3 is activated by a wide range of cytokines, including the gp130 family of cytokines and γ_c-dependent cytokines. Mice deficient in this STAT protein exhibit fetal lethality [51]. However, use of the Cre-loxP system to create conditional knockouts has revealed a number of functions for Stat3 [52]. Mice lacking Stat3 in T-cells exhibit a defect in lymphocytes in IL-2-induced IL-2 receptor α chain gene transcription, which is associated with a defect in proliferation to IL-2. Additionally, these lymphocytes do not normally respond to IL-6. Stat3-deficient macrophages and neutrophils do not respond to IL-10. Thus, within the immune system, Stat3 is vital for normal responses to IL-2, IL-6, and IL-10, and likely other cytokines as well. Stat3 has also been demonstrated to be vital for normal hair cycle and wound healing processes as well as for the normal involution of the mammary epithelium [52].

Stat4 Deficiency

Stat4 is activated primarily if not solely by IL-12. In accord with the role of IL-12 in the induction of T helper 1 (Th1) $CD4^+$ T-cells, Stat4-deficient mice exhibit a marked defect in the ability of IL-12 to induce Th1 T-cell development. Furthermore, in these mice, IL-12 is unable to enhance the function of NK cells. Thus, Stat4 appears to be uniquely and obligately required for the actions of IL-12 [53, 54].

Stat5 Deficiency

Stat5a and Stat5b are two very related STAT proteins. There has been considerable discussion as to whether these STATs are redundant, overlapping, or distinctive in their actions. In support of at least partially distinctive actions, it is interesting that Stat5a and Stat5b are each well conserved across species. For example, human and murine Stat5a are more similar to each other than they are to the Stat5b protein within the same species. Similar observations exist regarding Stat5b. In addition, different phenotypes have been reported for Stat5a-deficient and Stat5b-deficient mice. Stat5a-deficient mice exhibit a marked defect in prolactin signaling: in particular, there is defective proliferation of the lobulo-alveolar ductal epithelium in the mammary gland during pregnancy [55]. This finding is consistent with the original identification of Stat5a as a prolactin-induced "mammary gland factor". In contrast, Stat5b-deficient mice instead have a major defect perhaps related to growth hormone signaling, causing a loss of normal sexually dimorphic growth; the phenotype is reminiscent of Laron's dwarfism in humans [56]. Within the lympho-hematopoietic systems, both Stat5a and Stat5b-deficient mice exhibit modest decreases in the number of T-cells and in IL-2-mediated induction of the IL-2 receptor α chain gene, a finding that is explained based on the existence of an IL-2-response element in the 5' regulatory region of that gene [57, 58]. However, there are major differences as well. Signaling in response to Flt3 ligand is defective in the Stat5a but not Stat5b knockout mice. Although both Stat5a and Stat5b knockout mice exhibit defects related to natural killer development and cytolytic activity, the defect in the Stat5b knockout mice is more profound. In these mice, a defect in IL-2Rβ expression has been identified. Moreover, IL-2- and IL-15-boosting of perforin expression is particularly defective in the Stat5b knockout mice [58]. Interestingly, like the IL-2Rα gene, the perforin gene has been shown to have Stat5-responsive elements. Stat5a and Stat5b have also been shown to be vital for the recruitment of eosinophils and T-cells into the airways in an antigen-induced asthma model [59]. This indicates a vital role for Stat5 proteins for physiological signaling by IL-5. Interestingly, not only was IL-5 responsiveness diminished, but production of IL-4 and IL-5 were diminished in response to anti-CD3 in both Stat5a and Stat5b-deficient mice. Thus, the defective eosinophil recruitment can be attributed both to defective production and responsiveness to IL-5 in these animals.

Stat5a/Stat5b double knockout mice have also been generated. Although T cells develop, there is a marked defect in the proliferation of these cells; this defect corresponded to a loss of induction of cyclin D2, cyclin D3, and cdk6 following stimulation with anti-CD3 [60]. Thus, there appear to be at least certain defects related to signaling *via* the T-cell receptor and/or IL-2. One of the most dramatic findings in the Stat5a/Stat5b double knockout mice is an abrogation of development of NK cells. Interestingly, IL-15 is the cytokine that appears to be most responsible for NK-cell development. The receptor for IL-15 is tripartite, being comprised of IL-15Rα, IL-2Rβ, and γ_c (each of which when deleted in mice also results in the abrogation of NK-cell development). Given that Stat5a and Stat5b have docking sites on IL-2Rβ, we can hypothesize that the defective NK-cell development in the Stat5a/Stat5b double knockout phenotype is due to a normal requirement for IL-15-mediated IL-2Rβ-dependent activation of Stat5a and Stat5b for development of this lineage.

Stat6 Deficiency

Stat6 is activated by IL-4 and IL-13, the receptors for which share the IL-4 receptor α chain, which serves as a docking site for Stat6. IL-4 plays a critical role in the differentiation of the T-helper 2 (Th2) subpopulation of $CD4^+$ T cells. Consistent with this observation, Stat6-deficient mice exhibit a marked defect in their responsiveness to IL-4 and IL-13, and there is defective differentiation of Th2 cells [61, 62]. There is also defective IgE class switching as well as defects related to antigen-induced eosinophil recruitment. Overall, the data from the Stat4 and Stat6 knockout mice confirm the at least partially "opposing" functions of Stat4 and Stat6 - the former is essential for the differentiation of Th1 cells, whereas the latter is essential for the differentiation of Th2 cells.

STATS IN TRANSFORMATION

STAT proteins have been found to be constitutively activated in settings of malignant or oncogene-mediated transformation. For example, Stat3 was originally reported to be constitutively activated in Src-transformed cells [63], and more recent data have suggested that Stat3 can function as an oncogene [64]. STAT proteins have also been shown to be constitutively activated in cells transformed by human T-cell lymphotropic virus, type I (HTLV-I) [65, 66] and v-Abl [67] and Epstein-Barr virus [68]. Together, these findings have indicated that STATs may participate in cellular transformation. In addition, humans with certain forms of leukemia have been shown to have constitutively activated STAT proteins. For example, Stat1 and Stat5 were found to be activated in certain cases of acute lymphocytic leukemia whereas Stat1, Stat3 and Stat5 are activated in some cases of acute myelocytic leukemia [68-70]. In malignancies where the JAK-STAT pathway is activated, it is conceivable that its manipulation could have therapeutic benefit.

CONCLUSIONS

The JAK-STAT signaling pathway has been intensively studied as a rapid membrane to nuclear signaling pathway that is activated following stimulation with interferons or cytokines. A tremendous amount is now understood regarding how this signaling pathway contributes to the specificity of cytokine signaling, thanks in part to the range of knockout models in cell lines and in mice as well as certain humans with genetic diseases. A broad challenge is to identify the genes that are regulated by each STAT protein. Moreover, the discovery of negative regulation of the JAK-STAT pathways *via* the PIAS proteins and the CIS/SOCS/JAB/SSI family of inhibitors has been extremely important in our understanding the JAK-STAT pathways. Our understanding of this pathway along with other pathways that are activated by cytokines together will eventually allow a more integrated understanding of the molecular basis of cytokine action.

REFERENCES

1. **Leonard, WJ.** 1999 Type I cytokines and interferons and their receptors. In: Paul, W.E. (Ed.), Fundamental Immunology Fourth Edition. Lippincott Raven Publishers, Philadelphia, PA, pp. 741-774.
2. **Darnell, JE Jr**. 1997 STATs and gene regulation. Science 277:1630-1635.
3. **Leonard WJ O'Shea JJ**. 1998 Jaks and STATs: Biological implications. Annu Rev Immunol 16:293-322.
4. **Please see multiple reviews of STAT proteins by multiple authors.** 2000 Oncogene 19:2466-2656.
5. **Argetsinger LS, Campbell GS, Yang X, Witthuhn BA, Silvennoinen O, Ihle JN, Carter-Su C**. 1993 Identification of JAK2 as a growth hormone receptor-associated tyrosine kinase. Cell 74:237-244.
6. **Witthuhn BA, Quelle FW, Silvennoinen O, Yi T, Tang B, Miura O, Ihle JN**. 1993 JAK2 associates with the erythropoietin receptor and is tyrosine phosphorylated and activated following stimulation with erythropoietin. Cell 74:227-236.
7. **Copeland NG, Gilbert DJ, Schindler C, Zhong Z, Wen Z, Darnell JE Jr, Mui, AL, Miyajima A, Quelle FW, Ihle JN, Jenkins NA**. 1995 Distribution of the mammalian Stat gene family in mouse chromosomes. Genomics 29:225-228.
8. **Lin J-X, Mietz, J, Modi WS, John S, Leonard WJ**. 1996 Cloning of human Stat5B. Reconstitution of interleukin-2-induced Stat5A and Stat5B DNA binding activity in COS-7 cells. J. Biol. Chem. 271:10738-10744.
9. **Darnell JE Jr, Kerr IM, Stark GR**. 1994 Jak-STAT pathways and transcriptional activation in response to IFNs and other extracellular signaling proteins. Science 264:1415-1421.
10. **Lew DJ, Decker T, Strehlow I, Darnell JE Jr.** 1991 Overlapping elements in the guanylate-binding protein gene promoter mediate transcriptional induction by alpha and gamma interferons. Mol Cell Biol 11:182-191.
11. **Horvath CM, Darnell JE Jr**. 1997 The state of the STATs: recent developments in the study of signal transduction to the nucleus. Curr Opin Cell Biol 9:233-239.
12. **Horvath CM, Wen Z, Darnell JE Jr**. 1995 A STAT protein domain that determines DNA sequence recognition suggests a novel DNA-binding domain. Genes Dev 9:984-994.
13. **Xu X, Sun Y-L, Hoey T**. 1996 Cooperative DNA binding and sequence-selective recognition conferred by the STAT amino terminal domain. Science 273:794-797.
14. **Soldaini E, John S, Moro S, Bollenbacher J, Schindler U, Leonard WJ**. 2000 DNA binding site selection of dimeric and tetrameric Stat5 proteins reveals a large repertoire of divergent tetrameric Stat5a binding sites. Mol Cell Biol 20:389-401.
15. **Schindler U, Wu P, Rothe M, Brasseur M, McKnight SL**. 1995 Components of a Stat recognition code: evidence for two layers of molecular selectivity. Immunity 2:689-697.
16. **Goh KC, Haque SJ, Williams BR**. 1999 p38 MAP kinase is required for STAT1 serine phosphorylation and transcriptional activation induced by interferons. EMBO J 18:5601-5608.
17. **Wen Z, Zhong Z, Darnell JE Jr**. 1995 Maximal activation of transcription by Stat1 and Stat3 requires both tyrosine and serine phosphorylation. Cell 82:241-250.
18. **Ng J, Cantrell D**. 1997 STAT3 is a serine kinase target in T lymphocytes. J Biol Chem 272:24542-24549.

19. **Zhang X, Blenis J, Li HC, Schindler C, Chen-Kiang S**. 1995 Requirement of serine phosphorylation for formation of STA-promoter complexes. Science 267:1990-1994.
20. **Zhang JJ, Zhao Y, Chait BT, Lathem WW, Ritzi M, Knippers R, Darnell JE Jr**. 1998 Ser727-dependent recruitment of MCM5 by Stat1alpha in IFN-gamma-induced transcriptional activation. EMBO J 17:6963-6971.
21. **Deckeer T, Kovarik P.** 2000. Serine phosphorylation of STATs. Oncogene 19:2628-2637.
22. **Bhattacharya S, Eckner R, Grossman S, Oldread E, Arany Z, D'Andrea A, Livingston DM**. 1996 Cooperation of Stat2 and p300/CBP in signalling induced by interferon α. Nature 383:344-347.
23. **Zhang JJ, Vinkemeier U, Gu W, Chakravarti D, Horvath CM, Darnell JE Jr**. 1996 Two contact regions between Stat1 and CBP/p300 in interferon gamma signaling. Proc Natl Acad Sci U S A. 93:15092-15096.
24. **Pfitzner E, Jahne R, Wissler M, Stoecklin E, Groner B**. 1998 P300/CREB-binding protein enhances the prolactin-mediated transcriptional induction through direct interaction with the transaction domain of Stat5, but does not participate in the Stat5-mediated suppression of the glucocorticoid response. Mol Endocrinol 12:1582-1593.
25. **Zhu M, John S, Berg M, Leonard WJ**. 1999 Functional association of Nmi with Stat5 and Stat1 in IL-2- and IFNgamma-mediated signaling. Cell 96:121-130.
26. **Look DC, Pelletier MR, Tidwell RM, Roswit WT, Holtzman MJ**. 1995 Stat1 depends on transcriptional synergy with Sp1. J Biol Chem 270:30264-30267.
27. **Shaefer TS, Sanders LK, Nathans D**. 1995 Cooperative transcriptional activity of Jun and Stat3β, a short form of Stat3. Proc Natl Acad Sci USA 92:9097-9101.
28. **Vinkemeier U, Cohen SL, Moarefi I, Chait BT, Kuriyan J, Darnell JE Jr**. 1996 DNA binding of in vitro activated Stat1a, Stat1b and truncated Stat1: interaction between NH2-terminal domains stabilizes binding of two dimers to tandem DNA sites. EMBO J 15:5616-5626.
29. **Vinkemeier U, Moarefi I, Darnell JE Jr, Kuriyan J**. 1998 Structure of the amino-terminal protein interaction domain of STAT-4. Science 279:1048-1052.
30. **John S, Vinkemeier U, Soldaini E, Darnell JE Jr, Leonard WJ**. 1999 The significance of tetramerization in promoter recruitment by Stat5. Mol Cell Biol 19:1910-1918.
31. **Kim TK, Maniatis T**. 1996 Regulation of interferon-gamma-activated STAT1 by the ubiquitin-proteasome pathway. Science 273:1717-1719.
32. **Imada K, Leonard WJ**. 2000 The Jak-STAT pathway. Molec Immunol 37:1-11.
33. **Shuai K**. 2000 Modulation of STAT signaling by STAT-interacting proteins. Oncogene 19:2638-2644.
34. **Yasukawa H, Sasaki A, Yoshimura A**. 2000 Negative regulation of cytokine signaling pathways. Annu Rev Immunol 18:143-164.
35. **Kovanen PE, Leonard WJ**. 1999 Cytokine signaling: inhibitors keep cytokines in check. Curr Biol 9:R899-902.
36. **Rodig S, Meraz MA, White JM, Lampe PA, Riley JK, Arthur CD, King KL, Sheehan KCF, Yin L, Pennica D, Johnson EM Jr, Schreiber RD**. 1998 Disruption of the Jak1 gene demonstrates obligatory and nonreduntant roles of the Jaks in cytokine-induced biologic responses. Cell 93:373-383.
37. **Neubauer H, Cumano A, Müller M, Wu H, Huffstadt U, Pfeffer K.** 1998 Jak2 deficiency defines an essential developmental checkpoint in definitive hematopoiesis. Cell 93:397-409.
38. **Parganas E, Wang D, Stravopodis D, Topham DJ, Marine J-C, Teglund S, Vanin E, Bodner S, Colamonici OR, van Deursen JM, Grosveld G, Ihle JN**. 1998 Jak2 is essential for signaling through a variety of cytokine receptors. Cell 93:385-395.
39. **Noguchi M, Yi H, Rosenblatt HM, Filipovich AH, Adelstein S, Modi WS, McBride OW, Leonard WJ**. 1993 Interleukin-2 receptor γ chain mutation results in X-linked severe combined immunodeficiency in humans. Cell 73:147-157.

40. **Macchi P, Villa A, Gillani S, Sacco MG, Frattini A, Porta F, Ugazio AG, Johnston JA, Candotti F, O'Shea JJ, Vezzoni P, Notarangelo LD**. 1995 Mutations of Jak-3 gene in patients with autosomal severe combined immune deficiency (SCID). Nature 377:65-68.
41. **Russell SM, Tayebi N, Nakajima H, Riedy MC, Roberts JL, Aman MJ, Migone T-S, Noguchi M, Markert ML, Buckley RH, O'Shea JJ, Leonard WJ**. 1995 Mutation of Jak3 in a patient with SCID: Essential role of Jak3 in lymphoid development. Science 270:797-800.
42. **DiSanto JP, Müller W, Guy-Grand D, Fischer A, Rajewsky K**. 1995 Lymphoid development in mice with a targeted deletion of the interleukin-2 receptor γ chain. Proc Natl Acad Sci USA 92:377-381.
43. **Cao X, Shores EW, Hu-Li J, Anver MR, Kelsall BL, Russell SM, Drago J, Noguchi M, Grinberg A, Bloom ET, Paul WE, Katz SI, Love PE, Leonard WJ**. 1995 Defective lymphoid development in mice lacking expression of the common cytokine receptor γ chain. Immunity 2:223-238.
44. **Nosaka T, van Deursen JM, Tripp RA, Thierfelder WE, Witthuhn BA, McMickle AP, Doherty PC, Grosveld GC, Ihle JN**. 1995 Defective lymphoid development in mice lacking Jak3. Science 270:800-802.
45. **Thomis DC, Gurniak CB, Tivol E, Sharpe AH, Berg LJ**. 1995. Defects in B lymphocyte maturation and T lymphocyte activation in mice lacking Jak3. Science 270:794-797.
46. **Shimoda K, Kato K, Aoki K, Matsuda T, Miyammoto A, Shibamori M, Yamashita M, Numata A, Takase K, Kobayashi S, Shibata S, Asano Y, Gondo H, Sekiguchhi K, Nakayama K, Nakayama T, Okamura T, Okamura S, Niho Y, Nakayama K.** 2000 Tyk2 plays a restricted role in IFNα signaling, although it is required for IL-12-mediated T cell function. Immunity 13:561-571.
47. **Karaghiosoff M, Neubauer H, Lassnig C, Kovarik P, Schindler H, Pircher H, McCoy B, Bogdan C, Decker T, Brem G, Pfeffer K, Muller M**. 2000 Partial impairment of cytokine responses in Tyk2-deficient mice. Immunity 13:549-560.
48. **Durbin JE, Hackenmiller R, Simon MC, Levy DE.** 1996 Targeted disruption of the mouse Stat1 gene results in compromised innate immunity to viral disease. Cell 84:443-450.
49. **Meraz MA, White JM, Sheehan KC, Bach EA, Rodig SJ, Dighe AS, Kaplan DH, Riley JK, Greenlund AC, Campbell D, Carver-Moore K, DuBois RN, Clark R, Aguet M, Schreiber RD.** 1996 Targeted disruption of the Stat1 gene in mice reveals unexpected physiologic specificity in the JAK-STAT signaling pathway. Cell 84:431-442.
50. **Park C, Li S, Cha E, Schindler C**. 2000 Immune response in Stat2 knockout mice. Immunity 13:795-804.
51. **Takeda K, Noguchi K, Shi W, Tanaka T, Matsumoto M, Yoshida N, Kishimoto T, Akira S.** 1997 Targeted disruption of the mouse Stat3 gene leads to early embryonic lethality. Proc Natl Acad Sci U S A 94:3801-3804.
52. **Akira S**. 2000 Roles of STAT3 define by tissue-specific gene targeting. Oncogene 19:2607-2611.
53. **Kaplan MH, Sun YL, Hoey T, Grusby MJ**. 1996 Impaired IL-12 responses and enhanced development of Th2 cells in Stat4-deficient mice. Nature 382:174-177.
54. **Thierfelder WE, van Deursen JM, Yamamoto K, Tripp RA, Sarawar SR, Carson RT, Sangster MY, Vignali DA, Doherty PC, Grosveld GC, Ihle JN**. 1996 Requirement for Stat4 in interleukin-12-mediated responses of natural killer and T cells. Nature 382:171-174.
55. **Liu X, Robinson GW, Gouilleux F, Groner B, Hennighausen L**. 1995. Cloning and expression of Stat5 and an additional homologue (Stat5b) involved in prolactin signal transduction in mouse mammary tissue. Proc Natl Acad Sci USA 92:8831-8835.

56. **Udy GB, Towers RP, Snell RG, Wilkins RJ, Park S-H, Ram PA, Waxman DJ, Davey HW**. 1997 Requirement of STAT5b for sexual dimorphism of body growth rates and liver gene expression. Proc Natl Acad Sci USA 94:7239-7244.
57. **Nakajima H, Liu XW, Wynshaw-Boris A, Rosenthal LA, Imada K Finbloom DS, Hennighausen L, Leonard WJ**. 1997 An indirect effect of Stat5a in IL-2-induced proliferation: a critical role for Stat5a in IL-2-mediated IL-2 receptor α chain induction. Immunity 7:691-701.
58. **Imada K, Bloom ET, Nakajima H, Horvath-Arcidiacono JA, Udy GB, Davey HW, Leonard WJ**. 1998 Stat5b is essential for natural killer cell-mediated proliferation and cytolytic activity. J Exp Med 188:2067-2074.
59. **Kagami S-I, Nakajima H, Kumano K, Suzuki K, Suto A, Imada K, Davey HW, Saito Y, Takatsu K, Leonard WJ, Iwamoto I.** 1999. Both Stat5a and Stat5b are required for antigen-induced eosinophil and T-cell recruitment into the tissue. Blood 95:1370-1377.
60. **Moriggl R, Topham DJ, Teglund S, Sexl V, McKay C, Wang D, Hoffmeyer A, van Deursen J, Sangster MY, Bunting KD, Grosveld GC, Ihle JN**. 1999 Stat5 is required for IL-2-induced cell cycle progression of peripheral T cells. Immunity 10:249-259.
61. **Kaplan MH, Schindler U, Smiley ST, Grusby MJ.** 1996 Stat6 is required for mediating responses to IL-4 and for development of Th2 cells. Immunity 4:313-319.
62. **Shimoda K, van Deursen J, Sangster MY, Sarawar SR, Carson RT, Tripp RA, Chu C, Quelle FW, Nosaka T, Vignali DA, Doherty PC, Grosveld G, Paul WE, Ihle JN**. 1996 Lack of IL-4-induced Th2 response and IgE class switching in mice with disrupted Stat6 gene. Nature 380:630-633.
63. **Yu CL, Meyer DJ, Campbell GS, Larner AC, Carter-Su C, Schwartz J, Jove R**. 1995 Enhanced DNA-binding activity of a Stat3-related protein in cells transformed by the Src oncoprotein. Science 269:81-83.
64. **Bromberg JF, Wrzeszczynska MH, Devgan G, Zhao Y, Pestell RG, Albanese C, Darnell JE Jr**. 1999 Stat3 as an oncogene. Cell 98:295-303.
65. **Migone T-S, Lin J-X, Cereseto A, Mulloy JC, O'Shea JJ, Franchini G, Leonard WJ**. 1995 Constitutively activated Jak-STAT pathway in T cells transformed with HTLV-I. Science 269:79-81.
66. **Takemoto S, Mulloy JC, Cereseto A, Migone T-S, Patel BK, Matsuoka M, Yamaguchi K, Takatsuki K, Kamihira S, White JD, Leonard WJ, Waldmann T, Franchini G**. 1997 Proliferation of adult T cell leukemia/lymphoma cells is associated with the constitutive activation of JAK/STAT proteins. Proc Natl Acad Sci U S A 94:13897-13902.
67. **Danial NN, Pernis A, Rothman PB**. 1995 Jak-STAT signaling induced by the v-abl oncogene. Science 269:1875-1877.
68. **Weber-Nordt RM, Egen C, Wehinger J, Ludwig W, Gouilleux-Gruart V, Mertelsmann R, Finke J.** 1996 Constitutive activation of STAT proteins in primary lymphoid and myeloid leukemia cells and in Epstein-Barr virus (EBV)-related lymphoma cell lines. Blood 88:809-816.
69. **Gouilleux-Gruart V, Gouilleux F, Desaint C, Claisse JF, Capiod JC, Delobel J, Weber-Nordt R, Dusanter-Fourt I, Dreyfus F, Groner B, Prin L**. 1996 STAT-related transcription factors are constitutively activated in peripheral blood cells from acute leukemia patients. Blood 87:1692-1697.
70. **Chai SK, Nichols GL, Rothman P**. 1997 Constitutive activation of JAKs and STATs in BCR-Abl-expressing cell lines and peripheral blood cells derived from leukemic patients. J Immunol 159:4720-4728.
71. **Asao H, Okuyama C, Kumaki S, Ishii N, Tsuchiya S, Foster D, Sugamura K.** 2001 The common γ-chain is an indispensable subunit of the IL-21 receptor complex. J Immunol. 167:1-5.
72. **Ozaki K, Kikly K, Michalovich D, Young PR, Leonard WJ**. 2000 Cloning of a type I cytokine receptor most related to the IL-2 receptor β chain. Proc Natl Acad Sci U S A. 97:11439-44.

7

Prolactin and Growth Hormone Receptors

Signal Transduction and Crosstalk

Li-yuan Yu-Lee[1*] and Sébastien Jeay[2]

[1]*Departments of Medicine, Molecular & Cellular Biology, and Immunology, and Cell & Molecular Biology Program, Baylor College of Medicine, Houston, Texas, USA*
[2] *INSERM Unit 344, Molecular Endocrinology, Faculté de Médecine Necker, 156 rue de Vaugirard, 75730 Paris Cedex 15, France*

INTRODUCTION

The pituitary hormones prolactin (PRL) and growth hormone (GH) share structural similarities with the helix bundle helix peptide hormones, many of which are cytokines [1]. The receptors for PRL (PRLR) and GH (GHR) also share structural as well as functional similarities with members of the Class I cytokine receptor or hematopoietin cytokine receptor superfamily. The diverse biological functions of PRL and GH as endocrine hormones as well as autocrine/paracrine cytokines will be discussed. This chapter focuses on the structure/function of the PRLR, the kinase cascades that are activated by both PRL and GH, the known signaling molecules that mediate their functions, and examples of how PRL signaling regulates the expression of target genes. How PRL crosstalks with other cytokine signaling pathway will also be introduced. For more in-depth discussion on PRL and its actions, the readers are referred to Ref. [2] and chapters within. For excellent reviews on GH, its receptor and signaling pathways, the readers are referred to Ref. [3-6].

[*] yulee@bcm.tmc.edu

PRL ACTIONS AND TARGET TISSUES

PRL is a peptide hormone that is synthesized and secreted primarily by lactotrophs in the anterior pituitary. PRL is also expressed in a number of extra-pituitary sites [7], ranging from neurons in the brain to epithelial cells of secretory glands to cells of the immune system [8,9]. PRL exerts profound effects on a wide range of tissues, with over 300 effects described in vertebrates [9]. PRL regulates differentiation of the mammary gland, ovary, male sex accessory organs, submaxillary and lacrimal glands, pancreas and liver [1]. PRL regulates proliferation in diverse tissues, including pigeon crop sac epithelium, pancreatic β cells, astrocytes, anterior pituitary cells and T lymphocytes [1,2]. PRL also exerts anti-apoptotic effects in lymphocytes undergoing glucocorticoid-induced programmed cell death. In mammals, PRL is primarily involved in reproduction and lactation. In the ovary, PRL participates in the differentiation of the corpus luteum and regulates progesterone synthesis. In the mammary gland, PRL regulates the growth and differentiation of the mammary epithelium and the synthesis of milk proteins [2,10,11]. Knockout (KO) mice lacking either the PRL or the PRLR gene revealed additional PRL targets, including the brain (maternal behavior), uterus (implantation) and bone (osteoblasts) [2,12].

Clinically, high PRL levels have been associated with certain autoimmune diseases [13]. Either higher or lower than normal PRL levels have been shown to compromise the immune responses in animal models [14]. In tissue culture, PRL has been shown to modulate the proliferation and effector functions of T cells, B cells, macrophages and natural killer cells [15]. *In vivo*, thymocytes and splenocytes synthesize and secrete PRL, which may act as autocrine or paracrine growth factors [8,9]. How pituitary and extra-pituitary PRL modulate the growth and differentiation of target tissues also depends on the cell type and its stage of differentiation.

PRLR SIGNAL TRANSDUCTION

PRLR Structure/Function

The diverse activities of PRL are mediated by the PRLR which is found on a wide variety of cell types. Although there is only one PRLR gene, a long (90 – 95 kDa) and a short (42 kDa) form of the PRLR exist, the latter resulting from differential splicing of 3' end exons encoding the cytoplasmic domain (Fig. 1) [16]. In addition, a third naturally occurring intermediate PRLR form

(65 kDa), found in Nb2 T lymphoma cells, results from an in-frame truncation in the intracytoplasmic domain (ICD), generating a shortened receptor tail. A truncated PRLR resembling the Nb2 form has been cloned from human mammary tumors but its functional status remains to be determined [2,17].

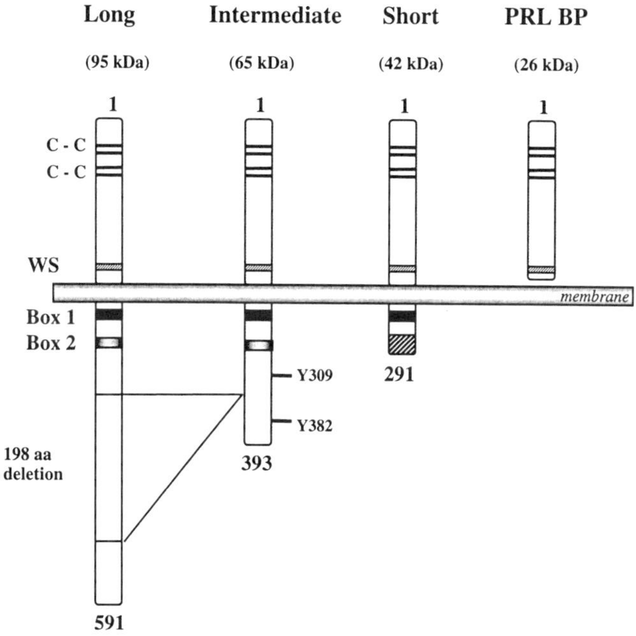

Figure 1. Schematic representation of the rat PRLR isoforms. The extracellular domain (1 – 210 aa) of the PRLR contains a conserved cysteine doublet (C-C) and a WS motif also found in other Type I cytokine receptors. The intracellular domain (235 – 591 aa) of the PRLR contains a conserved Box 1 or proline-rich motif that interacts with JAK2, and an acidic Box 2 motif. An in-frame deletion (323 – 520 aa) generates a truncated intermediate form. The last exon of the short PRLR is unique. A membrane truncation generates a soluble PRL binding protein (PRLBP). Tyrosine residues Y309 and Y382 are highlighted in the intermediate Nb2 receptor (see text for discussion). The equivalent Y residues in the long PRLR isoform are not shown.

Upon ligand binding, the PRLR undergoes homodimerization, which generates the active signaling complex. Interestingly, the half-life of PRL binding to the PRLR is estimated to be about 0.6 min while GH binding to the GHR is stable for about 30 min [18]. This difference in the rate of ligand/receptor interaction suggests intrinsic differences in signaling capacities of the two closely-related receptor systems. The Nb2 PRLR is more potent than the long PRLR in both mitogenic [19] and lactogenic [12] signaling. Although the signaling capacity of the short PRLR is unclear, its steady-state levels change with the physiologic states in some tissues, leading to the suggestion that the short PRLR modulates long or Nb2 PRLR signaling by engaging them in heterodimer complex formation [2,12]. In the Nb2 T cells, which express 12,000 PRLR/cell, only 30% occupancy of surface PRLR is needed to elicit maximal biological response [18].

PRLR Motifs

Several ICD motifs in the PRLR are conserved throughout the cytokine receptor superfamily and are important for signal transduction (see Chapter 6). These include the 8 amino acid (aa) Box 1 proline-rich sequence (I-F-P-P-V-P-X-P, where X is non consensus aa) proximal to the transmembrane domain, which is thought to interact with the receptor-associated protein tyrosine kinase (PTK) JAK2 [16]. Although this relationship has been determined by receptor mutagenesis and transfection analyses, it is still unclear how JAK2 physically interacts with the Box 1 region of the PRLR. Other regions such as the acidic Box 2 sequences and specific tyrosine residues are also important for PRLR signaling. Multiple tyrosines in the long PRLR ICD and only the last tyrosine Y382 in the Nb2 PRLR, or its equivalent in the long PRLR, are important for mediating differentiated functions [16], while both Y309 and Y382 in the Nb2 PRLR are needed for signaling to an immediate early response gene, interferon regulatory factor-1 (IRF-1) [20] (Fig. 1). One function of phosphorylated receptor tyrosine residues is to provide a "docking site" for the recruitment of SH2-containing proteins, including Stats, phosphatases and adaptors [16]. Other as yet undefined PRLR ICD regions interact either constitutively or inducibly with other kinases and cytoplasmic factors that affect PRLR signaling in target tissues [2,21].

JAK/Stat Pathway Activated by PRL

Our understanding of how PRL delivers diverse signals to target cells is facilitated by the elucidation of the "JAK/Stat pathway" [22] as the prototype signaling pathway used by all hematopoietin/cytokine receptors (see Chapter 6; Fig. 2). Briefly, PRL binding to the PRLR leads to receptor dimerization or oligomerization, which leads to the activation of PRLR-associated JAK2 [16]. Activated JAK2 then phosphorylates downstream targets, including the PRLR ICD itself, a family of latent transcription factors called Stats (signal transducers and activators of transcription), and other SH2-containing signaling molecules [9,19]. Activated Stats form homo- or heteromeric complexes, translocate into the nucleus, bind conserved DNA elements called interferon (IFN) gamma activated sequence (GAS) and regulate target gene transcription.

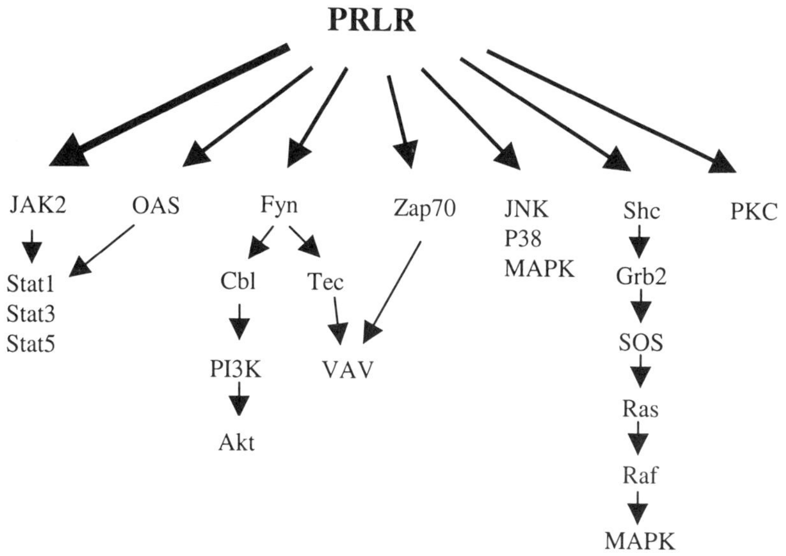

Figure 2. PRL-inducible JAK/Stat and parallel kinase signaling pathways. One of the main signaling pathways activated by PRL stimulation is the JAK/Stat pathway (bold arrow). However, other parallel kinase and signaling cascades are also activated. JAK2, Fyn and OAS have been shown to be constitutively associated with the PRLR, while the other molecules are found in the PRLR complex after ligand binding. Although the kinase cascades are depicted in parallel, extensive cross talk between the JAK/Stat and other pathways is well documented. How cells respond to PRL is likely to be determined by a combination of the various kinase pathways activated in a cell type specific manner.

JAK2 and Parallel Kinase Pathways Activated by PRL

One of the first molecules to be activated in the PRLR signaling pathway is the PTK JAK2. It is interesting to note that JAK2 is prebound to the inactive PRLR monomer. Mutagenesis and *in vitro* binding analyses have shown that the amino terminus of the 120 kDa JAK2 is involved in binding to the PRLR, while the carboxyl terminus is critical for JAK2 tyrosine kinase activity [4]. In addition to the JAK/Stat signaling pathway (see Chapter 6), other PTKs such as Fyn, Src, ZAP-70, Tec, as well as serine/threonine kinases such as Ras, Raf, PI3 kinase, MAPK, JNK and protein kinase C [2], have all been suggested to play a role in PRLR signaling in different target cells (Fig. 2). While more is known about the JAK/Stat signaling pathway, activation of other parallel kinase cascades in coordination with the JAK/Stat signaling pathway elicits specific patterns of gene expression in the various PRL responsive cells and tissues. The pleiotropic actions of PRL on cellular proliferation, differentiation or apoptosis will likely be determined by the activation and interactions of these parallel kinase cascades.

PRLR Signaling Molecules

Stat Factors

Stat factors are one of the earliest mediators of signaling from cytokine receptors [22-25]. Seven mammalian Stat genes, Stat1, 2, 3, 4, 5a, 5b and 6, have been identified. Stat factors are in general 750 – 800 amino acids in size and contain distinct functional domains, including a coiled-coiled domain, DNA binding domain, linker domain, SH2 domain, a critical tyrosine residue that is important for dimerization, nuclear translocation and DNA binding, and a carboxyl terminus transactivation domain (see also Chapter 6). Additional serine [23] and tyrosine [26] residues in the carboxyl terminus of Stat1, Stat3 and Stat5 further modify transactivation potential. Stat1, 3, 5a, 5b and 6 have naturally occurring splice variants with truncations in the carboxyl terminus, generating dominant negative β isoforms [27,28]. Stat factors utilize various domains to interact/crosstalk with a diverse set of proteins, both in the cytoplasm and nucleus, to mediate target gene transcription.

Stats Interact with Cytoplasmic Proteins

Stat factors reside basally in the cytoplasm of unstimulated cells. Within minutes of cytokine stimulation, Stats are recruited into the receptor complex and, through their SH2 domain, interact with phosphorylated tyrosine residues on the receptor tail and in turn become tyrosine phosphorylated, presumably by JAK PTK. Stats may also interact directly with JAK itself or serve as adapters to couple PI3 kinase to receptors [29]. The amino terminus of Stats contains a region that can undergo Stat/Stat tetramer formation [30]. The coiled-coil domain of Stats (except Stat2) can complex with the cytoplasmic N-myc interacting protein (Nmi) [22], and the resulting Stat/Nmi complex translocates into the nucleus to enhance Stat transactivation potentials. Other studies have isolated various Stat interacting proteins, including StIP1 (Stat3 interacting protein) which interacts with both JAK2 and Stat3 [31] and PIAS (protein inhibitors of activated Stats) [24]. Stat1 also interacts with the nuclear transport importin α/β complex for import into the nucleus [32]. These studies show that in the cytoplasm, Stats acquire signal-transducing capability *via* interactions with cytokine receptors, JAK and other factors to mediate transcriptional responses.

Stats Interact with Nuclear Proteins

Activated Stat complexes translocate within minutes into the nucleus [22]. Once in the nucleus, Stats can interact with other nuclear proteins, bind to cognate DNA elements (ISRE or GAS), and participate in gene transcription (activators of transcription function). Stats exhibit different affinities for different binding sites. However, higher order Stat/Stat tetramers can activate transcription from weak binding sites [30]. The transactivation potentials of Stats are further modulated by interactions with other proteins, such as p48 (a member of the IRF family), IRF-1, c-jun, Sp1, Src, nuclear hormone receptors, MCM5 and BRCA1 [11,24,25,33]. Although full-length proteins carry out the majority of Stat functions, both Stat3β and Stat5aβ can activate gene transcription by complexing with other transcription factors and utilizing their transactivation domain for activity [27,28]. These studies illustrate that the functional activities of Stats can be modulated by interaction with other DNA binding proteins at target promoters.

Coactivators

A recurrent theme in transcriptional regulation is the extensive protein/protein interactions that occur at the level of the target promoters. A class of nuclear factors called coactivators mediates the functional integration of various transcription factors at target genes [34]. The most extensively characterized coactivators are the CBP/p300 proteins which have several interesting properties. First, CBP/p300 do not bind DNA directly, but they act as a scaffold not only to integrate the activities of multiple DNA binding proteins, but also to facilitate interactions with other coactivators as well as with components of the basal transcription machinery. Second, CBP/p300 exhibit intrinsic histone acetyltransferase (HAT) activities, which acetylate histones and lead to remodeling of chromatin and transcriptional activation of genes. The functional activities of Stats are enhanced by their interactions with coactivators. Indeed, Stat1 can recruit multiple coactivators for transcriptional activation. Three regions in CBP/p300 interact with Stat1 [25,35], where one of these regions also interacts with Stat5 [36]. It is interesting to speculate that Stat5 competition with Stat1 for binding to CBP/p300 forms one basis for competitive action between these two Stats at target promoters (see below). Thus, coactivators can integrate the activities of multiple Stats or Stats with non-Stat proteins in regulation of gene transcription. On the other hand, coactivators can also become targets of competitive binding by different Stats or with non-Stat proteins which results in inhibition of gene transcription. In this manner, Stats can inhibit the functions of other transcription factors by "squelching" limiting pools of CBP/p300 as one mechanism of inhibition of promoter activity [34,37].

PRL SIGNALING TO TARGET GENES

To understand PRL signaling to nuclear targets, the expression of two PRL-inducible genes has been extensively characterized, both of which are primarily activated by PRL inducible Stats. The milk protein β-casein gene is a marker of PRL-inducible differentiated functions in mammary epithelial cells [2,11]. The IRF-1 gene is a PRL-inducible immediate-early response gene that is activated during mitogenic stimulation in T cells [8,38]. For a list of PRL-inducible genes in Nb2 T cells obtained by various differential screening techniques, the readers are referred to [39].

PRL Inducible Stats

PRL stimulates the rapid tyrosine phosphorylation of Stat1, Stat3 and Stat5 in many cell types [40-42]. Stat5a was first cloned as a PRL-inducible mammary gland factor (MGF), which is followed by the cloning of the closely-related Stat5b [43,44]. PRL stimulates primarily Stat5 to regulate transcription of the milk protein genes β-casein, αs1-casein, β-lactoglobulin, and whey acidic protein [11,42]. PRL stimulates Stat1 to regulate transcription of the IRF-1 gene [20]. Unexpectedly, PRL inducible Stat5 inhibits IRF-1 promoter activity [37,44].

PRL Inducible Genes

PRL Stimulates β-casein Gene Transcription

In regulating the transcription of the β-casein milk protein gene, PRL does not act alone but works in concert with two other hormones, insulin and hydrocortisone. The combined actions of the three lactogenic hormones are integrated at the level of composite response elements (-70/-230 bp) that contain binding sites for Stat5, C/EBPβ and the glucocorticoid receptor (GR) [2,11]. In the absence of the lactogenic hormones, the β-casein promoter is repressed by the transcription factor YY1 (ying yang 1). This repression is relieved upon PRL stimulation, as PRL-inducible Stat5 binding to a GAS element (-82/-176 bp) weakens YY1 interaction at the β-casein promoter. The functional activity of Stat5 is further enhanced by its interaction with GR which increases Stat5 tyrosine phosphorylation and DNA binding, and could account in part for the synergistic effects of glucocorticoids and PRL on β-casein gene transcription [11]. GR also enhances transcriptional activity of C/EBPβ which binds to several sites upstream of the GAS element [45]. Thus, GR is critical for cooperative transactivation by Stat5 and C/EBPβ at the β-casein promoter. The actions of these factors at the β-casein composite elements are further integrated by their interactions with coactivators such as p300 [2,36]. Thus, PRL signaling to the β-casein promoter involves a relief of repression by PRL-inducible Stat5, whose activity is enhanced by cooperative interactions with GR, GR with C/EBPβ, and presumably their combined interactions with coactivators. Interestingly, Stat5 KO mice show that Stat5a plays a major role in mediating PRL signaling to milk protein genes, while Stat5b is important for GH signaling to male-specific liver gene expression [46].

PRL Stimulates IRF-1 Gene Transcription

PRL stimulates the expression of a panel of immediate-early genes in Nb2 T cells, as part of a mitogenic signaling pathway (see also Ref. [39]). IRF-1, one of the most PRL responsive genes, is itself a multifunctional transcription factor that regulates the expression of a host of genes important for mediating immune responses [8]. Additionally, mutations and/or deletions in IRF-1 are correlated with a high incidence of leukemias and myelodysplasia, suggesting a role of IRF-1 in tumor suppression [47]. In Nb2 T lymphoma cells, PRL stimulates the biphasic transcription of the IRF-1 gene, with a transient but dramatic 25-fold induction during early G1 (1 hr) and a second peak of induction at the G1/S transition phase (8 – 10 hr) of the cell cycle [38].

Positive mediators: Stat1, CBP/p300 and Sp1. The rapid IRF-1 gene transcriptional response to PRL stimulation during G1 is positively mediated by at least three factors: PRL-inducible Stat1 binding to a GAS element at -110/-120 bp [38], constitutive Sp1 binding at -200 bp [48], and protein-protein interaction between Stat1 and the coactivator CBP/p300 [37] (Fig. 3A). Our working hypothesis is that upon PRL stimulation, activated Stat1 binds to the IRF-1 GAS and together with the pre-bound Sp1 forms an enhanceosome (assembly of transcription factors) [49], which recruits coactivators such as CBP/p300 and CRSP (cofactor required for Sp1) [50], as well as the general transcription machinery, for transcriptional activation of the IRF-1 promoter. Using chromatin immunoprecipitation assays (ChIP) to identify DNA that are found in "active" chromatin, we show that the IRF-1 promoter is associated with acetylated H4 histones in response to PRL stimulation, thus linking HAT activities and chromatin remodeling at the IRF-1 promoter with PRL stimulation of IRF-1 gene transcription *in vivo* [51]. Further ChIP analysis would elucidate how Stat1, Sp1 and CBP/p300, together with chromatin remodeling and/or cell cycle regulated factors, coordinate PRL activation of IRF-1 gene transcription *in vivo*.

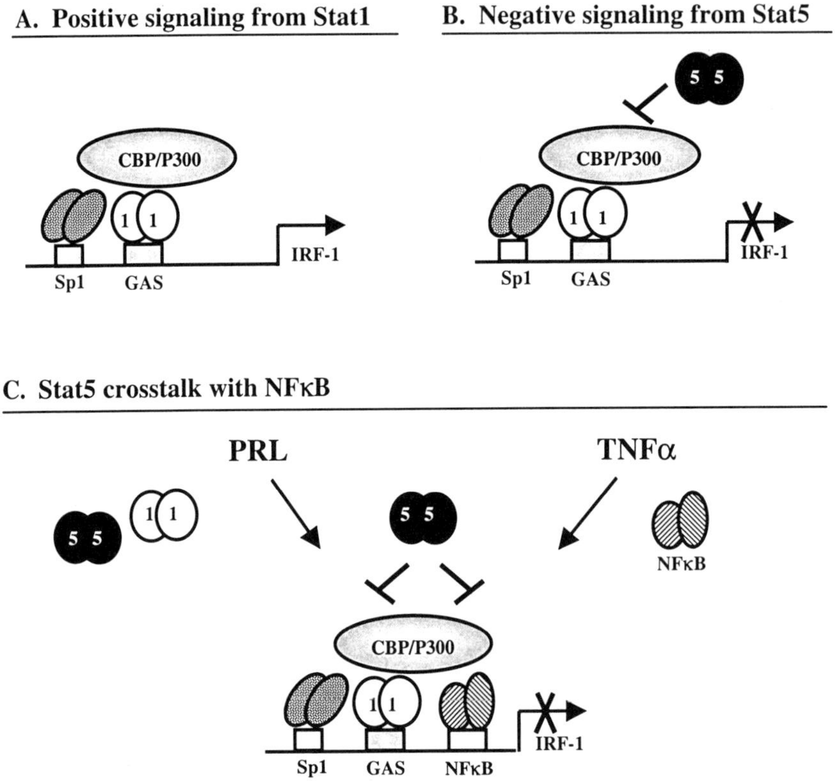

Figure 3. Mechanisms of PRL regulation of IRF-1 gene transcription: Stat signaling and crosstalk. Black and white circles represent Stats 5 and 1, respectively. See text for discussion.

Negative mediators: Stat5 and corepressors. Interestingly, although Stat5 does not need to bind to the IRF-1 GAS, Stat5 is detected as a minor component in the PRL-inducible IRF-1 GAS complex during G1 PRL stimulation of Nb2 T cells [41]. Surprisingly, the functional consequence of Stat5 interaction at the IRF-1 promoter is one of transcriptional repression [37,44]. Both the amino and carboxyl termini but not the DNA binding activity of Stat5b are required for inhibition, suggesting that Stat5b inhibits the IRF-1 promoter without binding to DNA. Our working hypothesis is that Stat5b is not competing with Stat1 for binding to the IRF-1 promoter, but that Stat5b is competing for a factor *via* protein/protein interactions ("squelching") to inhibit PRL signaling to the IRF-1 promoter [19] (Fig. 3B). One target of Stat5b inhibition at the IRF-1 promoter is the coactivator p300 [37]. Thus, PRL signaling to the IRF-1 promoter involves Stat1 binding to the IRF-1 GAS and cooperative interactions between Stat1, Sp1 and coactivators to promote IRF-1 transcription. PRL signaling to Stat5 appears to squelch limiting amounts of coactivators, leading to IRF-1 promoter inactivation which likely contributes to IRF-1 gene down-regulation. Recent studies show that Stat5b can interact with corepressor proteins in a transient overexpression assay (data not shown). Thus, two distinct mechanisms may mediate transcriptional repression by Stat5: Stat5 can squelch coactivators and/or recruit corepressors at target promoters to shut down gene transcription. Stat5a/5b [52] KO mice showed that the absence of Stat5 resulted in the "increased expression" or "derepression" of certain genes, thus confirming that Stat5 can act as transcriptional repressors *in vivo*.

Dichotomous Stat functions. Intriguingly, the functional consequence of PRL-inducible Stat5 interaction with CBP/p300 is distinct depending on the target promoter, that is, transcriptional upregulation at the β-casein promoter but transcriptional repression at the IRF-1 promoter. Such dichotomous Stat functions are reflected in the developing mammary gland where activated Stat1 in the virgin and early pregnant gland (highly proliferative) is correlated with high IRF-1 expression while activated Stat5 in the lactating gland (fully differentiated) is associated with the complete absence of IRF-1 expression but maximal β-casein expression (data not shown). While the mechanistic details of how Stats function in such opposite ways at different promoters are still unclear, these studies show that transcriptional regulation by Stats is a complex process. Stats can act as transcriptional activators or transcriptional repressors, depending on the target promoter, the complement of coactivators, corepressors and other DNA binding proteins which are recruited into the promoter-specific "enhanceosome", and the stage of differentiation of the target cell.

Signals that Down-Regulate PRLR Signaling

Up to now, the focus has been on how PRL activates the JAK/Stat signaling pathway, *via* a well-orchestrated series of phosphorylation events. However, most robust cytokine signaling is transient. How then is signaling turned off? Two types of SH2-containing protein tyrosine phosphatases (PTP) [12,53] have been implicated in turning off cytokine signaling, SHP-1 which is found primarily in hematopoietic cells and SHP-2 which is ubiquitous (see Chapter 2). PTPs are recruited into the receptor complex *via* their SH2 domains, become activated by tyrosine phosphorylation, and are thought to dephosphorylate JAKs and/or the cytokine receptor, thereby shutting down the signaling process [12]. However, not all PTPs are involved in receptor down-regulation, as SHP-2 is necessary for initiating PRLR [12] and GHR [54] signaling. A cycle of JAK2/SHP-1 tyrosine phosphorylation/de-phosphorylation events appears to tightly regulate Stat5b activation and inactivation in response to GHR signaling in the liver [46]. Further, PTPs have been suggested to directly inactivate Stats in the nucleus [55]. Using the GHR or PRLR cytoplasmic domains as baits, molecules that attenuate GH or PRL signaling have been identified. These include Grb10 which down-regulates JAK2 signaling [6] and 2'5'-oligoadenylate synthetase (OAS) which reduces Stat1 tyrosine phosphorylation and DNA binding [21]. Interestingly, a group of molecules whose expression is induced by PRL feedbacks to inhibit further signaling from the PRLR. These include SOCS (suppressors of cytokine signaling) which bind to and inhibit JAKS [56], and PIAS which bind to and inhibit Stats [24](see Chapter 6). Interestingly, not all SOCS proteins inhibit receptor signaling as SOCS2, which binds directly to the PRLR, potentiates receptor signaling [2,6].

SIGNAL CROSSTALK

Through their extensive interactions with cytoplasmic as well as nuclear factors (see above and Ref. [24]), Stats crosstalk positively as well as negatively with other factors to regulate target gene expression. As one example of such crosstalk, Stats interact physically as well as functionally with another group of signaling molecules which also translocate from the cytoplasm into the nucleus upon activation, NFκB (nuclear factor of κB) [57]. Stat1 synergizes with NFκB while Stat5b antagonizes NFκB signaling to the IRF-1 promoter [37] (Fig. 3C). Interestingly, Stat5b appears to antagonize TNFα mediated NFκB signaling at NFκB target promoters, again through squelching of limiting coactivators. These observations provide a mechanistic understanding of how PRL may antagonize TNFα signaling, by mobilizing Stat5 to inhibit NFκB signaling. This paradigm may underlie one way in which PRL protects against GR-mediated apoptosis or TGFβ mediated immunosuppression in various disease states.

16K PRL

Members of the PRL/GH/placental lactogen (PL)/proliferin (PLF) family have an additional interesting property, that is, regulation of angiogenesis [58,59]. Full-length PRL, GH, PL and PLF display angiogenic activities, stimulating blood vessel formation from the vasculature [60]. In the placenta, PLF promotes angiogenesis during the first half of pregnancy while later in gestation PLF related protein (PRP) is anti-angiogenic which prevents further placental neovascularization. Recently, PRP is also shown to inhibit tumor growth by preventing neovascularization of tumors [59]. An amino-terminal proteolytic cleavage fragment of the 23 kDa PRL or 16K PRL, as well as 16K GH and 16K PL, exhibits anti-angiogenic activity in blocking new capillary formation from existing blood vessels [58], presumably by inducing programmed cell death (apoptosis) in endothelial cells [61]. 16K PRL appears to be made by the hypothalamo-neurohypophyseal system and both 16K PRL and 16K GH can be found in rodent and human serum. Recent studies also show that pulmonary fibroblasts can synthesize and secrete bioactive 16K PRL to regulate NO production, as part of a proinflammatory pathway [62]. 16K PRL also stimulates PAI-1 (plasminogen activator inhibitor-1) expression but inhibits MAPK activation and endothelial cell proliferation, presumably by blocking the Ras/Raf pathway [33,60]. It is interesting that 16K PRL does not work through the PRLR, but appears to bind to a high affinity site distinct from the PRLR [58,62]. Thus, full-length PRL and the PRLR versus 16K PRL and its binding protein display opposing activities on angiogenic and/or inflammatory responses. What the 16K hormones are binding to and whether 16K PRL, 16K GH and 16K PL also have anti-tumor activity are areas of active investigation.

GROWTH HORMONE

GH Target Tissues Activated by GH

GH is synthesized and secreted primarily by somatotrophic cells in the anterior pituitary. Its secretion is regulated by two pituitary neuropeptides: growth hormone releasing hormone (GHRH) which stimulates GH production, and somatostatin which exerts a negative control on GH production. In addition, extra-pituitary sites of GH production have been recently discovered such as mammary gland, placenta [63] and lymphoid tissues [64,65], suggesting an autocrine/paracrine action of GH on these tissues.

As introduced in this Chapter, PRL and GH belong to a family of polypeptide hormones which share structural similarities [1]. However, members of the PRL/GH/placental lactogen (PL)/proliferin (PLF) family have different biological effects. GH exerts multiple actions on a wide range of target tissues, *via* two different modes of action: either GH acts directly on its target tissues, or GH effects are mediated through a local production of insulin-like growth factor-1 (IGF-1; see Chapter 5). IGF-1 expression is directly controlled by GH, but signaling pathway(s) activated by GH and leading to expression of IGF-1 is (are) still unknown.

One of the most important effect of GH is its ability to stimulate growth of skeleton and soft tissues. GH also regulates proliferation in a wide variety of cells, including chondrocytes, fibroblasts, T lymphocytes, myoblasts and adipocytes. These effects were shown to be indirect, mediated by a local production of IGF-1. GH also directly controls differentiation in diverse tissues including muscle and adipose tissue. In addition to its proliferative and differentiative effects, GH is also able to directly regulate metabolism of proteins, carbohydrates, and lipids.

GHR Signal Transduction

JAK/Stat Pathway Activated by GH

As diverse activities of PRL are mediated by PRLR, biological effects of GH are transduced by its own receptor: the GHR. PRLR and GHR belong to the class I superfamily of cytokine receptors, sharing structural and functional similarities [6], and whose transduction processes utilize associated tyrosine kinases of the JAK family and members of the Stat proteins [66]. Indeed, by the same mechanism of action used by PRL, GH binding to its receptor promotes GHR dimerization which leads to the recruitment of the tyrosine kinase JAK2 (JAK2 does not seem to be pre-associated to the GHR, in contrast to what has been previously described for PRLR [4]). Activation of JAK2 by both GH and PRL leads to phosphorylation and activation of the same Stat proteins: Stat1, Stat3, Stat5a and Stat5b. Activation of the JAK/Stat pathway by GH leads to homo- or heterodimerization, then translocation of Stats into the nucleus, and activation of the transcription of GH target genes. For example, Stat1 and Stat3 are able to recognize the SIE (c-sis inducible element) present in the promoter of c-*fos* gene [67]. Also, Stat5 activated by GH is able to bind the GHRE2 element in the promoter of *spi* 2.1 gene [68], and LHRE element of the β-casein gene promoter [69].

Among the various regulators known to control the activation of GH-induced JAK/Stat pathway (e.g. SOCS and SH2 domain-containing phosphatases (SHPs); see Ref. 6 for a review), an additional signaling molecule for GH, called SH2-Bβ, has been identified, which binds to JAK2 and is phosphorylated on tyrosine in response to GH stimulation [70]. SH2-Bβ is a splicing variant of the SH2 domain-containing protein SH2-B, and in 3T3-F442A preadipocytes it has been shown to be a potent cytoplasmic activator of JAK2 as well as of Stat3 and Stat5b, in response to GH (Fig. 4) [70,71]. Furthermore, SH2-Bβ has been shown to be required for GH-induced actin reorganization by a mechanism discrete from the action of SH2-Bβ as a stimulator of JAK2 kinase activity [72]. The same group has also reported two different binding sites for JAK2 on SH2-Bβ [73]. Moreover, a correlation between differential binding to and regulation of JAK2 by the two binding sites of SH2-Bβ has been shown: when JAK2 is bound to the C-terminal containing SH2 region of SH2-Bβ, SH2-Bβ acts as a potent activator of JAK2 and some Stat proteins [71,73], whereas when the JAK2/SH2-Bβ interaction occurs on the N-terminal region of SH2-Bβ, SH2-Bβ acts as a negative regulator of JAK2 [73]. Thus, SH2-Bβ represents an important dual regulator of the JAK/Stat pathway activated by GH.

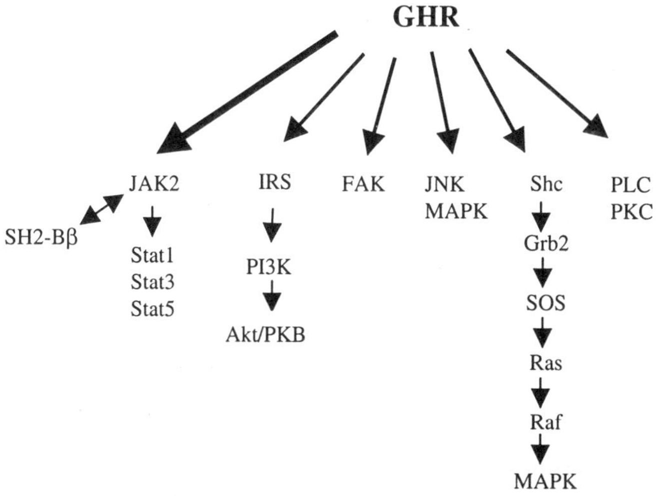

Figure 4. GH-inducible JAK/Stat and parallel kinase signaling pathways. One of the main signaling pathways activated by GH stimulation is the Jak/Stat pathway (bold arrow). The GH-activated Jak/Stat pathway is regulated by several signaling molecules, including SH2-Bβ, which seems to exert a dual positive/negative regulation of JAK2, depending on its interaction with the kinase. However, other parallel kinase and signaling cascades are also activated. JAK2, IRS (insulin receptor substrate) and Shc are found in the GHR complex after ligand binding. How cells respond to GH is likely to be determined by a combination of the various kinase pathways activated in a cell type specific manner.

Likewise, because of its other potential protein-protein interaction domains and its recruitment and phosphorylation by kinases that are not activated by SH2-Bβ, SH2-Bβ would likely mediate other, more specific actions of GH.

In addition to these *in vitro* investigations, studies of Stat5a KO mice indicate that Stat5a is required for mammary gland development and lactation, suggesting an important role of Stat5a in mediating PRL biological effects *in vivo* [74]. Studies of Stat5b KO mice have shown the same morphological characteristics as the GHR KO mice (i.e. short stature, sexual dismorphism), indicating an important role of Stat5b in GH actions *in vivo* [46,75,76]. GH and PRL effects seem then to be mainly related to the JAK/Stat pathway *in vitro* and *in vivo* [77].

Parallel Kinase Pathways Activated by GH

In addition to the JAK/Stat signaling pathway, other PTKs such as FAK (focal adhesion kinase), Ras and Raf, as well as serine/threonine kinases such as PI3 kinase, Akt/protein kinase B, MAPK, JNK, phospholipase C (PLC) and protein kinase C (PKC) have been found to be activated by GH and to play a role in GH actions, in different cellular models (Fig. 4). For example, PI3 kinase activation by GH is related to the ability of GH to regulate lipogenesis and lypolysis in adipocytes [78]. Also, activation of PLC and PKC is associated with GH-dependent activation of calcium channels and increase of intracellular calcium concentration in different type of cells such as adipocytes [79], lymphoid cells [80] and CHO cells [81]. Finally, activation of MAPK (see Chapter 3) by GH seems to be important for activation of numerous transcription factors such as Elk-1, c-Jun, c-Myc and Stat3, principally involved in the regulation of proliferation.

As is true for PRL, more is known about the JAK/Stat pathway in GH signaling, but parallel pathways may also have an important role in GH effects, with interactions between intracellular proteins and crosstalk between different pathways leading to specific patterns of gene expression in various GH responsive cells and tissues. As an example, signaling molecules other than those in the JAK/Stat pathway could be involved in mediating both anti-apoptotic and proliferative effects of GH in pro-B Ba/F3 cells. As discussed above, GH is involved in the regulation of proliferation in a wide range of cells and tissues. However, little is known about the signaling pathway(s) acivated in response to the proliferative effects of GH. In 1995, an *in vitro* study by Wang & Wood suggested the involvement of c-*fos* gene transcription in the proliferative effect of GH in lymphoid Ba/F3 cells [67]. The increase in c-*fos* transcription was dependent upon the activation of the JAK/Stat pathway in Ba/F3 cells, suggesting a role of JAK/Stat in the proliferative effect of GH. Furthermore, an effect of GH on cell survival was observed in these cells [67].

A recent study characterized signaling molecules that could be involved in both anti-apoptotic and proliferative effects of GH in Ba/F3 cells. Ba/F3 cells were shown to produce small quantities of endogenous GH, which were sufficient to promote Ba/F3 cell survival, without inducing proliferation, when cells were deprived of serum and growth factors [82]. Using this convenient cellular model, a new GH signaling pathway, dependent upon activation of NFκB and expression of Bcl-2, was shown to be crucial for the GH anti-apoptotic effect [82] (Fig. 5). Moreover, although locally-produced GH was only able to induce partial survival in these cells, addition of exogenous GH promoted survival and also proliferation of Ba/F3 cells [67,82,83] (Fig. 5). Studies of cyclin expression induced by GH showed that cyclins D levels (which are expressed in surviving cells) were dependent on the activation of the NFκB/Bcl-2 pathway, whereas cyclin A and E levels (which are expressed only in proliferative cells) required the activation of another GH-induced

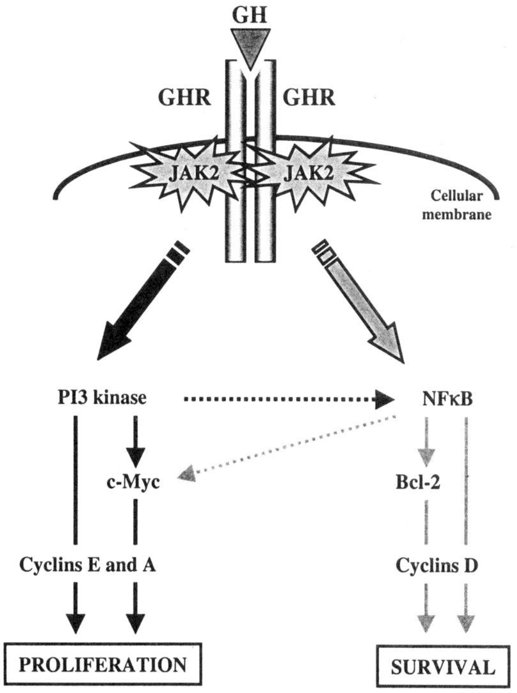

Figure 5. Schematic representation of the signaling pathways for GH effects in Ba/F3 cells. GH binding promotes JAK2 activation, which, in turn, induces simultaneous activation of both PI3 kinase (black arrows) and NFκB (gray arrows) signaling pathways. PI3 kinase activation accounts for the proliferative response, probably through c-Myc activation and/or expression. NFκB activation is responsible for survival, regulating anti-apoptotic proteins such as Bcl-2. Expression of cyclins D is probably dependent on the NFκB induction pathway, whereas cyclins E and A expression depends on PI3 kinase. A link between these two main pathways might exist, as NFκB activation can be partially controlled by PI3 kinase activation, and inhibition of NFκB provokes a down-regulation of c-Myc expression (dotted arrows).

signaling pathway, which is dependent on the activation of PI3 kinase and on the expression of the oncoprotein c-Myc [83] (Fig. 5). These observations suggest that anti-apoptotic and proliferative effects of GH are two distinct effects, mediated by separate signaling pathways: NFκB/Bcl-2 and PI3 kinase/c-Myc pathways, respectively. Furthermore, GH is able to regulate cell cycle of Ba/F3 cells directly, independent of a local production of IGF-1 [84].

In summary, although the JAK/Stat pathway is well characterized in PRL and GH signaling, some other known or unknown signaling pathways may be important in mediating some PRL and GH effects, depending on the target cell and tissue types that are involved.

CONCLUSIONS AND FUTURE QUESTIONS

Our studies thus highlight novel features of PRL- and GH-inducible signaling mediators and pathways: *i)* Stats act as either transcriptional activators or repressors depending on target promoters and cell type; *ii)* Different Stats can be recruited over the cell cycle to regulate gene transcription; *iii)* Stats work in concert with non-Stat proteins to regulate gene transcription. To mediate such diverse activity, multiple Stat complexes may exist within cells. Indeed, recent studies show that Stats are found in large cytoplasmic complexes (1–4 MDa) or "statosome" [85], which contain accessory molecules that can present Stats to receptor complex as well as chaperone activated Stats to the nucleus. In addition to cytoplasmic statosomes, once in the nucleus, Stats interact with enzymes that either open or close chromatin and with multiple nuclear factors to regulate gene expression [33]; *iv)* Cytokine signals cross talk via protein/protein interactions at the level of target genes; and *v)* Parallel signaling pathways, in collaboration with the JAK/Stat pathway, coordinate the diverse functions of PRL and GH. Understanding signal crosstalk may elucidate how PRL antagonizes TNFα (NFκB) and TGFβ (Smads) mediated signaling in various inflammatory and immunosuppressive diseases. Signal crosstalk also underlies how GH mediates proliferative as well as anti-apoptotic cell survival responses. Future research will focus on sorting out which PRL- or GH-inducible signaling pathways are involved in regulating specific patterns of gene expression in different target tissues, under normal versus disease states.

REFERENCES

1. **Horseman ND, Yu-Lee L-y.** 1994 Transcriptional regulation by the helix bundle peptide hormones: Growth hormone, prolactin, and hematopoietic cytokines. Endocr Rev 15:627-649.
2. **Horseman, N. D.** ed. 2001 Prolactin. pp1-416. Boston, Kluwer Academic Publishers.
3. **Herrington J, Smit LS, Schwartz J, Carter-Su C.** 2000 The role of Stat proteins in growth hormone signaling. Oncogene 19:2585-2597.
4. **Carter-Su C, Rui L, Herrington J.** 2000 Role of the tyrosine kinase Jak2 in signal transduction by growth hormone. Pediatr Neurol 14:550-557.
5. **Carter-Su C, Rui L, Stofega MR.** 2000 SH2-B and SIRP: Jak2 binding proteins that modulate the actions of growth hormone. Recent Prog Horm Res 55:293-311.
6. **Finidori J**. 2000 Regulators of growth hormone signaling. Vitam Horm 59:71-97 .
7. **Ben-Jonathan N, Mershon J, Allen D, Steinmetz R.** 1996 Extrapituitary prolactin: Distribution, regulation, functions and clinical aspects. Endocrine Rev 17:639-669.
8. **Yu-Lee L-y.** 1997 Molecular actions of prolactin in the immune system. Proc Soc Exp Biol Med 215:35-52.
9. **Bole-Feysot C, Goffin V, Edery M, Binart N, Kelly PA.** 1998 Prolactin (PRL) and its receptor: actions, signal transduction pathways and phenotypes observed in PRL receptor knockout mice. Endocr Rev 19:225-268.
10. **Brisken C, Kaur S, Chavarria TE, Binart N, Sutherland RL, Weinberg RA, Kelly PA, Ormandy CJ.** 1999 Prolactin controls mammary gland development via direct and indirect mechanisms. Dev Biol 210:96-106.
11. **Rosen JM, Wyszomierski SL, Hadsell D.** 1999 Regulation of milk protein gene expression. Annu Rev Nutr 19:407-436.
12. **Goffin V, Binart N, Clement-Lacroix P, Bouchard B, Bole-Feysot C, Edery M, Lucas BK, Touraine P, Pezet A, Maaskant R, Pichard C, Helloco C, Baran N, Favre H, Bernichtein S, Allamando A, Ormandy C, Kelly PA.** 1999 From the molecular biology of prolactin and its receptor to the lessons learned from knockout mice models. Genet Anal Biomol Engin 15:189-201.
13. **Walker SE, McMurray RW, Houri JM, Allen SH, Keisler D, Sharp GC, Schlechte JA.** 1998 Effects of prolactin in stimulating disease activity in systemic lupus erythematosus. Ann New York Acad Sci 840:762-772.
14. **Elbourne KB, Keisler D, McMurray RW.** 1998 Differential effects of estrogen and prolactin on autoimmune disease in the NZB/NZW F1 mouse model of systemic lupus erythematosus. Lupus 7:420-427.
15. **Matera L, Mori M, Geuna M.** 2000 Prolactin in autoimmunity and antitumor defense. J Neuroimmunol 109:47-55.
16. **Goffin V, Kelly PA.** 1997 The prolactin/growth hormone receptor family: Structure/function relationships. J Mamm Gland Biol Neoplasia 2:7-17.
17. **Clevenger CV, Chang WP, Ngo W, Pasha TLM, Montone KT, Tomaszweski JE.** 1995 Expression of prolactin and prolactin receptor in human breast carcinoma. Am J Pathol 146:695-705 .
18. **Gertler A.** 1997 Recombinant analogues of prolactin, growth hormone, and placental lactogen: correlations between physical structure, binding characteristics, and activity. J Mamm Gland Biol Neoplasia 2:69-80.
19. **Yu-Lee L-y, Luo G, Moutoussamy S, Finidori J.** 1998 Prolactin and growth hormone signal transduction in lymphohemopoietic cells. Cell Mol Life Sci 54:1067-1075.

20. **Wang Y, O'Neal KD, Yu-Lee L-y.** 1997 Multiple prolactin receptor cytoplasmic residues and Stat1 mediate prolactin signaling to the IRF-1 promoter. Mol Endocrinol 11:1353-1364.
21. **McAveney KM, Book ML, Ling P, Horvath G, Chebath J, Yu-Lee L-y.** 2000 Association of 2',5'-oligoadenylate synthetase with the prolactin receptor: Alteration in prolactin-inducible Stat1 signaling to the IRF-1 promoter. Mol Endocrinol 14:295-306.
22. **Schindler C.** 1999 Cytokine and Jak-Stat signaling. Exp Cell Res 253:7-14.
23. **Bromberg J, Darnell JE, Jr.** 2000 The role of STATs in transcriptional control and their impact on cellular function. Oncogene 19:2468-2473.
24. **Shuai K.** 2000 Modulation of Stat signaling by Stat-interacting proteins. Oncogene 19:2683-2644.
25. **Horvath CM.** 2000 Stat proteins and transcriptional responses to extracellular signals. Trends Biochem Sci 25:496-502.
26. **Kazansky AV, Kabotyanski E, Wyszomierski SL, Yel J, Rosen JM.** 1999 Differential effects of prolactin and src/abl kinase on the nuclear translocation of Stat5b and Stat5a. J Biol Chem 274:22484-22492.
27. **Moriggl R, Fouilleux-Gruart V, Jahne R, Berchtold S, Gartmann C, Liu X, Hennighausen L, Sotiropoulos A, Groner B, Gouilleux F.** 1996 Deletion of the carboxyl-terminal transactivation domain of MGF-Stat5 results in sustained DNA binding and a dominant negative phenotype. Mol Cell Biol 16:5691-5700.
28. **Shaefer TS, Sanders LK, Park OK, Nathans D.** 1997 Functional differences between Stat3α and Stat3β. Mol Cell Biol 17:5307-5316.
29. **Pfeffer LM, Mullersman JE, Pfeffer SR, Murti A, Yang CH.** 1997 Stat3 as an adapter to couple phosphatidylinositol 3-kinase to the IFNAR1 chain of the type I interferon receptor. Science 276:1418-1420.
30. **John S, Vinkemeier U, ldaini E, Darnell JE, Jr., Leonard WJ.** 1999 The significance of tetramerization in promoter recruitment by Stat5. Mol Cell Biol 19:1910-1918.
31. **Collum RG, Brutsaer S, Lee G, Schindler C.** 2000 A Stat3 interacting protein (StIP1) regulates cytokine signal transduction. Proc Natl Acad Sci 97:10120-10125.
32. **Sekimoto T, Imamoto N, Makajima K, Hirano T, Yoneda Y.** 1997 Extracellular signal-dependent nuclear import of Stat1 is mediated by nuclear pore targeting complex formation with NPI-1, but not Rch1. EMBO J 16:7067-7077.
33. **Chatterjee-Kishore M, Wright KL, Ting JPY, Stark GR.** 2000 How Stat1 mediates constitutive gene expression: a complex of unphosphorylated Stat1 and IRF1 supports transcription of the LMP2 gene. EMBO J 19:4111-4122.
34. **Glass CK, Rose DW, Rosenfeld MG.** 1997 Nuclear receptor coactivators. Curr Op Cell Biol 9:222-232.
35. **Kurokawa R, Kalafus D, Ogliastro MH, Kioussi C, Xu L, Torchia J, Rosenfeld MG, Glass CK.** 1998 Differential use of CREB binding protein-coactivator complexes. Science 279:700-703.
36. **Pfitzner E, Jahne R, Wissler M, Stoecklin E, Groner B.** 1998 p300/CREB-binding protein enhances the prolactin-mediated transcriptional induction through direct interaction with the transactivation domain of Stat5, but does not participate in the Stat5-mediated suppression of the glucocorticoid response. Mol Endocrinol 12:1582-1593.
37. **Luo G, Yu-Lee L-y.** 2000 Stat5 inhibits NFκB-mediating signaling. Mol Endocrinol 14:114-123.
38. **Stevens AM, Wang Y, Sieger KA, Lu H, Yu-Lee L-y.** 1995 Biphasic transcriptional regulation of the interferon regulatory factor-1 gene by prolactin: Involvement of gamma-interferon activated sequence and Stat-related proteins. Mol Endocrinol 9:513-525.

39. **Bole-Feysot C, Perret E, Roustan P, Bouchard B, Kelly PA.** 2000 Analysis of prolactin-modulated gene expression profiles during the Nb2 cell cycle using differential screening techniques. Genome Biol 1:research0008.1-0008.15.
40. **DaSilva L, Rui H, Erwin RA, Howard OMZ, Kriken RA, Malabarba MG, Hackett RH, Larner AC, Farrar WL.** 1996 Prolactin recruits Stat1, Stat3, and Stat5 independent of conserved receptor tyrosines Tyr402, Tyr479,Tyr515 and Tyr 580. Mol Cell Endocrinol 117:131-140.
41. **Wang Y, Yu-Lee L-y.** 1996 Multiple Stat complexes interact at the IRF-1 GAS in prolactin-stimulated Nb2 T cells. Mol Cell Endocrinol 121:19-28.
42. **Gouilleux F, Wakao H, Mundt M, Groner B.** 1994 Prolactin induces phosphorylation of Tyr694 of Stat5 (MGF), a prerequisite for DNA binding and induction of transcription. EMBO J 13:4361-4369.
43. **Wakao H, Gouilleux F, Groner B.** 1994 Mammary gland factor (MGF) is a novel member of the cytokine regulated transcription factor gene family and confers the prolactin response. EMBO J 13:2182-2191.
44. **Luo G, Yu-Lee L-y.** 1997 Transcriptional inhibition by Stat5: Differential activities at growth-related versus differentiation-specific promoters . J Biol Chem 272:26841-26849.
45. **Wyszomierski SL, Rosen JM.** 2001 Cooperative effects of STAT5 (Signal Transducer and Activator of Transcription 5) and C/EBPß (CCAAT/Enhancer-Binding Protein-ß) on ß-casein gene transcription are mediated by the glucocorticoid receptor. Mol Endocrinol 15: 15(2):228-240
46. **Waxman DJ.** 2000 Growth hormone pulse-activated STAT5 signaling: a unique regulatory mechanism governing sexual dimorphism of liver gene expression. Novartis Found Symp 227:61-74.
47. **Tanaka N, Taniguchi T.** 2000 The interferon regulatory factors and oncogenesis. Semin Cancer Biol 10:73-81.
48. **McAlexander MB, Yu-Lee L-y.** 2001 Sp1 is required for prolactin activation of the interferon regulatory factor 1 gene. Mol Cell Endocrinol, in press
49. **Carey M.** 1998 The enhanceosome and transcriptional synergy. Cell 92:5-8.
50. **Ryu S, Zhou S, Ladurner AG, Tijan R.** 1999 The transcriptional cofactor complex CRSP is required for activity of the enhancer-binding protein Sp1. Nature 397:446-450.
51. **McAlexander MB, Yu-Lee L-y.** 2001 Prolactin activation of IRF-1 transcription involves changes in histone acetylation. FEBS Lett 488:91-94.
52. **Teglund S, McKay C, Schuetz E, van Deursen JM, Stravopodis D, Wang D, Brown M, Bodner S, Grosveld G, Ihle JN.** 1998 Stat5a and Stat5b proteins have essential and nonessential, or redundant, roles in cytokine responses. Cell 93:841-850.
53. **Hunter T**. 1995 Protein kinases and phosphatases: The yin and yang of protein phosphorylation and signaling. Cell 80:225-236.
54. **Kim S-O, Jiang J, Yi W, Feng GS, Frank SJ.** 1998 Involvement of the Src homology 2-containing tyrosine phosphatase SHP-2 in growth hormone signaling. J Biol Chem 273:2344-2354.
55. **Haspel RL, Darnell JE, Jr.** 1999 A nuclear tyrosine phosphatase is required for the inactivation of Stat1. Proc Natl Acad Sci 96:10188-10193.
56. **Naka T, Fujimoto M, Kishimoto T.** 1999 Negative regulation of cytokine signaling: Stat-induced Stat inhibitor. Trends Biochem Sci 24:394-398.
57. **Perkins ND.** 2000 The Rel/NFκB family: friend and foe. Trends Biochem Sci 25:434-440.
58. **Clapp C, Martinez de la Escalera G.** 1997 Prolactins: Novel regulators of angiogenesis. News Physiol Sci 12:231-237.
59. **Bengtson NW, Linzer DIH.** 2000 Inhibition of tumor growth by the antiangiogenic placental hormone, proliferin-related protein. Mol Endocrinol 14:1934-1943.

60. **Struman I, Bentzien F, Lee H, Mainfroid V, Angelo G, Goffin V, Weiner RI, Martial JA.** 1999 Opposing actions of intact and N-terminal fragments of the human prolactin/growth hormone family members on angiogenesis: An efficient mechanism of the regulation of angiogenesis. Proc Natl Acad Sci 96:1246-1251.
61. **Martini J, Piot C, Humeau LM, Stuman I, Martial JA, Weiner RI.** 2000 The antiangiogenic factor 16K PRL induces programmed cell death in endothelial cells by caspase activation. Mol Endocrinol 14:1536-1549.
62. **Corbacho A, Nava G, Eiserich J, Noris G, Macotela Y, Struman I, Martinez de la Escalera G, Freeman B, Clapp C.** 2000 Proteolytic cleavage confers nitric oxide synthase inducing activity upon prolactin. J Biol Chem 275:13183-13186.
63. **Mol JA, Henzen-Logmans SC, Hageman P, Misdorp W, Blankenstein MA, Rijnberk A.** 1995 Expression of the gene encoding growth hormone in the human mammary gland. J Clin Endocrinol Metab 80:3094-3096.
64. **Hattori N, Shimatsu A, Sugita M, Kumagai S, Imura H.** 1990 Immunoreactive growth hormone (GH) secretion by human lymphocytes: augmented release by exogenous GH. Biochem Biophys Res Commun 168:396-401.
65. **de Mello-Coelho V, Gagnerault MC, Souberbielle JC, Strasburger CJ, Savino W, Dardenne M, Postel-Vinay MC.** 1998 Growth hormone and its receptor are expressed in human thymic cells. Endocrinology 139:3837-3842.
66. **Heim MH.** 1999 The Jak-STAT pathway: cytokine signalling from the receptor to the nucleus. J Recept Signal Transduct Res 19:75-120.
67. **Wang YD, Wood WI.** 1995 Amino acids of the human growth hormone receptor that are required for proliferation and Jak-STAT signaling. Mol Endocrinol 9:303-311.
68. **Wang YD, Wong K, Wood WI.** 1995 Intracellular tyrosine residues of the human growth hormone receptor are not required for the signaling of proliferation or Jak-STAT activation. J Biol Chem 270:7021-7024.
69. **Gouilleux F, Pallard C, Dusanter-Fourt I, Wakao H, Haldosen LA, Norstedt G, Levy D, Groner B.** 1995 Prolactin, growth hormone, erythropoietin, and granulocyte-macrophage colony stimulating factor induce MGF-Stat5 DNA binding activity. EMBO J 14:2005-2013.
70. **Rui L, Mathews LS, Hotta K, Gustafson TA, Carter-Su C.** 1997 Identification of SH2-Bβ as a substrate of the tyrosine kinase JAK2 involved in growth hormone signaling. Mol Cell Biol 17:6633-6644.
71. **Rui L, Carter-Su C.** 1999 Identification of SH2-Bβ as a potent cytoplasmic activator of the tyrosine kinase Janus kinase 2. Proc Natl Acad Sci USA 96:7172-7177.
72. **Herrington J, Diakonova M, Rui L, Gunter DR, Carter-Su C.** 2000 SH2-B is required for growth hormone-induced actin reorganization. J Biol Chem 275:13126-13133.
73. **Rui L, Gunter DR, Herrington J, Carter-Su C.** 2000 Differential binding to and regulation of JAK2 by the SH2 domain and N-terminal region of SH2-Bβ. Mol Cell Biol 20:3168-3177.
74. **Liu X, Robinson GW, Wagner KU, Garrett L, Wynshaw-Boris A, Hennighausen L.** 1997 Stat5a is mandatory for adult mammary gland development and lactogenesis. Genes Dev 11:179-186.
75. **Waxman DJ, Ram PA, Park SH, Choi HK.** 1995 Intermittent plasma growth hormone triggers tyrosine phosphorylation and nuclear translocation of a liver-expressed, Stat 5-related DNA binding protein. Proposed role as an intracellular regulator of male-specific liver gene transcription. J Biol Chem 270:13262-13270.
76. **Udy GB, Towers RP, Snell RG, Wilkins RJ, Park SH, Ram PA, Waxman DJ, Davey HW.** 1997 Requirement of STAT5b for sexual dimorphism of body growth rates and liver gene expression. Proc Natl Acad Sci USA 94:7239-7244.

77. **Teglung S, McKay C, Schuetz E, van Deursen JM, Stravopodis D, Wang D, Brown M, Bodner S, Grosveld G, Ihle JN.** 1998 Stat5a and Stat5b proteins have essential and nonessential, or redundant, roles in cytokine responses. Cell 93:841-850.
78. **Ridderstrale M, Tornqvist H.** 1994 PI-3-kinase inhibitor Wortmannin blocks the insulin-like effects of growth hormone in isolated rat adipocytes. Biochem Biophys Res Commun 203:306-310.
79. **Schwarz Y, Yamaguchi H, Goodman HM.** 1992 Growth hormone increases intracellular free calcium in rat adipocytes: correlation with actions on carbohydrate metabolism. Endocrinology 131:772:778.
80. **Ilondo MM, De Meyts P, Bouchelouche P.** 1994 Human growth hormone increases cytosolic free calcium in cultured human IM-9 lymphocytes: a novel mechanism of growth hormone transmembrane signalling. Biochem Biophys Res Commun 202:391-397.
81. **Billestrup N, Bouchelouche P, Allevato G, Ilondo M, Nielsen JH.** 1995 Growth hormone receptor C-terminal domains required for growth hormone-induced intracellular free Ca2+ oscillations and gene transcription. Proc Natl Acad Sci USA 92:2725-2729.
82. **Jeay S, Sonenshein GE, Postel-Vinay MC, Baixeras E.** 2000 Growth hormone prevents apoptosis through activation of nuclear factor-kB in interleukin-3-dependent Ba/F3 cell line. Mol Endocrinol 14:650-661.
83. **Jeay S, Sonenshein GE, Kelly PA, Postel-Vinay MC, Baixeras E.** 2001 Growth hormone exerts antiapoptotic and proliferative effects through two different pathways involving nuclear factor-kB and phosphatidylinositol 3-kinase. Endocrinology 142:147-156
84. **Baixeras E, Jeay S, Kelly PA, Postel-Vinay MC.** 2001 The proliferative and anti-apoptotic actions of growth hormone and insulin-like growth factor-1 are mediated through distinct signaling pathways in the pro-B Ba/F3 cells. Endocrinology 142:2968-2977.
85. **Sehgal PB.** 2000 Stat-signaling through the cytoplasmic compartment: Consideration of a new paradigm. Cell Signal 12:525-535.

8

Erythropoietin, Thrombopoietin and Leptin Receptors

Fabrice Gouilleux*

Laboratory of Immunology, School of Medicine, University of Picardie-Jules Verne, 3, rue des Louvels, 80036, Amiens, France

INTRODUCTION

Erythropoietin (Epo), Thrombopoietin (Tpo) and leptin are hormones with distinct physiological properties. While the first two regulate survival, growth and differentiation of erythroid and megakaryocytic progenitors respectively, leptin is crucial for mammalian body weight regulation[1-3]. Receptors for these ligands have been isolated and well- characterized; they belong to the class I cytokine receptor family that includes most interleukin receptors involved in hematopoiesis as well as those for prolactin and growth hormone (see Chapter 7) which play an important role in metabolism and reproduction [4]. This family of receptors shares structural similarities both in their extracellular and intracellular domains and is characterized by the absence of an intrinsic tyrosine kinase. Nevertheless, ligand binding to this type of receptor induces the tyrosine phosphorylation of many cellular substrates including the receptor itself, leading to the activation of distinct signaling pathways. This review will focus first on the structure and biological properties of Epo, Tpo, leptin and their receptors. The second part will summarize the signal transduction pathways induced by these three ligands and their roles in cell proliferation, differentiation and survival.

---

* fabrice.gouilleux@sa.u-picardie.fr

STRUCTURE AND BIOLOGICAL PROPERTIES OF LIGANDS AND RECEPTORS

Epo, Tpo and Leptin Ligands

Epo. Epo was purified at the end of the 1970s and its cDNA cloned several years after[5]. This cDNA was one of the first hematopoietic growth factor to be cloned and encoded a primary translation product of 166 amino acids. This hormone is highly glycosylated. Its molecular mass is ~34 kDa, although the protein only accounts for 18 kDa. Epo is highly conserved among different mammalian species with 82% homology between human and mouse. The three dimensional structure of the protein has been determined and shows that Epo adopts a four helix bundle fold found in most cytokines, which is essential for functional activity [6]. During adult life, Epo is expressed mainly in kidney and to a lesser extent in liver. In contrast, the liver is the main site of Epo production during the embryonic life. In brain, Epo expression has been detected in astrocytes [1]. Mice deficient in Epo die by embryonic day 13.5 due to a lack of definitive erythropoieis in the fetal liver [7]. However, the presence of erythroid progenitors CFU-E and BFU-E is unaffected, indicating that Epo is not necessary for the commitment to the erythroid lineage but affects the terminal differentiation of erythroid progenitors.

Tpo. The c-Mpl receptor was an orphan receptor for a long time. Inhibition of megakaryocytic colony formation *in* vitro by introduction of c-mpl anti-sense oligonucleotides in CD34+ cells suggested that c-Mpl encoded a receptor for a megakaryocyte lineage specific growth factor [8]. Different strategies for cloning the putative c-Mpl ligand were used, ranging from its purification using different chromatography techniques and partial amino acid sequence analysis allowing cDNA cloning, or the use of cDNA libraries prepared from cells expressing c-Mpl and selection based on their growth independency. The cDNAs isolated from different species encoded protein products with high degree of identity (69-77%)[2]. The predicted primary translation product of human Tpo cDNA contains 335 amino acid with a peptide signal of 21 amino-acids. The protein can be separated in two domains, one of 153 amino acids which is highly conserved among different species sharing 23% identity with Epo, and the second one with 181 amino acids particularly rich in serine, threonine and proline residues and highly glycosylated. A truncated form of Tpo containing only the Epo like domain was shown to be active *in vitro.* In addition, a presumed splicing variant of Tpo (Tpo2) with a four amino acid deletion within the Epo like domain is unable to confer growth of cells expressing the c-Mpl receptor indicating that the Epo like domain is crucial for Tpo activity [9]. Indeed this domain contains four cystein residues that form disulfide bridges required for activity. This amino-terminal domain therefore adopts a helical conformation described for Epo and other cytokines. In mouse and human, Tpo is expressed mainly in liver and kidney. However, Tpo mRNA was also detected in skeletal muscle, brain, spleen and bone marrow. Hepatocytes and hepatoma cell lines express also Tpo mRNA. Mice lacking Tpo are severely thrombocytopenic with 80 to 90% reduction in both platelets and megakaryocytes demonstrating the essential physiological role of Tpo in megakaryocytopoiesis and the production of platelets *in vivo* [10].

Leptin (Ob). Among the loci identified in mice where extreme obesity mutations occur, *obese* (*ob*) and *diabetes* (*db*) were the most intensively studied [11,12]. The parabiosis (cross-circulation) experiments of an *ob/ob* mice and a lean wild-type control resulted in *ob/ob* mouse weight normalization indicating that a circulating factor responsible for the weight regulation was absent in *ob/ob* mice. Purification of this factor was unsuccessful and a positional cloning strategy was used to isolate the *ob* gene and to identify its encoded 167 kDa protein, leptin (Ob) [13]. The mature polypeptide is secreted by adipose tissue and is found in the blood of various mammalian species including mice and humans. The predicted tertiary structure suggests that leptin adopts a helical cytokine like structure similar to Interleukin-2 (IL–2) and Growth Hormone (GH)[14]. Administration of recombinant leptin to *ob/ob* mice results in food intake reduction and weight loss. Leptin mRNA and protein expression are regulated in humans and rodents by changes in percentage of body fat but also by changes in food intake. Expression is up-regulated when the body fat increased and down-regulated when the body weight is reduced. Beside its metabolic and endocrine functions, leptin also plays a regulatory role in immunity, inflammation and hematopoiesis. Alterations in immune and inflammatory responses are present in leptin deficient mice [15].

Epo/Tpo/leptin Receptors

EpoR. Epo, Tpo, and leptin receptors belong to the type 1 super-family of single transmembrane cytokine receptors (see Fig. 1). The Epo receptor (EpoR) is the founding member of this family and is composed of a unique chain of 66 kDa [16,17]. The extracellular part of the receptor contains two fibronectin III like sub-domains. Within the distal subdomain (called D1), two pairs of cystein residues form a disulfide bridge while the membrane proximal sub-domain (called D2) contains the conserved motif WSXWS known to be important for Epo receptor folding [18]. The cytoplasmic domain (236 amino acids) can be divided into two functional regions. The membrane proximal region contains two domains called Box1 and Box2 which are conserved in other cytokine receptors and are necessary for JAK kinase activation and Epo-induced mitogenesis [19-21]. The membrane distal region includes eight tyrosine residues which are phosphorylated upon Epo stimulation [22].

The physiological importance of EpoR has been clearly demonstrated by gene *knockout* studies in mice. Like *Epo*$^{-/-}$ mice, *EpoR*$^{-/-}$ mice die at embryonic day 12.5 due to severe anemia. EpoR is essential for definitive erythropoiesis in fetal liver but is not required for the development of primitive yolk-sac derived erythrocytes. No defects were apparent in non-erythroid cell lineages in which EpoR is expressed like megakaryocytes or endothelial cells [7,23].

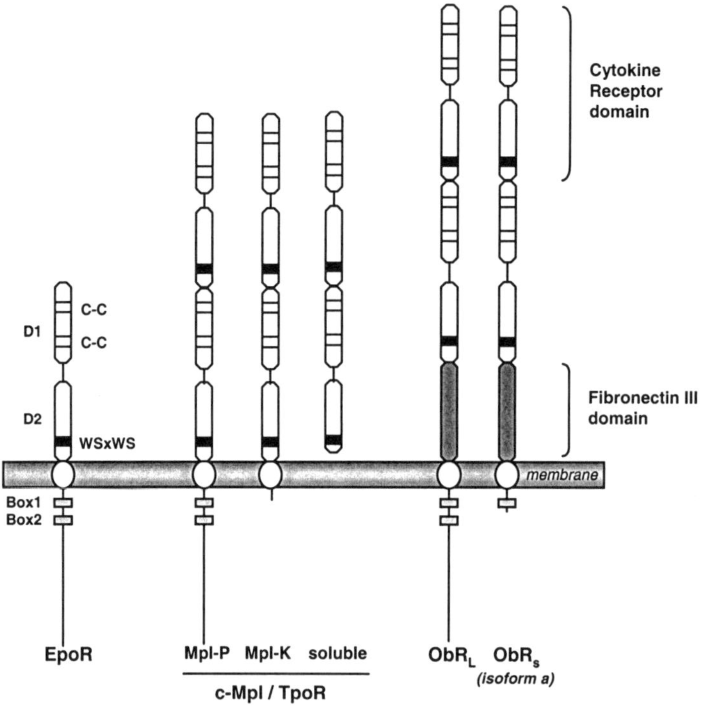

Figure1. Shematic representation of EpoR, c-Mpl (TpoR) and O-R. The extracellular parts contain cytokine receptor domains with two characteristic features: two pairs of disulphide-linked cysteines and the WS motif. In the cytoplasmic domains, Box1 and Box2 are the only conserved motifs found in these receptors. The cytoplasmic region of EpoR, c-Mpl and ObR contains also several tyrosine residues which are phosphorylated after ligand binding to the receptor.

TpoR. The Tpo receptor was discovered as the product of the *c-mpl* gene, the cellular homologue of the oncogene v-*mpl*, the transforming principle of the murine myeloproliferative leukaemia virus (MPLV)[24]. Human and murine *c-mpl* cDNAs were isolated and sequenced and display approximately 81% identity at the amino acid sequence level. Both human and murine sources show the presence of multiple isoforms of c-Mpl produced as a result of alternative splicing [25,26]. Three major isoforms of c-Mpl have been described, Mpl-P encoding the full length Tpo receptor, Mpl-K which lacks most of the cytoplasmic region and is probably unable to transmit signal, and a soluble form of c-Mpl lacking the transmembrane and the cytoplasmic domains. The extracellular portion of c-Mpl (Mpl-P) is composed of a duplicated 200 amino acid domain containing four spaced cysteine residues and a WSxWS motif adjacent to the transmembrane region, a feature shared with the common β-subunit of the IL-3, IL-5 and GM-CSF receptors [27]. Similarly to Epo receptor, the cytoplasmic part of c-Mpl can be divided into two domains: the membrane proximal domain containing the Box1 and Box2 crucial for Tpo-induced JAK kinase activation and cell proliferation and the membrane distal domain containing several tyrosine residues which are phosphorylated upon Tpo stimulation and are necessary for the recruitment of various signaling molecules[28,29]. Mice deficient in c-Mpl display a similar phenotype to $Tpo^{-/-}$ mice and develop a thrombocytopenia. However, residual platelet production (5-10%) is observed in these *knockout* mice and it is sufficient to prevent spontaneous bleeding and allows a normal lifespan. This implies that Tpo-independent mechanisms of platelet production occurs *in vivo*. In addition to the deficit in megakaryocytes and platelets, progenitor cells of all hematopoietic lineages are produced in reduced numbers in $Tpo^{-/-}$ and $c\text{-}Mpl^{-/-}$ mice indicating an important role of Tpo signaling in hematopoietic stem cell regulation [10, 30]. Recently, point mutations or deletions in the coding region of c-Mpl have been found in a patient with congenital amegakaryocytic thrombocytopenia [31].

ObR. The leptin receptor (ObR) was first identified by expression cloning techniques. Sequencing the original murine ObR cDNA isolated from the expression cloning screen revealed a single membrane-spanning receptor with homology to members of the class I cytokine receptor family, the closest relatives of ObR being the gp130-transducing chain of the interleukin (IL-) 6 receptor, the granulocyte colony stimulating factor (G-CSF) receptor and the leukaemia inhibitory factor (LIF) receptor α chain [32]. The predicted extracellular region of ObR was ~816 amino acids while the intracellular part contained 34 amino acids, suggesting that this protein might not have the ability to transmit signals (ObRa). However, further screening and analysis of cDNA libraries allowed identification of several isoforms in both mice and humans, including a long form with an intracellular domain of 303 amino acids which is expressed preferentially in the hypothalamus (ObRb). Differences in receptor forms arising from alternative splicing results in ObR intracellular domains with differing length and composition (Fig. 1). These ObR isoforms have been therefore divided in two classes: the long form (ObR_L or ObRb), and the various short forms that have been isolated to date (ObR_S, including ObRa, -c, -d, -f)[33]. A soluble form of ObR (ObRe) has also been described and is believed to modulate circulating leptin levels [34]. The large extracellular region of ObR includes two cytokine binding domains, each containing a single copy of the characteristic WSxWS motif and a fibronectin type III domain adjacent to the transmembrane region. Deletion/substitution experiments revealed that only the second cytokine binding domain is necessary for leptin binding and receptor activation. The intracellular region of the long form of ObR contains the classical Box1 and Box2 domains crucial for JAK kinase activation and several tyrosine residues which are phosphorylated upon leptin stimulation indicating that the long form is capable of transducing a signal. In contrast, the original short form OBRa retains only Box1 and is deficient in signal transducing capacities [35-37].

The complete insensitivity of *db* mice to leptin administration and the identical phenotype of *ob* and *db* mice suggested that the *db* locus encoded the leptin receptor. Indeed, the leptin receptor maps to chromosome 4 of the mouse in a region that contain the *db/db* mutation. The *db* mutation consists of a point mutation (GT transversion) within an exon encoding the extreme C-terminus and the 3' untranslated region of the short form of leptin receptor resulting in the generation of a new splice donor site and inappropriate splicing of the long form of the ObR. As a consequence, the ObR_L has a truncated intracellular domain similar to the predominant isoform OBR_S [38,39].

SIGNALING PATHWAYS INDUCED BY EPO, TPO AND LEPTIN

Epo, Tpo and Leptin Receptor Activation: Early Events in Signal Transduction

The first observation that a mutant EpoR (R129C) covalently dimerized by interchain disulfide bonds is constitutively active, provides strong evidence that receptor homodimerization is a crucial step in signal transduction [40]. Additional studies have demonstrated that the mutations E132C and E133C in the extracellular domain of Epo receptor also cause Epo receptor dimerisation via disulphide bond formation and render it constitutively active [41]. Recent data from crystallographic studies have shown that unliganded Epo receptors exist as a pre-formed dimer. This dimerisation is mediated by the D1-D2 intervening regions but the cytoplasmic portions of each paired monomers are so far apart that signal transduction cannot be initiated. In the presence of Epo, these pre-formed dimers are brought into functional proximity and allow initiation of signal transduction [42]. Similarly it is believed that the Tpo receptor forms homodimers and mutants of c-Mpl that constitutively associate display also ligand independent activity [43]. Given the relative structural similarities between EpoR and c-Mpl, it is possible that pre-formed c-Mpl dimers exist in absence of ligand. Leptin receptors form also homodimers, both in the presence or absence of ligands. Each leptin receptor binds one molecule of leptin resulting in a tetrameric complex. Activation of the receptor is thought to result from a ligand-induced conformational change [15].

One of the earliest responses detected within cells after ligand-induced receptor activation is a transient increase in tyrosine phosphorylation of cellular proteins including the receptor. Since EpoR, c-Mpl and ObR like other hematopoietic cytokine receptors, lack intrinsic tyrosine kinase activity, these receptors must interact with and activate protein tyrosine kinases allowing signal transduction. Numerous studies over the last ten years demonstrated that the JAK tyrosine kinases are crucial for this obligatory step. The JAK kinases are usually associated with the membrane proximal domain of the cytoplasmic tail of the cytokine receptors. Presumably ligand-induced conformational change brings the associated JAK kinases into proximity and enables them to transphosphorylate and activate each other. Once activated, the JAK kinases then phosphorylate tyrosine residues within the cytoplasmic portion of the receptors which then provide docking sites for the recruitment of SH2-domain (Src Homology domain 2) containing molecules (Fig. 2). These effectors are next eventually phosphorylated and activate their respective signaling pathways [44,45].

The JAK/STAT Signaling Pathway

The JAK tyrosine kinases include four family members: Jak1, Jak2, Jak3 and Tyk2 (see Chapter 6). Jak2 is commonly activated after ligand binding to EpoR, c-Mpl and ObR [46-50]. In addition activation of Tyk2 during Tpo signaling has been also reported in cells expressing c-Mpl [51]. The absolute requirement of Jak2 in Epo or Tpo-induced signal transduction has been demonstrated in mice deficient in Jak2. *Jak2*$^{-/-}$ mice die as embryos between day 13 and 15 from severe anemia with a phenotype similar to *EpoR*$^{-/-}$ mice. However, both the total number of fetal liver cells and the number of erythroid progenitor cells are greatly reduced suggesting that Jak2 is required at an earlier stage of erythropoietic development than is the EpoR. A possible explanation for this difference could be related to the effect of Tpo signaling. Indeed, hematopoietic progenitors from fetal livers of *Jak2-null* mice fail to respond to Tpo. In addition, Jak2 but not Tyk2 is probably essential for most Tpo-induced cellular functions because the genetic deletion of Jak2 in mice resulted in the complete absence of megakaryocyte development [52-54]. Since Jak2 deficient mice are not viable, it is difficult to predict its physiological role in leptin signaling events. The use of mice with conditional deletion of Jak2 in the hypothalamus will resolved this important question. The proximal region of the cytosolic domain in EpoR, c-Mpl and ObR_L receptors as in many cytokines receptors contains two conserved segments identified as Box1 and Box2 (Fig. 1). Deletion or mutation of Box1 or Box2 or of the region between Box1 and Box2 in EpoR and c-Mpl impairs activation of Jak2 and disrupts all measurable signal transduction and mitogenic stimuli, although there are some exceptions to the rule [19,21,28,55].

Among the substrates activated by the JAK kinases are members of the Signal Transducer and Activator of Transcription (STAT) protein family. STAT proteins are latent cytoplasmic transcription factors which are activated by tyrosine phosphorylation following ligand binding to the receptors. Activated STATs form homodimers or heterodimers through a reciprocal interaction between the SH2 domain and the phosphorylated tyrosine residue present in the C-terminal region of these proteins. The complex then moves to the nucleus, where it binds to specific DNA elements and activates the transcription of target genes (Fig. 2).

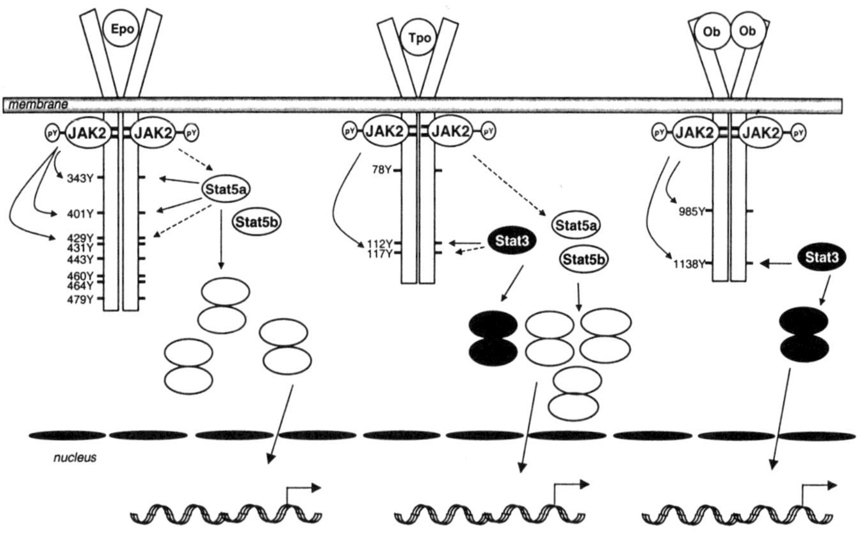

Figure2. The JAK/STAT pathway in EpoR, TpoR and ObR signaling (see details in the text)

Seven mammalian members have been isolated (Stats 1, 2, 3, 4, 5A, 5B and 6)[45]. Activation of STAT proteins by Epo, Tpo and leptin have been extensively described in the literature. It has been reported that Stat1, Stat3 and Stat5 (5A and 5B) are activated following Epo or Tpo stimulation of different cell lines [56, 57-59]. In addition to Stat1, Stat3 and Stat5, Stat6 is also activated by leptin in cells expressing the long form of ObR [36,60,61]. However, the patterns of activated STAT observed for these ligands *in vitro* vary from one cell type to another and do not always follow the *in vivo* situation. Stat5A and Stat5B are the main STATs activated by Epo in normal primary erythroid cells while Stat3 and, to a lesser extent, Stat5 are activated in megakaryocytes or platelets in response to Tpo [56,62]. Leptin injection specifically induces activation of Stat3 in the hypothalamus of wild type but not *db/db* mice [63]. Before their activation, STAT proteins are recruited to phosphorylated tyrosine residues of the activated receptors. Tyrosine residues which are necessary for Stat5 or Stat3 activation have been identified in the cytoplasmic tail of EpoR, c-Mpl and ObR. In the EpoR, mutation of the eight tyrosine residues revealed that among phosphorylated tyrosine residues, Y343 or Y401 are independently sufficient for maximal activation of Stat5, while Y429 and Y431 can partially activate Stat5 [64-66]. Interestingly , low level activation of Stat5 can occur in the absence of any tyrosine residues on the EpoR suggesting that Stat5 may be activated independently of the receptor probably through a direct interaction with Jak2 [67]. The use of cell lines expressing receptor mutants or dominant negative forms of Stat5 allowed analysis of the importance of Stat5 in Epo-stimulated cell proliferation and differentiation. Conflicting results emerged from these studies probably reflecting cell type specific differences. Some investigators have suggested a correlation between Stat5 activation and Epo-stimulated cell proliferation and/or survival while others have demonstrated a correlation with haemoglobin synthesis and differentiation [65,68-71]. The most surprising result came from the analysis of a combined *Stat5a/5b* knock-out mouse. Hematopoiesis in these mice appears to be normal including normal number of red cells and normal bone marrow erythroid progenitors (CFU-E)[72]. However, a recent report demonstrated that Stat5 is essential for the high erythropoietic rate in the fetal liver of *Stat5a/5b*$^{-/-}$ mice. *Stat5a/5b*$^{-/-}$ embryos are severely anemic and erythroid progenitors are present in low numbers, show higher levels of apoptosis and are less responsive to Epo [73]. These features may be explained by an important role of Stat5 in Epo-mediated inhibition of apoptosis. Indeed, Stat5 mediates the transcriptional induction of the anti-apoptotic *Bcl-x* gene in erythroid cells [73].

Tyrosine residues in the cytosolic portion of c-Mpl necessary for STAT activation have also been identified. Two phosphorylated tyrosine residues Y112 and Y117 are capable of Stat3 recruitment and phosphorylation, although the effect of Y117 on Stat3 activation is less pronounced. Similarly, Y112, plays a partial role in Stat5 phosphorylation. Like EpoR, Stat5 activation occurs in absence of any cytoplasmic tyrosine residues. Mutations of Y112 and Y117 showed that both residues contribute to, but are not essential for Tpo dependent cell proliferation [29]. The role of Stat5 and Stat3 in megakaryocytic development and platelet formation remains unclear. In a human megakaryoblastic leukemia cell line, Stat5 is believed to mediate Tpo induction of the cyclin-dependent kinase (Cdk) inhibitor, p21 thereby contributing to Tpo-stimulated differentiation while in MO7e cells, constitutive activation of Stat5 has been correlated with ligand-independent growth [74,75]. However, *Stat5a/5b*$^{-/-}$ mice show no obvious defect in the number of megakaryocytes, their precursors or their progeny. The role of Stat3 in Tpo signaling cannot be analyzed because *Stat3*$^{-/-}$ mice are not viable, even at the earliest stages of embryonic development.

Leptin induces activation of Stat3 in cells expressing the long but not the short form of ObR and this requires the unique tyrosine residue Y1138 in the carboxyl-terminus of ObR_L [50] . For the same reason, the lethality of *Stat3*$^{-/-}$ mice does not allow to assess the physiological role of Stat3 in leptin signaling.

Other Tyrosine Kinases Associated with Epo/Tpo/Leptin Receptors

Other tyrosine kinases have been shown to be activated during Epo signaling, including c-Fes/Fps, Syk, Tec and Lyn [76-79]. Lyn, a member of the Src tyrosine kinase (see Chapter 1), is involved in Epo-induced differentiation of the fetal liver cell line J2E [79]. Lyn interacts with the phosphorylated Y464 and Y479 of Epo receptor and with phosphorylated Jak2 through its SH2 domain. The kinase domain of Lyn binds also to the EPOR membrane proximal 91 amino-acids. In addition Lyn was demonstrated to phosphorylate EpoR and Stat5 and to interact with the hemopoietin-specific protein HS1 which is involved in Epo induced proliferation and differentiation of J2E cells [80,81]. Because mice deficient in Lyn do not show obvious hematopoietic defect, it is unclear whether Lyn is essential for EpoR signaling. Tec is a member of the Tec/Btk (Bruton's tyrosine kinase) family of tyrosine kinases which are expressed primarily in hematopoietic cells. In FD-M6 cells stimulated with Epo, Tec has been shown to co-immunoprecipitate with the nucleotide exchange factor Vav and the small adaptor Grb2 [78]. Phosphorylation of Tec in cells stimulated with Tpo has been also reported [82] .

Besides complex formation with cytosolic tyrosine kinases, functional and/or physical interaction with distinct receptor tyrosine kinases have been described. c-Kit which binds Stem Cell Factor (SCF) is co-expressed with EpoR at BFU-E and CFU-E stages and SCF acts at each stage to enhance erythroid colony size and red cell production. Biochemical analysis in different erythroid cell lines have shown that SCF binding to c-Kit induces tyrosine phosphorylation of EpoR and that c-Kit associates physically with the extended Box2 of the cytosolic domain of EpoR [83, 84]. SCF stimulation of c-Kit also promotes EpoR and Stat5 expression and enhances the Stat5-dependent expression of target genes [85,86].

A functional interaction between EpoR and IGF-I (Insulin Growth Factor-I; see Chapter 5) receptor has also been reported [87].

Signaling Through the Phospholipid-Modifying Enzymes: PI 3-kinase, PLCγ and SHIP

PI 3-kinase. Phosphatidyl-inositol 3-kinase (PI 3-kinase) activation following cytokine stimulation of hematopoietic cells has been shown to play an important role in proliferation and survival. PI 3-kinase is composed of a p85 (regulatory) and a p110 (catalytic) subunit. Membrane localization and activation of PI 3-kinase is required to generate phosphatidylinositol 3,4-biphosphate and phosphatidylinositol 3,4,5-triphosphate, two lipid products that are needed to activate various isoforms of the serine/threonine kinases PKC and Akt (also named PKB). Activated Akt is known to induce phosphorylation of Bad, a Bcl2 family member, and FKHRL1, a member of the forkhead transcription factor family, thereby generating distinct anti-apoptotic signals [88]. Activation of Akt and phosphorylation of its downstream targets Bad and FKHRL1 have been observed in erythroid progenitors and cell lines following Epo stimulation and LY294002, a specific PI 3-kinase inhibitor, blocks the phosphorylation of Akt and induces apoptosis of human erythroid progenitors, indicating that PI 3-kinase delivers an anti-apoptotic signal in these human primary cells [89,90] (Fig. 3).

Epo-dependent association of EpoR with PI 3-kinase has been reported and several mechanisms have been shown to activate the PI 3-kinase in Epo stimulated cells [91]. Phosphorylation of Y479 of the EpoR is involved in binding the PI 3-kinase and studies in $EpoR^{-/-}$ mice have demonstrated that a receptor containing only Y479 supports an almost normal number of fetal CFU-Es and BFU-Es [92]. However, mutant receptor lacking this tyrosine residue are capable to support normal erythroid differentiation. In addition, deletion or mutation of tyrosine 479 does not abrogate PI 3-kinase activation following Epo stimulation of different cell lines indicating that alternate pathways for PI 3-kinase activation without direct binding to Epo receptor exist. Indeed, several mechanisms have been described. One involves the nucleotide exchange factor Vav and the other, the adaptor protein IRS-2 (Insulin-Receptor-Substrate 2). These two proteins are associated with the Epo receptor, are tyrosine phosphorylated in response to Epo stimulation and provide a binding site for PI 3-kinase [93,94]. More recently, the IRS-related proteins Gab-1 (Grb2 Associated Binder 1) and/or Gab-2 have been shown to be tyrosine phosphorylated in Epo-stimulated cell lines and primary erythroid progenitors and to recruit PI 3-kinase [95,96].

Tpo stimulation of c-Mpl receptor bearing cell lines, megakaryocytes and platelets also leads to activation of the PI 3-kinase and to the generation of an anti-apoptotic signal in human megakaryoblasts [97-99]. In contrast to EpoR, no direct association of p85 with activated c-Mpl receptor has been detected so far and recent data have shown that IRS-2 and Gab2 in c-Mpl expressing Ba/F3 cells are tyrosine phosphorylated following Tpo stimulation, recruits p85 and enhance the enzymatic activity of PI 3-kinase. In addition, the tyrosine phosphatase SHP-2 (see Chapter 2) which is thought to act as a positive effector of signaling, associates with Gab2 and PI 3-kinase in these cells. Disruption of SHP-2/Gab2 association by expression of a dominant negative form of SHP-2 decreases PI 3-kinase activity and reduces Akt phosphorylation. Tyrosine phosphorylation of SHP-2 and its association with p85 have been also detected in primary murine megakaryocytes stimulated with Tpo, indicating the important role of SHP-2 as a regulator of PI 3-kinase activation in c-Mpl signaling. Neither Gab2 or IRS-2 are phosphorylated in murine megakaryocytes, but instead a ~100kDa tyrosine phosphorylated protein immunologically unrelated to Gab proteins co-immunoprecipitates with p85 [100]. In contrast, in human megakaryocytes, Tpo induces the phosphorylation of Gab1 [101]. Besides the role of PI 3-kinase activation in Tpo-induced cell proliferation and survival, it has been reported that Tpo enhances platelet activation *in vitro* and potentiates collagen receptor signaling through a PI 3-kinase dependent pathway [98].

Leptin mimicks insulin effects on glucose transport and glycogen synthesis through a PI 3-kinase dependent pathway in C2C12 myotubes. This activation requires Jak2 and IRS-2. Leptin stimulates tyrosine phosphorylation of Jak2 and IRS-2 and IRS-2, associated PI 3-kinase activity has been found in C2C12 cells[102]. Leptin-induced K(ATP) channel activation in the rat CRI-G1 insulinoma cell line also seems to require PI 3-kinase activity [103].

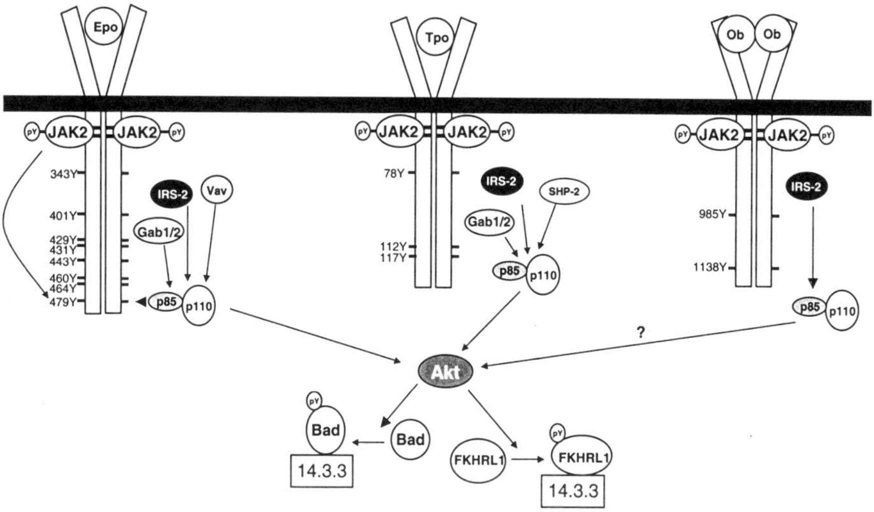

Figure 3: EpoR, c-Mpl, ObR signaling through PI 3-kinase. PI 3-kinase induces activation of Akt which in turn phosphorylates Bad and FKHRL1. Phosphorylated Bad and FKHRL1 are sequestered in the cytoplasm by the protein 14.3.3 resulting in the inhibition of their pro-apoptotic function.

PLC-γ and Ship. Other phospholipid-modifying enzymes that have been found to be associated with EpoR include phospholipase C-γ (PLC-γ) and phosphatidylinositol 3,4,5 triphosphate 5-phosphatase (Ship). In UT-7 cells, Epo rapidly induces the tyrosine phosphorylation of PLC-γ1 that is followed by diacylglycerol and 1,4,5 trisphosphate formation. These two second messengers activate PKC and increase intracellular calcium, respectively, indicating that PLC-γ1 activation could contribute to mitogenic signaling through the EpoR [104]. In FDCP-1 cells engineered to express EpoR, Epo induces the tyrosine phosphorylation of PLC-γ2, a process that contributes to glycosylphosphatidylinositol hydrolysis [105].

Ship is also tyrosine phosphorylated following Epo stimulation and associates with EpoR [106,107]. Ship catalyzes the hydrolysis of inositol 3,4,5-trisphosphate and inositol 1,3,4,5-tetraphosphate to generate inositol 3,4-biphosphate and inositol 1,3,4,-trisphosphate respectively. Two distinct *ship* genes have been isolated. *Ship1* is expressed exclusively in hematopoietic tissues while *ship2* has a wider tissue distribution. In Ba/F3 cells expressing various tyrosine mutants of EpoR, it has been recently reported that Ship1 is recruited to phosphorylated Y401, Y429 and Y431 through its SH2 domain, phosphorylated Y401 appearing to be the major site of Ship1 binding. In these studies, tyrosine phosphorylation of Ship1 induces its association with the

SH2-containing adaptor Shc and the constitutive binding to the Grb2 adaptor protein, thereby linking Ship1 to activation of the Ras/MAP kinase pathway [108]. However, Ship1 recruitment to EpoR does not directly correlate with activation of MAP kinase. Cells expressing EpoR mutants containing only Y343 are also capable to activate Erks following Epo stimulation but Ship1 recruitment is abrogated in these conditions [109]. Association of Ship with PI 3-kinase, Gab-1 or IRS-2 following Epo stimulation has been also described [94,95]. *Ship1*$^{-/-}$ mice exhibit a shortened life-span and a reduced number of mature, but not primitive erythroid progenitors suggesting that Ship1 might be involved in a late stage of erythroid maturation [110].

In Ba/F3 cells engineered to express c-Mpl, Tpo induces the tyrosine phosphorylation of Ship. The residue Y112 of the cytoplasmic tail of c-Mpl is both a site of phosphorylation and a critical residue for the Tpo dependent phosphorylation of Ship. Ship could either interact directly with phopshorylated Y112 *via* its SH2 domain or it may be recruited indirectly though its association with the adaptor Shc [29].

Activation of MAP Kinases Through Distinct Signaling Cascades

Activation of the Ras/MAP kinase (Mitogen Activated Protein kinase; see Chapter 3) pathway is supposed to be initiated by the binding of the small linker Grb-2 to activated receptor complexes. The guanidine nucleotide exchange factor Sos (Son of Sevenless) pre-associated to Grb-2 localizes to the membrane and removes GDP from inactive Ras allowing GTP to bind and activate Ras. Ras mediated membrane translocation of the serine-kinase Raf results in increased Raf kinase activity and the induction of a proteine kinase cascade involving the MAP kinase kinase (MEK1/2) and their downstream targets MAP kinases: Erk-1 and Erk-2 (Extracellular signal Related Kinases) [111].

EpoR. Grb-2 is recruited to the EpoR *via* the adaptor Shc [112,113](Fig. 4). The small adaptor proteins Shc are widely expressed as a 52 kDa and a 46 kDa form and contain a N-terminal phosphotyrosine-binding domain (PTB; see Chapter 1), a central effector domain and a C-terminal SH2 domain. Shc seems not to interact directly with the EpoR but is recruited to Ship-1 or Jak2 [108,113]. Following Epo stimulation, Shc is tyrosine phosphorylated, thereby providing docking sites for the SH2 domain of Grb-2. Grb-2 binds also to the scaffolding protein Gab-1 or the tyrosine phosphatase SHP-2 in Epo-stimulated cells [95] and it has been proposed that in CTLL-2 cells expressing EpoR, Epo induces activation of Raf1/MEK/MAP kinase *via* a Shc independent pathway [114].

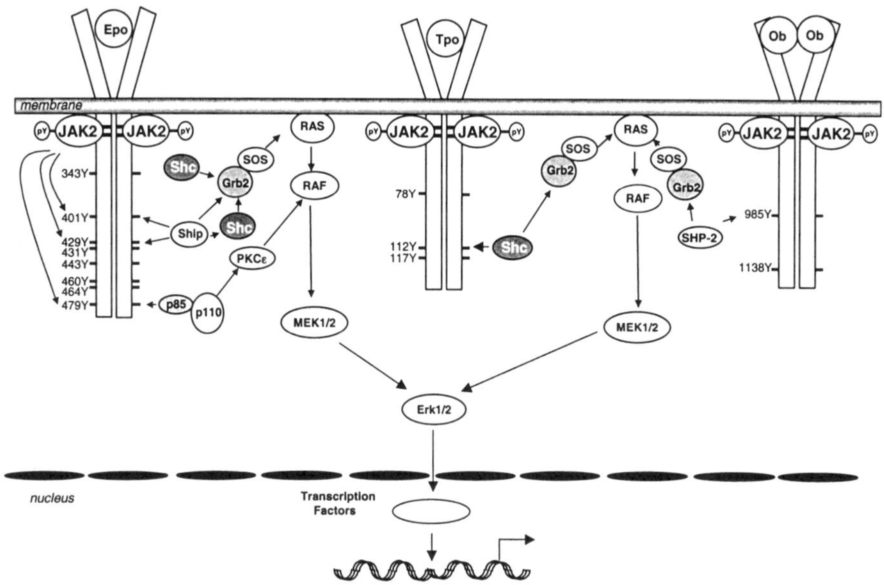

Figure 4. Activation of Erk1/2 through multiple signaling cascades (see details in the text).

An alternate mechanism for Ras/MAP kinase activation involving the SH2/SH3 containing adaptor protein CrkL has been also reported. In 32D cells expressing the EpoR, CrkL is tyrosine phosphorylated following Epo stimulation and associates with SHP-2, Ship-1 and Shc. In these cells, CrkL regulates the Epo dependent activation of the Ras/MAP kinase pathway through its interaction with the guanine nucleotide exchange factor C3G [115]. Ras independent activation of MAP kinase has also been described. Binding of PI 3-kinase to phosphorylated Y479 of EpoR results in MAP kinase activation *via* a PKCε dependent pathway [92]. The role of MAP kinase/Erk activation in EpoR signaling remains unclear. In CTLL-2 cells expressing EpoR or in HCD57 erythroid cells, Erk activity is required for the mitogenic effect of Epo[116,117].

TpoR. The distal cytoplasmic domain of c-Mpl has been shown to be crucial for Ras/MAP kinase activation. Shc is recruited to phosphorylated Y112 present in this domain, is tyrosine phosphorylated following Tpo stimulation and associates with Grb-2 [29](Fig. 4). Substitution of Y112 by phenylalanine within this c-Mpl domain is sufficient to abolish the normal differentiation response of WEHI3B-D+ and M1 cells expressing c-Mpl and the Tpo dependent phosphorylation of Shc, its association with Grb2 or the receptor and the induction of c-*fos* gene expression [118]. In 32D cells bearing c-Mpl receptors, introduction of a dominant negative form of Shc also inhibits Tpo-mediated differentiation [119]. Interestingly, Tpo is capable of inducing transformation of FRE rat fibroblasts engineered to express c-Mpl; Y112 and Shc phosphorylation have been shown to be crucial for the transforming capacity of c-Mpl [120]. In contrast to these different data, distinct C-terminal regions that do not include Y112 were also identified as necessary for Tpo dependent differentiation. In UT-7 cells expressing c-Mpl, a subdomain between cytoplasmic residues 71 and 94 has been shown to be responsible for megakaryocytic differentiation and prolonged activation of MAP kinase/Erk following Tpo stimulation [121,122]. In c-Mpl expressing F36P cells, prolonged activation of Ras plays also a central role in Tpo mediated differentiation [123]. In these different studies, the use of dominant negative or constitutively active forms of Ras or MAP kinase or the use of a chemical specific inhibitor of MEK1/2 further showed the requirement of Ras and MAP kinase in Tpo-induced differentiation. Although cell type specific differences exist that could explain the use of distinct cytoplasmic domains and signaling cascades, activation of the Ras/MAP kinase pathway makes an important contribution to differentiation signal triggered by Tpo. This was further evaluated in normal human and murine progenitors induced to differentiate into megakaryocytes by Tpo. In these primary cells, blockade of Erk1/2 activation reduces or delays the generation of megakaryocytes [124,125]. However, the role of the Ras/MAP kinase pathway in megakaryocytic differentiation *in vivo* remains uncertain. A recent report showed that mice expressing a mutant c-Mpl, without the cytoplasmic distal region have normal numbers of megakaryocytes and platelets compared to wild-type mice [126]. Thus, it is possible that activation of MAP kinase/Erk is not required or not essential for megakaryocytic differentiation *in vivo,* or that other cytokines may compensate for thrombopoietin *in vivo* activation of Erks.

ObR. Leptin stimulation of cells expressing ObR induces activation of MAP kinase/Erk [50,127,128]. The tyrosine residue Y985 in the cytoplasmic region of ObR plays a crucial role in this activation and mediates the recruitment of the tyrosine phosphatase SHP-2 which is tyrosine phosphorylated following leptin stimulation and associates with Grb2[50, 129](Fig. 4). The role of Erks activation in leptin receptor signaling remains largely undefined. In C2C12 muscle cells, activation of Erk2 appears to play a role in glucose uptake induced by leptin stimulation[130]. Leptin also stimulates the proliferation of Min6 cells and C3H10T1/2 cells through activation of MAP kinase[127,128].

The MAP kinase family is divided into four groups: Erk1 and -2 described above, the c-Jun amino terminal kinases (JNK), the p38 MAP kinase (p38) and

Erk5/Bmk1. Epo and Tpo have been reported to activate JNKs in different cell line models and in certain instances, p38 MAP kinase [131-133]. In SKT6 cells, Epo dependent activation of JNK and p38 have been shown to be important for Epo induced hemoglobinization while in HCD57 cells activation of p38 and JNKs is associated with Epo-induced proliferation [117,134]. The role of JNK and p38 in EpoR remains to be established. Recent data demonstrated that whereas some embryos from *p38*$^{-/-}$ mice die between day 11.5 and 12.5, those that develop past this stage are anemic owing to a failure of definitive erythropoiesis. A role for p38 in the regulation of Epo gene expression and stress-induced erythropoiesis rather than a signaling effector of EpoR has been proposed to explain the phenotype of these *p38*$^{-/-}$ mice[135].

Additional Signaling Adaptors and Effectors

Crk, c-Cbl, APS. Additional molecular adaptors that have been demonstrated to associate with Epo receptor complexes include Crk family members, c-Cbl, and APS. Crk proteins are the cellular homologues of v-crk which was originally described as an oncogene from the avian retroviruses CT10 and ASV-1. Three Crk protein variants are expressed in hematopoietic cells: CrkI, CrkII and CrkL. CrkII and CrkL proteins contain one SH2 and two SH3 domains (see Chapter 1) while CrkI contains only one SH3 domain [136]. CrkL has been shown to be tyrosine phosphorylated in response to Epo stimulation of 32D cells and to associate with SHP-2 and Shc. Epo-induced tyrosine phosphorylation of CrkL is dependent on the membrane proximal EpoR cytoplasmic region involved in the activation of Jak2 as well as the C-terminal 145 amino acid region which is required for tyrosine phosphorylation of SHP-2 and Shc [137]. In these cells, CrkL is constitutively associated through its SH3 domains with the guanine nucleotide exchange factor C3G and is believed to modulate activity of Ras and/or Rap1, a Ras family member involved in the regulation of beta1 integrin-mediated hematopoietic cell adhesion [115,138]. In Epo-stimulated human erythroid progenitors, CrkL associates also with DNA bound Stat5 and is thought to modulate its transcriptional activity [139]. Similarly, CrkL becomes tyrosine phosphorylated in human platelets in response to Tpo stimulation and associates with Stat5 through an interaction between the SH2 domain of CrkL and the phosphorylated tyrosines of Stat5 [140]. Finally CrkL and CrkII have been found to interact with the proto-oncogene c-Cbl following Epo stimulation of different cell lines [141]. c-Cbl was originally described as the transforming oncogene of the CAS NS-1 retrovirus. p120Cbl, the product of c-*cbl* is a scaffold protein containing a ring finger motif, a proline-rich region, conserved P(Y) sites and a PTB domain [142]. In Epo-stimulated UT-7 cells, c-Cbl is phosphorylated and associates with Grb2. This association is constitutive and probably involves an interaction between the proline-rich region of c-Cbl and the SH3 domain of Grb2 [143]. Similar results were obtained in the megakaryocytic cell line M-07E and in human platelets following Tpo stimulation [144-146]. An association between c-Cbl and PI 3-kinase has also been described in Tpo-stimulated cells [97]. Past studies have linked c-Cbl to negative regulation of growth factor signaling and to protein turn-over. Recent evidence showed that c-Cbl recruits components of the ubiquitin-conjugation pathway and acts as an ubiquitin-protein ligase suggesting that c-Cbl may also negatively regulate EpoR or TpoR signaling [142].

APS. APS is a novel adaptor protein containing a Pleckstrin Homology (PH) domain plus one SH2 domain. Upon Epo exposure, it is tyrosine phosphorylated at its C-terminus and associates with c-Cbl. Forced expression of APS in an Epo-dependent hematopoietic cell line results in reduced activation of Stat5 apparently by its binding to phosphorylated Y343 of EpoR. Interestingly, co-expression of c-Cbl and APS is necessary to inhibit Stat5 activation in 293 cells [147].

PKC. Other candidate intracellular regulatory molecules are members of the protein kinase C (PKC) family of serine/threonine kinases. It has been proposed that PKC plays a role in the lineage determination of erythroid and megakaryocytic differentiation [148]. Other studies in human and murine erythroleukemic cell lines have shown that the modulation of PKCδ and PKCε or both play an important role in erythroid maturation indicating that distinct isoforms of PKCs may be involved in EpoR or TpoR signaling [149,150]. As mentioned above, PKCε is believed to link PI 3-kinase and MAP kinase and to regulate Epo dependent induction of c-myc and proliferation [92,151]. In contrast, down regulation of PKCα isoform expression inhibits Epo induced erythroid differentiation of human CD34+ progenitors cells [152]. More recently, it has been shown that PKCα controls erythropoietin receptor signaling and is important for Epo-dependent phosphorylation of EpoR and of various effectors [153]. A role for PKCs in Tpo-mediated induction of c-myc and proliferation in UT-7 cells expressing c-Mpl and in the activation of Erks in human platelets has also been reported [154-156]. However ,the precise role of each PKC isoform in TpoR signaling remains to be investigated.

Positive and Negative Control of Cytokine Receptor Signaling by the Tyrosine Phosphatases SHP-2 and SHP-1

SHP-1 and SHP-2 are two structurally related tyrosine phosphatases (see Chapter 2) containing two SH2 domains which bind directly at Epo receptor sites pY429/pY431 and pY401 respectively (Fig. 5). SHP-1 (also called HCP or PTP1C) is mutated in motheaten mice and is associated with hematopoietic cell hypertrophy, including an overexpansion of splenic CFU-E [157]. Cells expressing the EpoR mutant Y429F do not activate SHP-1 and exhibit prolonged Epo-induced activation of Jak2 in different hematopoietic cell lines, and overexpression of SHP-1 inhibits Epo-induced differentiation and suppression of apoptosis in J2E erythroleukemic cells, indicating that SHP-1 is an important negative regulator of EpoR signaling and is responsible for Jak2 dephosphorylation and inactivation [158,159]. Familial and congenital polycythemia, an autosomal dominant benign erythrocytosis, has been reported to be associated with mutations of the EpoR. Several mutations cause deletion of the last 59-83 C-terminal amino acids, which removes a negative regulatory domain including the binding sites of SHP-1, Y429 and Y431. Cell lines expressing these truncated receptors are hypersensitive to Epo and shows prolonged activation of Jak2 [160]. In contrast to SHP-1, SHP-2 (also called Syp or PTP1D) is believed to positively regulate proliferation; thus, mutation of Y401 of EpoR abolishes proliferative activity [161]. As mentioned above, SHP-2 is tyrosine phosphorylated and associates with Grb2, Ship, PI 3-kinase, Gab1, Shc, CrkL and Cbl in different cell lines following Epo stimulation [95,137,162]. SHP-2 is also involved in the activation of PI 3-kinase and MAP kinase in Tpo and Leptin-induced signaling events [50, 100,129,163]. It appears therefore that in addition to its phosphatase activity, SHP-2 plays a pivotal role as an adaptor molecule in the activation of distinct signaling pathways elicited by Epo, Tpo and leptin.

Negative Regulation of EpoR, c-Mpl and ObR Signaling by CIS/SOCS Family Members

Members of the supressor of cytokine signaling (SOCS) family were discovered as negative regulators of cytokine signaling by inhibition of the JAK/STAT pathway (see Chapter 6). Eight structurally related family members have been isolated: Cis-1 and SOCS-1 to-7, each of which contains a conserved C-terminal domain of 40 to 50 amino acids called the SOCS box and a central SH2 domain [164]. The N-terminal region is the most divergent part of Cis-1/SOCS proteins and is believed to be important for the ability of SOCS proteins to inhibit tyrosine kinases and/or other signaling effectors.

The gene encoding Cis-1, the founding member of this family, was isolated as a Stat5 target [165]. Epo, Tpo and other cytokines like IL-2, IL-3, GM-CSF rapidly induce *cis* gene expression and Epo-induced transcription is mediated in part by Stat5 binding sites present in the human and murine *cis* gene promoters [166]. In the Epo receptor system, Cis-1 binds to phosphorylated Y401, a major Stat5 binding site and has been proposed to blunt proliferative signaling by inhibiting Stat5 activation [167,168]. Cis-1 is also ubiquinated and is involved in the proteasome degradation of the EpoR[169]. In FDCP-2 cells bearing c-Mpl, Tpo induces Cis-1 expression and its binding to tyrosine phosphorylated c-Mpl, a process involving probably the SH2 domain of Cis-1. As is true for EpoR, Cis-1 could be important for c-Mpl degradation in this cell line [170]. The physiological role of Cis-1 in EpoR and TpoR remains unclear. As in *Stat5a/5b*$^{-/-}$ mice, transgenic mice constitutively expressing Cis-1 from a β-actin gene promoter show defects in Prl-dependent mammary gland formation, T cell proliferation and growth hormone signaling, but adult erythropoiesis is apparently normal [168].

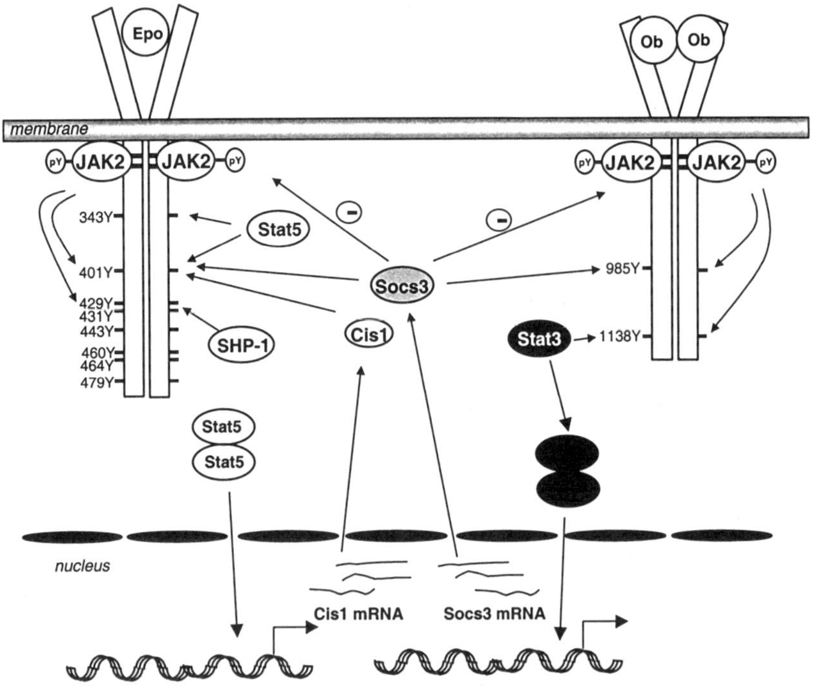

Figure 5. Negative regulation of EpoR and ObR signaling.(See details in the text)

Epo-induced SOCS-1, SOCS-2 and SOCS-3 expression has also been reported. SOCS-1, also called JAB (JAK binding protein), binds to Y1007 of Jak2 and inhibits Jak2 activity and tyrosine phosphorylation of EpoR in 293 cells [171]. In erythroid progenitors and in Ba/F3 cells expressing EpoR, SOCS-3 binds to phosphorylated Y401 of EpoR and to Jak2, and inhibits Epo-dependent proliferation and Stat5 activation. Binding of SOCS-3 to EpoR and Jak2 requires the N- and C-terminal regions adjacent to the SH2 domain. Deletion of the Y401-containing region of EpoR reduces the inhibitory effect of SOCS-3 [172]. Mice deficient in SOCS-3 die as embryos at days 12-16 and a marked erythrocytosis is associated with this embryonic lethality. Moreover, the *in vitro* proliferative capacity of progenitors is greatly increased. Transgene-mediated expression of SOCS-3 also results in embryonic lethality and blocks fetal erythropoiesis [173]. *SOCS-1*$^{-/-}$ mice were also generated but showed a completely different phenotype and develop a complex fatal neonatal disease due to an hyperresponsiveness to IFNγ [174]. Thus, SOCS-3 but not SOCS-1 plays a crucial role in negatively regulating fetal liver erythropoiesis indicating that the specific inhibitory role of SOCS-3 in erythropoiesis may be explained by its ability to bind to both EpoR and Jak2.

In Tpo-dependent hematopoietic cell lines, IFNα supresses Tpo-induced phosphorylation of c-Mpl, Stat5 and Tpo-induced megakaryocytopoiesis by induction of SOCS-1 [175]. However, the identity of SOCS proteins involved in the negative regulation of Tpo signaling *in vivo* remains to be determined.

Peripheral leptin administration rapidly induces SOCS-3 mRNA in the hypothalamus of *ob/ob* but not *db/db* mice, but has no effect on Cis-1, SOCS-1 and SOCS-2 expression. In addition, in Ay/a mice, a model of leptin-resistant murine obesity, an increase in SOCS-3 mRNA expression was observed in the arcuate and dorsomedial hypothalamic nuclei [176]. In cell lines expressing ObR, induction of SOCS-3 expression by leptin is mediated by Stat3 and expression of SOCS-3, but not Cis-1 or SOCS-2, blocks leptin-induced signal transduction [50, 177]. The use of cells expressing ObR tyrosine mutants demonstrated that SOCS-3 binds to phosphorylated Y985 which was previously shown to be a docking site for SHP-2. Mutation of Y985 or overexpression of SHP-2 blunts SOCS-3-mediated inhibition of ObR signaling indicating that SOCS-3 inhibits OBR signaling by binding to phosphorylated Y985 [178]. In conclusion, SOCS-3 is a leptin inducible inhibitor of leptin signaling and a potential mediator of central leptin resistance.

CONCLUSIONS

Since the discovery of Epo, Tpo, leptin and their receptors, remarkable progress has been made toward understanding, at the molecular level, the signaling mechanisms that underlie the physiological properties of these different cytokines. Epo, Tpo and leptin signal transduction pathways appear to be similar and most of the signaling molecules thus far identified as being activated by Epo, Tpo and leptin are also found in the repertoire of signaling pathways used by other cytokine receptors (see Chapter 6). Questions that concern the molecular mechanisms of cytokine specificity and the permissive versus instructive role of cytokines in cell differentiation are still under debate. The permissive model proposes that cytokine receptors provide only signals for survival and/or proliferation. Several data from the literature seem to favor the permissive role of Epo and Tpo on erythroid and megakaryocytic differentiation. The c-Mpl cytoplasmic domain can be replaced by that of granulocyte colony-stimulating factor receptor *in vivo* without a measurable effect on megakaryocyte differentiation and expression of prolactin receptor in *EpoR*$^{-/-}$ progenitors fully supports erythroid differentiation indicating that signals emanating from these receptors are qualitatively sufficient to mimic those required for the biological properties of Epo and Tpo [179, 180]. In addition, it became evident that one of the primary action of Tpo or Epo is to prevent induction of apoptosis and that some of the signaling pathways induced by these ligands, e.g PI 3-kinase, deliver an anti-apoptotic signal in erythroid progenitors and megakaryocytes. One other persistent question concerns the identification of the key components downstream the receptor and Jak2 which are necessary for Epo or Tpo-induced proliferation, survival and differentiation *in vivo*. Previous studies with Epo receptor mutants in which all the tyrosine residues of the cytoplasmic part of the receptor were mutated indicated that the tyrosine residues are required for erythroid colony formation from murine CFU-E [181,182]. Contrasting results were however obtained with similar mutants in the erythroid differentiation of human CD34+ progenitors [183]. EpoR mutants containing any one of the tyrosine residues Y343, Y429,Y431,Y443 are able to support the formation of significant numbers of murine CFU-E, although to a reduced level, while the EpoR mutant retaining only Y479 support the formation of an almost normal number of CFU-E. These results indicate that proliferation, differentiation or survival of erythroid progenitors are at least in part mediated by redundant signaling pathways [18].

The conclusions that can be drawn from signaling studies in cell lines show that the signaling pathways described in this review are activated through different mechanisms, can interact and probably regulate each other through different signaling adaptors or effectors and are able to act together to fully support proliferation, differentiation or survival. Activation of PI 3-kinase through its binding to phosphorylated Y479, for instance, results in MAP kinase activation. This may explain the relatively high activity of this receptor mutant retaining only Y479 on erythroid differentiation [66]. Activation of both Jak2/Stat5 and Ras pathways is also required for full erythroid differentiation [183]. In a similar vein, although mice expressing a c-Mpl mutant without the cytoplasmic membrane distal region have normal numbers of megakaryocytes and platelets, they exhibit a smaller increase in platelet numbers than wild-type mice in response to exogenous Tpo treatment indicating that signals emanating from the distal part of the receptor determines the quantitative but not the qualitative response to Tpo and that cooperation between the proximal and the distal domain of the c-Mpl cytoplasmic region is necessary for an appropriate acute response to Tpo[126]. Cooperation between different signaling pathways may therefore explain the final outcome that optimize the biological properties of Epo, Tpo and leptin. The use of mice with deletion or conditional deletion of one or more genes encoding various signaling effectors or adaptors will be helpful in the future to determine the downstream component(s) that typify the biological actions of Epo, Tpo and leptin.

REFERENCES

1. **Lacombe, C. and Mayeux, P.** 1998 Biology of erythropoietin. Haematologica 83 :724-32.
2. **Wendling, F.**1999 Thrombopoietin: its role from early hematopoiesis to platelet production. Haematologica 84: 158-66.
3. **White, D.W. and Tartaglia, L.A.** 1996 Leptin and OB-R: body weight regulation by a cytokine receptor. Cytokine Growth Factor Rev 7: 303-9.
4. **Bazan, J.F.** 1990 Structural design and molecular evolution of a cytokine receptor superfamily. Proc Natl Acad Sci U S A 87: 6934-8.
5. **Recny, M.A., Scoble, H.A. and Kim, Y.** 1987 Structural characterization of natural human urinary and recombinant DNA-derived erythropoietin. Identification of des-arginine 166 erythropoietin. J Biol Chem 262: 17156-63.
6. **Bazan, J.F.** 1990 Haemopoietic receptors and helical cytokines. Immunol Today 11: 350-4.
7. **Wu, H., Liu, X., Jaenisch, R. and Lodish, H.F.** 1995 Generation of committed erythroid BFU-E and CFU-E progenitors does not require erythropoietin or the erythropoietin receptor. Cell 83: 59-67.
8. **Methia, N., Louache, F., Vainchenker, W. and Wendling, F.** 1993 Oligodeoxynucleotides antisense to the proto-oncogene c-mpl specifically inhibit in vitro megakaryocytopoiesis. Blood 82: 1395-401.
9. **Wendling, F. and Vainchenker, W.** 1998 Thrombopoietin and its receptor. Eur Cytokine Netw 9: 221-31.
10. **Alexander, W.S.** 1999 Thrombopoietin and the c-Mpl receptor: insights from gene targeting. Int J Biochem Cell Biol 31: 1027-35.

11. **Ingalls, A.M., Dickie, M.M. and Snell, G.D**. 1996 Obese, a new mutation in the house mouse. Obes Res 4: 101.
12. **Hummel, K.P., Dickie, M.M. and Coleman, D.L**. 1966 Diabetes, a new mutation in the mouse. Science 153: 1127-8.
13. **Zhang, Y., Proenca, R., Maffei, M., Barone, M., Leopold, L. and Friedman, J.M.** 1994 Positional cloning of the mouse obese gene and its human homologue . Nature 372: 425-32.
14. **Madej, T., Boguski, M.S. and Bryant, S.H**. 1995 Threading analysis suggests that the obese gene product may be a helical cytokine. FEBS Lett 373: 13-8.
15. **Fantuzzi, G. and Faggioni, R**. 2000 Leptin in the regulation of immunity, inflammation, and hematopoiesis. J Leukoc Biol 68: 437-46.
16. **D'Andrea, A.D., Lodish, H.F. and Wong, G.G.** 1989 Expression cloning of the murine erythropoietin receptor. Cell 57: 277-85.
17. **Jones, S.S., D'Andrea, A.D., Haines, L.L. and Wong, G.G.** 1990 Human erythropoietin receptor: cloning, expression, and biologic characterization. Blood 76: 31-5.
18. **Constantinescu, S.N., Ghaffari, S. and Lodish, H.F.** 1999 The Erythropoietin Receptor: Structure, Activation and Intracellular Signal Transduction. Trends Endocrinol Metab 10: 18-23.
19. **Jiang, N., He, T.C., Miyajima, A. and Wojchowski, D.M**. 1996 The box1 domain of the erythropoietin receptor specifies Janus kinase 2 activation and functions mitogenically within an interleukin 2 beta-receptor chimera. J Biol Chem 271: 16472-6.
20. **Miura, O., D'Andrea, A., Kabat, D. and Ihle, J.N.** 1991 Induction of tyrosine phosphorylation by the erythropoietin receptor correlates with mitogenesis. Mol Cell Biol 11: 4895-902.
21. **Miura, O., Cleveland, J.L. and Ihle, J.N.** 1993 Inactivation of erythropoietin receptor function by point mutations in a region having homology with other cytokine receptors. Mol Cell Biol 13: 1788-95.
22. **Dusanter-Fourt, I., Casadevall, N., Lacombe, C., Muller, O., Billat, C., Fischer, S. and Mayeux, P**. 1992 Erythropoietin induces the tyrosine phosphorylation of its own receptor in human erythropoietin-responsive cells. J Biol Chem 267: 10670-5.
23. **Lin, C.S., Lim, S.K., D'Agati, V. and Costantini, F**. 1996 Differential effects of an erythropoietin receptor gene disruption on primitive and definitive erythropoiesis. Genes Dev 10: 154-64.
24. **Souyri, M., Vigon, I., Penciolelli, J.F., Heard, J.M., Tambourin, P. and Wendling, F**. 1990 A putative truncated cytokine receptor gene transduced by the myeloproliferative leukemia virus immortalizes hematopoietic progenitors. Cell 63: 1137-47.
25. **Skoda, R.C., Seldin, D.C., Chiang, M.K., Peichel, C.L., Vogt, T.F. and Leder, P**. 1993 Murine c-mpl: a member of the hematopoietic growth factor receptor superfamily that transduces a proliferative signal. Embo J 12: 2645-53.
26. **Vigon, I., Mornon, J.P., Cocault, L., Mitjavila, M.T., Tambourin, P., Gisselbrecht, S. and Souyri, M.** 1992 Molecular cloning and characterization of MPL, the human homolog of the v-mpl oncogene: identification of a member of the hematopoietic growth factor receptor superfamily. Proc Natl Acad Sci U S A 89: 5640-4.
27. **Souyri, M.** 1998 Mpl: from an acute myeloproliferative virus to the isolation of the long sought thrombopoietin. Semin Hematol 35: 222-31.
28. **Gurney, A.L., Wong, S.C., Henzel, W.J. and de Sauvage, F.J.** 1995 Distinct regions of c-Mpl cytoplasmic domain are coupled to the JAK-STAT signal transduction pathway and Shc phosphorylation. Proc Natl Acad Sci U S A 92: 5292-6.
29. **Drachman, J.G. and Kaushansky, K**. 1997 Dissecting the thrombopoietin receptor: functional elements of the Mpl cytoplasmic domain. Proc Natl Acad Sci U S A 94, 2350-5.
30. **Kaushansky, K.** 1999 Thrombopoietin and hematopoietic stem cell development. Ann N Y Acad Sci 872: 314-9.

31. **Ihara, K., Ishii, E., Eguchi, M., Takada, H., Suminoe, A., Good, R.A. and Hara, T.** 1999 Identification of mutations in the c-mpl gene in congenital amegakaryocytic thrombocytopenia. Proc Natl Acad Sci U S A 96: 3132-6.
32. **Tartaglia, L.A., Dembski, M., Weng, X., Deng, N., Culpepper, J., Devos, R., Richards, G.J., Campfield, L.A., Clark, F.T., Deeds, J. and et al.** 1995 Identification and expression cloning of a leptin receptor, OB-R. Cell 83: 1263-71.
33. **Tartaglia, L.A.** 1997 The leptin receptor. J Biol Chem 272: 6093-6.
34. **Huang, L., Wang, Z. and Li, C.** 2000 Modulation of Circulating Leptin Levels by Its Soluble Receptor. J Biol Chem .
35. **Baumann, H., Morella, K.K., White, D.W., Dembski, M., Bailon, P.S., Kim, H., Lai, C.F. and Tartaglia, L.A.** 1996 The full-length leptin receptor has signaling capabilities of interleukin 6-type cytokine receptors. Proc Natl Acad Sci U S A 93: 8374-8.
36. **Wang, Y., Kuropatwinski, K.K., White, D.W., Hawley, T.S., Hawley, R.G., Tartaglia, L.A. and Baumann, H.** 1997 Leptin receptor action in hepatic cells. J Biol Chem 272: 16216-23.
37. **Bjorbaek, C., Uotani, S., da Silva, B. and Flier, J.S.** 1997 Divergent signaling capacities of the long and short isoforms of the leptin receptor. J Biol Chem 272: 32686-95.
38. **Chen, H., Charlat, O., Tartaglia, L.A., Woolf, E.A., Weng, X., Ellis, S.J., Lakey, N.D., Culpepper, J., Moore, K.J., Breitbart, R.E., Duyk, G.M., Tepper, R.I. and Morgenstern, J.P.** 1996 Evidence that the diabetes gene encodes the leptin receptor: identification of a mutation in the leptin receptor gene in db/db mice. Cell 84: 491-5.
39. **Lee, G.H., Proenca, R., Montez, J.M., Carroll, K.M., Darvishzadeh, J.G., Lee, J.I. and Friedman, J.M.** 1996 Abnormal splicing of the leptin receptor in diabetic mice. Nature 379: 632-5.
40. **Watowich, S.S., Yoshimura, A., Longmore, G.D., Hilton, D.J., Yoshimura, Y. and Lodish, H.F.** 1992 Homodimerization and constitutive activation of the erythropoietin receptor. Proc Natl Acad Sci U S A 89: 2140-4.
41. **Watowich, S.S., Hilton, D.J. and Lodish, H.F.** 1994 Activation and inhibition of erythropoietin receptor function: role of receptor dimerization. Mol Cell Biol 14: 3535-49.
42. **Remy, I., Wilson, I.A. and Michnick, S.W.** 1999 Erythropoietin receptor activation by a ligand-induced conformation change. Science 283: 990-3.
43. **Alexander, W.S., Metcalf, D. and Dunn, A.R.** 1995 Point mutations within a dimer interface homology domain of c-Mpl induce constitutive receptor activity and tumorigenicity. Embo J 14: 5569-78.
44. **Ihle, J.N., Thierfelder, W., Teglund, S., Stravapodis, D., Wang, D., Feng, J. and Parganas, E.** 1998 Signaling by the cytokine receptor superfamily. Ann N Y Acad Sci 865: 1-9.
45. **Pellegrini, S. and Dusanter-Fourt, I.** 1997 The structure, regulation and function of the Janus kinases (JAKs) and the signal transducers and activators of transcription (STATs). Eur J Biochem 248: 615-33.
46. **Witthuhn, B.A., Quelle, F.W., Silvennoinen, O., Yi, T., Tang, B., Miura, O. and Ihle, J.N.** 1993 JAK2 associates with the erythropoietin receptor and is tyrosine phosphorylated and activated following stimulation with erythropoietin. Cell 74: 227-36.
47. **Drachman, J.G., Griffin, J.D. and Kaushansky, K.** 1995 The c-Mpl ligand (thrombopoietin) stimulates tyrosine phosphorylation of Jak2, Shc, and c-Mpl. J Biol Chem 270: 4979-82.
48. **Miyakawa, Y., Oda, A., Druker, B.J., Kato, T., Miyazaki, H., Handa, M. and Ikeda, Y.** 1995 Recombinant thrombopoietin induces rapid protein tyrosine phosphorylation of Janus kinase 2 and Shc in human blood platelets. Blood 86: 23-7.
49. **Ghilardi, N. and Skoda, R.C.** 1997 The leptin receptor activates janus kinase 2 and signals for proliferation in a factor-dependent cell line. Mol Endocrinol 11: 393-9.
50. **Banks, A.S., Davis, S.M., Bates, S.H. and Myers, M.G., Jr.** 2000 Activation of downstream signals by the long form of the leptin receptor. J Biol Chem 275: 14563-72.

51. **Sattler, M., Durstin, M.A., Frank, D.A., Okuda, K., Kaushansky, K., Salgia, R. and Griffin, J.D.** 1995 The thrombopoietin receptor c-MPL activates JAK2 and TYK2 tyrosine kinases. Exp Hematol 23: 1040-8.
52. **Parganas, E., Wang, D., Stravopodis, D., Topham, D.J., Marine, J.C., Teglund, S., Vanin, E.F., Bodner, S., Colamonici, O.R., van Deursen, J.M., Grosveld, G. and Ihle, J.N.** 1998 Jak2 is essential for signaling through a variety of cytokine receptors. Cell 93: 385-95.
53. **Neubauer, H., Cumano, A., Muller, M., Wu, H., Huffstadt, U. and Pfeffer, K.** 1998 Jak2 deficiency defines an essential developmental checkpoint in definitive hematopoiesis. Cell 93: 397-409.
54. **Drachman, J.G., Millett, K.M. and Kaushansky, K.** 1999 Thrombopoietin signal transduction requires functional JAK2, not TYK2. J Biol Chem 274: 13480-4.
55. **Dorsch, M., Fan, P.D., Danial, N.N., Rothman, P.B. and Goff, S.P.** 1997 The thrombopoietin receptor can mediate proliferation without activation of the Jak-STAT pathway. J Exp Med 186:1947-55.
56. **Oda, A., Sawada, K., Druker, B.J., Ozaki, K., Takano, H., Koizumi, K., Fukada, Y., Handa, M., Koike, T. and Ikeda, Y.** 1998 Erythropoietin induces tyrosine phosphorylation of Jak2, STAT5A, and STAT5B in primary cultured human erythroid precursors. Blood 92: 443-51.
57. **Miyakawa, Y., Oda, A., Druker, B.J., Miyazaki, H., Handa, M., Ohashi, H. and Ikeda, Y.** 1996 Thrombopoietin induces tyrosine phosphorylation of Stat3 and Stat5 in human blood platelets. Blood 87: 439-46.
58. **Pallard, C., Gouilleux, F., Benit, L., Cocault, L., Souyri, M., Levy, D., Groner, B., Gisselbrecht, S. and Dusanter-Fourt, I.** 1995 Thrombopoietin activates a STAT5-like factor in hematopoietic cells. Embo J 14: 2847-56.
59. **Schulze, H., Ballmaier, M., Welte, K. and Germeshausen, M.** 2000 Thrombopoietin induces the generation of distinct Stat1, Stat3, Stat5a and Stat5b homo- and heterodimeric complexes with different kinetics in human platelets. Exp Hematol 28: 294-304.
60. **Ghilardi, N., Ziegler, S., Wiestner, A., Stoffel, R., Heim, M.H. and Skoda, R.C.** 1996 Defective STAT signaling by the leptin receptor in diabetic mice. Proc Natl Acad Sci U S A 93: 6231-5.
61. **Rosenblum, C.I., Tota, M., Cully, D., Smith, T., Collum, R., Qureshi, S., Hess, J.F., Phillips, M.S., Hey, P.J., Vongs, A., Fong, T.M., Xu, L., Chen, H.Y., Smith, R.G., Schindler, C. and Van der Ploeg, L.H.** 1996 Functional STAT 1 and 3 signaling by the leptin receptor (OB-R); reduced expression of the rat fatty leptin receptor in transfected cells. Endocrinology 137: 5178-81.
62. **Drachman, J.G., Rojnuckarin, P. and Kaushansky, K.** 1999 Thrombopoietin signal transduction: studies from cell lines and primary cells. Methods 17: 238-49.
63. **Vaisse, C., Halaas, J.L., Horvath, C.M., Darnell, J.E., Jr., Stoffel, M. and Friedman, J.M.** 1996 Leptin activation of Stat3 in the hypothalamus of wild-type and ob/ob mice but not db/db mice. Nat Genet 14: 95-7.
64. **Damen, J.E., Wakao, H., Miyajima, A., Krosl, J., Humphries, R.K., Cutler, R.L. and Krystal, G.** 1995 Tyrosine 343 in the erythropoietin receptor positively regulates erythropoietin-induced cell proliferation and Stat5 activation. Embo J 14: 5557-68.
65. **Gobert, S., Chretien, S., Gouilleux, F., Muller, O., Pallard, C., Dusanter-Fourt, I., Groner, B., Lacombe, C., Gisselbrecht, S. and Mayeux, P.** 1996 Identification of tyrosine residues within the intracellular domain of the erythropoietin receptor crucial for STAT5 activation. Embo J 15: 2434-41.
66. **Klingmuller, U.** 1997 The role of tyrosine phosphorylation in proliferation and maturation of erythroid progenitor cells--signals emanating from the erythropoietin receptor. Eur J Biochem 249: 637-47.
67. **Fujitani, Y., Hibi, M., Fukada, T., Takahashi-Tezuka, M., Yoshida, H., Yamaguchi, T., Sugiyama, K., Yamanaka, Y., Nakajima, K. and Hirano, T.** 1997 An alternative

pathway for STAT activation that is mediated by the direct interaction between JAK and STAT. Oncogene 14: 751-61.
68. **Chretien, S., Varlet, P., Verdier, F., Gobert, S., Cartron, J.P., Gisselbrecht, S., Mayeux, P. and Lacombe, C.** 1996 Erythropoietin-induced erythroid differentiation of the human erythroleukemia cell line TF-1 correlates with impaired STAT5 activation. Embo J 15: 4174-81.
69. **Bittorf, T., Seiler, J., Ludtke, B., Buchse, T., Jaster, R. and Brock, J.** 2000 Activation of STAT5 during EPO-directed suppression of apoptosis. Cell Signal 12: 23-30.
70. **Iwatsuki, K., Endo, T., Misawa, H., Yokouchi, M., Matsumoto, A., Ohtsubo, M., Mori, K.J. and Yoshimura, A.** 1997 STAT5 activation correlates with erythropoietin receptor-mediated erythroid differentiation of an erythroleukemia cell line. J Biol Chem 272: 8149-52.
71. **Gregory, R.C., Jiang, N., Todokoro, K., Crouse, J., Pacifici, R.E. and Wojchowski, D.M.** 1998 Erythropoietin receptor and STAT5-specific pathways promote SKT6 cell hemoglobinization. Blood 92: 1104-18.
72. **Teglund, S., McKay, C., Schuetz, E., van Deursen, J.M., Stravopodis, D., Wang, D., Brown, M., Bodner, S., Grosveld, G. and Ihle, J.N.** 1998 Stat5a and Stat5b proteins have essential and nonessential, or redundant, roles in cytokine responses. Cell 93: 841-50.
73. **Socolovsky, M., Fallon, A.E., Wang, S., Brugnara, C. and Lodish, H.F.** 1999 Fetal anemia and apoptosis of red cell progenitors in Stat5a-/-5b-/- mice: a direct role for Stat5 in Bcl-X(L) induction. Cell 98: 181-91.
74. **Matsumura, I., Ishikawa, J., Nakajima, K., Oritani, K., Tomiyama, Y., Miyagawa, J., Kato, T., Miyazaki, H., Matsuzawa, Y. and Kanakura, Y.** 1997 Thrombopoietin-induced differentiation of a human megakaryoblastic leukemia cell line, CMK, involves transcriptional activation of p21(WAF1/Cip1) by STAT5. Mol Cell Biol 17: 2933-43.
75. **Liu, R.Y., Fan, C., Garcia, R., Jove, R. and Zuckerman, K.S.** 1999 Constitutive activation of the JAK2/STAT5 signal transduction pathway correlates with growth factor independence of megakaryocytic leukemic cell lines. Blood 93: 2369-79.
76. **Hanazono, Y., Chiba, S., Sasaki, K., Mano, H., Yazaki, Y. and Hirai, H.** 1993 Erythropoietin induces tyrosine phosphorylation and kinase activity of the c-fps/fes proto-oncogene product in human erythropoietin-responsive cells. Blood 81: 3193-6.
77. **Duprez, V., Blank, U., Chretien, S., Gisselbrecht, S. and Mayeux, P.** 1998 Physical and functional interaction between p72(syk) and erythropoietin receptor. J Biol Chem 273: 33985-90.
78. **Machide, M., Mano, H. and Todokoro, K.** 1995 Interleukin 3 and erythropoietin induce association of Vav with Tec kinase through Tec homology domain. Oncogene 11: 619-25.
79. **Tilbrook, P.A., Ingley, E., Williams, J.H., Hibbs, M.L. and Klinken, S.P.** 1997 Lyn tyrosine kinase is essential for erythropoietin-induced differentiation of J2E erythroid cells. Embo J 16: 1610-9.
80. **Chin, H., Arai, A., Wakao, H., Kamiyama, R., Miyasaka, N. and Miura, O.** 1998 Lyn physically associates with the erythropoietin receptor and may play a role in activation of the Stat5 pathway. Blood 91: 3734-45.
81. **Ingley, E., Sarna, M.K., Beaumont, J.G., Tilbrook, P.A., Tsai, S., Takemoto, Y., Williams, J.H. and Klinken, S.P.** 2000 HS1 interacts with Lyn and is critical for erythropoietin-induced differentiation of erythroid cells. J Biol Chem 275: 7887-93.
82. **Yamashita, Y.,Miyazato, A.,Shimizu, R.,Komatsu, N., Miura, Y., Ozawa, K., Mano, H.** 1997 Tec protein-tyrosine kinase is involved in the thrombopoietin/c-Mpl signaling pathway. Exp Hematol 25: 211-16.
83. **Wu, H., Klingmuller, U., Besmer, P. and Lodish, H.F.** 1995 Interaction of the erythropoietin and stem-cell-factor receptors. Nature 377 242-6.
84. **Wu, H., Klingmuller, U., Acurio, A., Hsiao, J.G. and Lodish, H.F.** 1997 Functional interaction of erythropoietin and stem cell factor receptors is essential for erythroid colony formation. Proc Natl Acad Sci U S A 94: 1806-10.

85. **Kapur, R. and Zhang, L.** 2001 A Novel Mechanism of Cooperation between c-Kit and Erythropoietin Receptor. Stem Cell Factor induces the expression of Stat5 and erythropoietin receptor, resulting in efficient proliferation and survival by erythropoietin. J Biol Chem 276: 1099-1106.
86. **Pircher, T.J., Geiger, J.N., Miller, C.P., Zhang, D., Gaines, P. and Wojchowski, D.M.** 2000 Integrative signaling by minimal EPO receptor forms and C-kit. J Biol Chem .
87. **Damen, J.E., Krosl, J., Morrison, D., Pelech, S. and Krystal, G.** 1998 The hyperresponsiveness of cells expressing truncated erythropoietin receptors is contingent on insulin-like growth factor-1 in fetal calf serum. Blood 92: 425-33.
88. **Datta, S.R., Brunet, A. and Greenberg, M.E.** 1999 Cellular survival: a play in three Akts. Genes Dev 13: 2905-27.
89. **Uddin, S., Kottegoda, S., Stigger, D., Platanias, L.C. and Wickrema, A.** 2000 Activation of the Akt/FKHRL1 pathway mediates the antiapoptotic effects of erythropoietin in primary human erythroid progenitors. Biochem Biophys Res Commun 275: 16-9.
90. **Haseyama, Y., Sawada, K., Oda, A., Koizumi, K., Takano, H., Tarumi, T., Nishio, M., Handa, M., Ikeda, Y. and Koike, T.** 1999 Phosphatidylinositol 3-kinase is involved in the protection of primary cultured human erythroid precursor cells from apoptosis. Blood 94: 1568-77.
91. **Miura, O., Nakamura, N., Ihle, J.N. and Aoki, N.** 1994 Erythropoietin-dependent association of phosphatidylinositol 3-kinase with tyrosine-phosphorylated erythropoietin receptor. J Biol Chem 269: 614-20.
92. **Klingmuller, U., Wu, H., Hsiao, J.G., Toker, A., Duckworth, B.C., Cantley, L.C. and Lodish, H.F.** 1997 Identification of a novel pathway important for proliferation and differentiation of primary erythroid progenitors. Proc Natl Acad Sci U S A 94: 3016-21.
93. **Shigematsu, H., Iwasaki, H., Otsuka, T., Ohno, Y., Arima, F. and Niho, Y.** 1997 Role of the vav proto-oncogene product (Vav) in erythropoietin-mediated cell proliferation and phosphatidylinositol 3-kinase activity. J Biol Chem 272: 14334-40.
94. **Verdier, F., Chretien, S., Billat, C., Gisselbrecht, S., Lacombe, C. and Mayeux, P.** 1997 Erythropoietin induces the tyrosine phosphorylation of insulin receptor substrate-2. An alternate pathway for erythropoietin-induced phosphatidylinositol 3-kinase activation. J Biol Chem 272: 26173-8.
95. **Lecoq-Lafon, C., Verdier, F., Fichelson, S., Chretien, S., Gisselbrecht, S., Lacombe, C. and Mayeux, P.** 1999 Erythropoietin induces the tyrosine phosphorylation of GAB1 and its association with SHC, SHP2, SHIP, and phosphatidylinositol 3-kinase. Blood 93: 2578-85.
96. **Wickrema, A., Uddin, S., Sharma, A., Chen, F., Alsayed, Y., Ahmad, S., Sawyer, S.T., Krystal, G., Yi, T., Nishada, K., Hibi, M., Hirano, T. and Platanias, L.C.** 1999 Engagement of Gab1 and Gab2 in erythropoietin signaling. J Biol Chem 274: 24469-74.
97. **Sattler, M., Salgia, R., Durstin, M.A., Prasad, K.V. and Griffin, J.D.** 1997 Thrombopoietin induces activation of the phosphatidylinositol-3' kinase pathway and formation of a complex containing p85PI3K and the protooncoprotein p120CBL. J Cell Physiol 171: 28-33.
98. **Oda, A., Miyakawa, Y., Druker, B.J., Ozaki, K., Ohashi, H., Kato, T., Miyazaki, H., Handa, M., Ikebuchi, K. and Ikeda, Y.** 1999 Thrombopoietin-induced signal transduction and potentiation of platelet activation. Thromb Haemost 82: 377-84.
99. **Majka, M., Ratajczak, J., Gewirtz, A.M. and Ratajczak, M.Z.** 2000 PI-3k-Akt axis inhibits apoptosis in normal human megakaryoblasts and is efficiently activated by thrombopoietin [In Process Citation]. Exp Hematol 28: 1492.
100. **Miyakawa, Y., Rojnuckarin, P., Habib, T. and Kaushansky, K.** 2000) Thrombopoietin induces PI3K activation through SHP2, Gab and IRS proteins in BaF3 cells and primary murine megakaryocytes. J Biol Chem .

101. **Bouscary, D., Lecoq-Lafon, C., Chretien, S., Zompi S., Fichelson S., Muller, O., Porteu F., Dusanter, I., Gisselbrecht, S., Mayeux, P. and Lacombe, C.** Role of Gab proteins in phosphatidyl-inositol 3-kinase activation by thrombopoietin (Tpo). Submitted
102. **Kellerer, M., Koch, M., Metzinger, E., Mushack, J., Capp, E. and Haring, H.U.** 1997 Leptin activates PI-3 kinase in C2C12 myotubes via janus kinase-2 (JAK-2) and insulin receptor substrate-2 (IRS-2) dependent pathways. Diabetologia 40: 1358-62.
103. **Harvey, J., McKay, N.G., Walker, K.S., Van der Kaay, J., Downes, C.P. and Ashford, M.L.** 2000 Essential role of phosphoinositide 3-kinase in leptin-induced K(ATP) channel activation in the rat CRI-G1 insulinoma cell line. J Biol Chem 275: 4660-9.
104. **Ren, H.Y., Komatsu, N., Shimizu, R., Okada, K. and Miura, Y.** 1994 Erythropoietin induces tyrosine phosphorylation and activation of phospholipase C-gamma 1 in a human erythropoietin-dependent cell line. J Biol Chem 269: 19633-8.
105. **Boudot, C., Petitfrere, E., Kadri, Z., Chretien, S., Mayeux, P., Haye, B. and Billat, C.** 1999 Erythropoietin induces glycosylphosphatidylinositol hydrolysis. Possible involvement of phospholipase c-gamma(2). J Biol Chem 274: 33966-72.
106. **Damen, J.E., Liu, L., Rosten, P., Humphries, R.K., Jefferson, A.B., Majerus, P.W. and Krystal, G.** 1996 The 145-kDa protein induced to associate with Shc by multiple cytokines is an inositol tetraphosphate and phosphatidylinositol 3,4,5-triphosphate 5-phosphatase. Proc Natl Acad Sci U S A 93: 1689-93.
107. **Liu, L., Damen, J.E., Cutler, R.L. and Krystal, G.** 1994 Multiple cytokines stimulate the binding of a common 145-kilodalton protein to Shc at the Grb2 recognition site of Shc. Mol Cell Biol 14: 6926-35.
108. **Mason, J.M., Beattie, B.K., Liu, Q., Dumont, D.J. and Barber, D.L.** 2000 The SH2 inositol 5-phosphatase Shipl is recruited in an SH2-dependent manner to the erythropoietin receptor. J Biol Chem 275: 4398-406.
109. **Bergelson, S., Klingmuller, U., Socolovsky, M., Hsiao, J.G. and Lodish, H.F.** 1998 Tyrosine residues within the intracellular domain of the erythropoietin receptor mediate activation of AP-1 transcription factors. J Biol Chem 273: 2396-401.
110. **Helgason, C.D., Damen, J.E., Rosten, P., Grewal, R., Sorensen, P., Chappel, S.M., Borowski, A., Jirik, F., Krystal, G. and Humphries, R.K.** 1998 Targeted disruption of SHIP leads to hemopoietic perturbations, lung pathology, and a shortened life span. Genes Dev 12: 1610-20.
111. **Marais, R. and Marshall, C.J.** 1996 Control of the ERK MAP kinase cascade by Ras and Raf. Cancer Surv 27: 101-25.
112. **Miura, Y., Miura, O., Ihle, J.N. and Aoki, N.** 1994 Activation of the mitogen-activated protein kinase pathway by the erythropoietin receptor. J Biol Chem 269: 29962-9.
113. **He, T.C., Jiang, N., Zhuang, H. and Wojchowski, D.M.** 1995 Erythropoietin-induced recruitment of Shc via a receptor phosphotyrosine-independent, Jak2-associated pathway. J Biol Chem 270: 11055-61.
114. **Barber, D.L., Corless, C.N., Xia, K., Roberts, T.M. and D'Andrea, A.D.** 1997 Erythropoietin activates Rafl by an Shc-independent pathway in CTLL-EPO-R cells. Blood 89: 55-64.
115. **Nosaka, Y., Arai, A., Miyasaka, N. and Miura, O.** 1999 CrkL mediates Ras-dependent activation of the Raf/ERK pathway through the guanine nucleotide exchange factor C3G in hematopoietic cells stimulated with erythropoietin or interleukin-3. J Biol Chem 274: 30154-62.
116. **Sakamoto, H., Kitamura, T. and Yoshimura, A.** 2000 Mitogen-activated Protein Kinase Plays an Essential Role in the Erythropoietin-dependent Proliferation of CTLL-2 Cells. J Biol Chem 275: 35857-35862.
117. **Jacobs-Helber, S.M., Ryan, J.J. and Sawyer, S.T.** 2000 JNK and p38 are activated by erythropoietin (EPO) but are not induced in apoptosis following EPO withdrawal in EPO-dependent HCD57 cells. Blood 96: 933-40.

118. **Alexander, W.S., Maurer, A.B., Novak, U. and Harrison-Smith, M.** 1996 Tyrosine-599 of the c-Mpl receptor is required for Shc phosphorylation and the induction of cellular differentiation. Embo J 15: 6531-40.
119. **Hill, R.J., Zozulya, S., Lu, Y.L., Hollenbach, P.W., Joyce-Shaikh, B., Bogenberger, J. and Gishizky, M.L.** 1996 Differentiation induced by the c-Mpl cytokine receptor is blocked by mutant Shc adaptor protein. Cell Growth Differ 7: 1125-34.
120. **Challier, C., Cocault, L., Flon, M., Pauchard, M., Porteu, F., Gisselbrecht, S. and Souyri, M.** 2000 A new feature of Mpl receptor: ligand-induced transforming activity in FRE rat fibroblasts. Oncogene 19: 2033-42.
121. **Porteu, F., Rouyez, M.C., Cocault, L., Benit, L., Charon, M., Picard, F., Gisselbrecht, S., Souyri, M. and Dusanter-Fourt, I.** 1996 Functional regions of the mouse thrombopoietin receptor cytoplasmic domain: evidence for a critical region which is involved in differentiation and can be complemented by erythropoietin. Mol Cell Biol 16: 2473-82.
122. **Rouyez, M.C., Boucheron, C., Gisselbrecht, S., Dusanter-Fourt, I. and Porteu, F.** 1997 Control of thrombopoietin-induced megakaryocytic differentiation by the mitogen-activated protein kinase pathway. Mol Cell Biol 17: 4991-5000.
123. **Matsumura, I., Nakajima, K., Wakao, H., Hattori, S., Hashimoto, K., Sugahara, H., Kato, T., Miyazaki, H., Hirano, T. and Kanakura, Y.** 1998 Involvement of prolonged ras activation in thrombopoietin-induced megakaryocytic differentiation of a human factor-dependent hematopoietic cell line. Mol Cell Biol 18: 4282-90.
124. **Rojnuckarin, P., Drachman, J.G. and Kaushansky, K.** 1999 Thrombopoietin-induced activation of the mitogen-activated protein kinase (MAPK) pathway in normal megakaryocytes: role in endomitosis. Blood 94: 1273-82.
125. **Fichelson, S., Freyssinier, J.M., Picard, F., Fontenay-Roupie, M., Guesnu, M., Cherai, M., Gisselbrecht, S. and Porteu, F.** 1999 Megakaryocyte growth and development factor-induced proliferation and differentiation are regulated by the mitogen-activated protein kinase pathway in primitive cord blood hematopoietic progenitors. Blood 94: 1601-13.
126. **Luoh, S.M., Stefanich, E., Solar, G., Steinmetz, H., Lipari, T., Pestina, T.I., Jackson, C.W. and de Sauvage, F.J.** 2000 Role of the distal half of the c-Mpl intracellular domain in control of platelet production by thrombopoietin in vivo. Mol Cell Biol 20: 507-15.
127. **Tanabe, K., Okuya, S., Tanizawa, Y., Matsutani, A. and Oka, Y.** 1997 Leptin induces proliferation of pancreatic beta cell line MIN6 through activation of mitogen-activated protein kinase. Biochem Biophys Res Commun 241: 765-8.
128. **Takahashi, Y., Okimura, Y., Mizuno, I., Iida, K., Takahashi, T., Kaji, H., Abe, H. and Chihara, K.** 1997 Leptin induces mitogen-activated protein kinase-dependent proliferation of C3H10T1/2 cells. J Biol Chem 272: 12897-900.
129. **Bjorbak, C., Buchholz, R.M., Davis, S.M., Bates, S.H., Pierroz, D.D., Gu, H., Neel, B.G., Myers, M.G., Jr. and Flier, J.S.** 2000 The role of SHP-2 in MAPK activation by leptin receptors. J Biol Chem .
130. **Berti, L. and Gammeltoft, S.** 1999 Leptin stimulates glucose uptake in C2C12 muscle cells by activation of ERK2. Mol Cell Endocrinol 157: 121-30.
131. **Nagata, Y., Moriguchi, T., Nishida, E. and Todokoro, K.** 1997 Activation of p38 MAP kinase pathway by erythropoietin and interleukin-3. Blood 90: 929-34.
132. **Nagata, Y., Nishida, E. and Todokoro, K.** 1997 Activation of JNK signaling pathway by erythropoietin, thrombopoietin, and interleukin-3. Blood 89: 2664-9.
133. **Ezumi, Y., Nishida, E., Uchiyama, T. and Takayama, H.** 1999 Thrombopoietin potentiates agonist-stimulated activation of p38 mitogen-activated protein kinase in human platelets. Biochem Biophys Res Commun 261: 58-63.
134. **Nagata, Y., Takahashi, N., Davis, R.J. and Todokoro, K.** 1998 Activation of p38 MAP kinase and JNK but not ERK is required for erythropoietin-induced erythroid differentiation. Blood 92: 1859-69.

135. **Tamura, K., Sudo, T., Senftleben, U., Dadak, A.M., Johnson, R. and Karin, M.** 2000 Requirement for p38alpha in erythropoietin expression: a role for stress kinases in erythropoiesis. Cell 102: 221-31.
136. **Feller, S.M., Posern, G., Voss, J., Kardinal, C., Sakkab, D., Zheng, J. and Knudsen, B.S.** 1998 Physiological signals and oncogenesis mediated through Crk family adapter proteins. J Cell Physiol 177: 535-52.
137. **Chin, H., Saito, T., Arai, A., Yamamoto, K., Kamiyama, R., Miyasaka, N. and Miura, O.** 1997 Erythropoietin and IL-3 induce tyrosine phosphorylation of CrkL and its association with Shc, SHP-2, and Cbl in hematopoietic cells. Biochem Biophys Res Commun 239: 412-7.
138. **Arai, A., Nosaka, Y., Kanda, E., Yamamoto, K., Miyasaka, N. and Miura, O.** 2000 Rap1 is activated by erythropoietin or interleukin-3 and is involved in regulation of {beta}1 integrin-mediated hematopoietic cell adhesion. J Biol Chem .
139. **Ota, J., Kimura, F., Sato, K., Wakimoto, N., Nakamura, Y., Nagata, N., Suzu, S., Yamada, M., Shimamura, S. and Motoyoshi, K.** 1998 Association of CrkL with STAT5 in hematopoietic cells stimulated by granulocyte-macrophage colony-stimulating factor or erythropoietin. Biochem Biophys Res Commun 252: 779-86.
140. **Oda, A., Wakao, H., Fujihara, M., Ozaki, K., Komatsu, N., Tanaka, S., Ikeda, H., Miyajima, A. and Ikebuchi, K.** 2000 Thrombopoietin and interleukin-2 induce association of CRK with STAT5 [In Process Citation]. Biochem Biophys Res Commun 278: 299-305.
141. **Barber, D.L., Mason, J.M., Fukazawa, T., Reedquist, K.A., Druker, B.J., Band, H. and D'Andrea, A.D.** 1997 Erythropoietin and interleukin-3 activate tyrosine phosphorylation of CBL and association with CRK adaptor proteins. Blood 89: 3166-74.
142. **Rudd, C.E. and Schneider, H.** 2000 Lymphocyte signaling: Cbl sets the threshold for autoimmunity. Curr Biol 10: R344-7.
143. **Hanazono, Y., Odai, H., Sasaki, K., Iwamatsu, A., Yazaki, Y. and Hirai, H.** 1996 Proto-oncogene products Vav and c-Cbl are involved in the signal transduction through Grb2/Ash in hematopoietic cells. Acta Haematol 95: 236-42.
144. **Sasaki, K., Odai, H., Hanazono, Y., Ueno, H., Ogawa, S., Langdon, W.Y., Tanaka, T., Miyagawa, K., Mitani, K., Yazaki, Y. and et al.** 1995 TPO/c-mpl ligand induces tyrosine phosphorylation of multiple cellular proteins including proto-oncogene products, Vav and c-Cbl, and Ras signaling molecules. Biochem Biophys Res Commun 216: 338-47.
145. **Brizzi, M.F., Dentelli, P., Lanfrancone, L., Rosso, A., Pelicci, P.G. and Pegoraro, L.** 1996 Discrete protein interactions with the Grb2/c-Cbl complex in SCF- and TPO-mediated myeloid cell proliferation. Oncogene 13: 2067-76.
146. **Oda, A., Ozaki, K., Druker, B.J., Miyakawa, Y., Miyazaki, H., Handa, M., Morita, H., Ohashi, H. and Ikeda, Y.** 1996 p120c-cbl is present in human blood platelets and is differentially involved in signaling by thrombopoietin and thrombin. Blood 88: 1330-8.
147. **Wakioka, T., Sasaki, A., Mitsui, K., Yokouchi, M., Inoue, A., Komiya, S. and Yoshimura, A.** 1999 APS, an adaptor protein containing Pleckstrin homology (PH) and Src homology-2 (SH2) domains inhibits the JAK-STAT pathway in collaboration with c-Cbl. Leukemia 13: 760-7.
148. **Lumelsky, N.L. and Schwartz, B.S.** 1997 Protein kinase C in erythroid and megakaryocytic differentiation: possible role in lineage determination. Biochim Biophys Acta 1358: 79-92.
149. **Patrone, M., Pessino, A., Passalacqua, M., Sparatore, B., Melloni, E. and Pontremoli, S.** 1996 Correlation between levels of delta protein kinase C and resistance to differentiation in murine erythroleukemia cells. Biochem Biophys Res Commun 220: 26-30.
150. **Patrone, M., Pessino, A., Passalacqua, M., Sparatore, B., Melloni, E. and Pontremoli, S.** 1994 Protein kinase C isoforms in murine erythroleukemia cells and their involvement in the differentiation process. FEBS Lett 344: 91-5.

151. **Li, Y., Davis, K.L. and Sytkowski, A.J.** 1996 Protein kinase C-epsilon is necessary for erythropoietin's up-regulation of c-myc and for factor-dependent DNA synthesis. Evidence for discrete signals for growth and differentiation. J Biol Chem 271: 27025-30.
152. **Myklebust, J.H., Smeland, E.B., Josefsen, D. and Sioud, M.** 2000 Protein kinase C-alpha isoform is involved in erythropoietin-induced erythroid differentiation of CD34(+) progenitor cells from human bone marrow. Blood 95: 510-8.
153. **von Lindern, M., Amelsvoort, M.P., van Dijk, T., Deiner, E., van Den Akker, E., van Emst-De Vries, S., Willems, P., Beug, H. and Lowenberg, B.** 2000 Protein kinase C alpha controls erythropoietin receptor signaling [In Process Citation]. J Biol Chem 275: 34719-27.
154. **Hong, Y., Dumenil, D., van der Loo, B., Goncalves, F., Vainchenker, W. and Erusalimsky, J.D.** 1998 Protein kinase C mediates the mitogenic action of thrombopoietin in c-Mpl-expressing UT-7 cells. Blood 91: 813-22.
155. **Kunitama, M., Shimizu, R., Yamada, M., Kato, T., Miyazaki, H., Okada, K., Miura, Y. and Komatsu, N.** 1997 Protein kinase C and c-myc gene activation pathways in thrombopoietin signal transduction. Biochem Biophys Res Commun 231: 290-4.
156. **Ezumi, Y., Uchiyama, T. and Takayama, H.** 1998 Thrombopoietin potentiates the protein-kinase-C-mediated activation of mitogen-activated protein kinase/ERK kinases and extracellular signal-regulated kinases in human platelets. Eur J Biochem 258: 976-85.
157. **Bignon, J.S. and Siminovitch, K.A.** 1994 Identification of PTP1C mutation as the genetic defect in motheaten and viable motheaten mice: a step toward defining the roles of protein tyrosine phosphatases in the regulation of hemopoietic cell differentiation and function. Clin Immunol Immunopathol 73: 168-79.
158. **Klingmuller, U., Lorenz, U., Cantley, L.C., Neel, B.G. and Lodish, H.F.** 1995 Specific recruitment of SH-PTP1 to the erythropoietin receptor causes inactivation of JAK2 and termination of proliferative signals. Cell 80: 729-38.
159. **Bittorf, T., Seiler, J., Zhang, Z., Jaster, R. and Brock, J.** 1999 SHP1 protein tyrosine phosphatase negatively modulates erythroid differentiation and suppression of apoptosis in J2E erythroleukemic cells. Biol Chem 380: 1201-9.
160. **Yoshimura, A. and Misawa, H.** 1998 Physiology and function of the erythropoietin receptor. Curr Opin Hematol 5: 171-6.
161. **Tauchi, T., Damen, J.E., Toyama, K., Feng, G.S., Broxmeyer, H.E. and Krystal, G.** 1996 Tyrosinc 425 within the activated erythropoietin receptor binds Syp, reduces the erythropoietin required for Syp tyrosine phosphorylation, and promotes mitogenesis. Blood 87: 4495-501.
162. **Tauchi, T., Feng, G.S., Shen, R., Hoatlin, M., Bagby, G.C., Jr., Kabat, D., Lu, L. and Broxmeyer, H.E.** 1995 Involvement of SH2-containing phosphotyrosine phosphatase Syp in erythropoietin receptor signal transduction pathways. J Biol Chem 270: 5631-5.
163. **Carpenter, L.R., Farruggella, T.J., Symes, A., Karow, M.L., Yancopoulos, G.D. and Stahl, N.** 1998 Enhancing leptin response by preventing SH2-containing phosphatase 2 interaction with Ob receptor. Proc Natl Acad Sci U S A 95: 6061-6.
164. **Yoshimura, A.** 1998 The CIS family: negative regulators of JAK-STAT signaling. Cytokine Growth Factor Rev 9: 197-204.
165. **Yoshimura, A., Ohkubo, T., Kiguchi, T., Jenkins, N.A., Gilbert, D.J., Copeland, N.G., Hara, T. and Miyajima, A.** 1995 A novel cytokine-inducible gene CIS encodes an SH2-containing protein that binds to tyrosine-phosphorylated interleukin 3 and erythropoietin receptors. Embo J 14: 2816-26.
166. **Verdier, F., Rabionet, R., Gouilleux, F., Beisenherz-Huss, C., Varlet, P., Muller, O., Mayeux, P., Lacombe, C., Gisselbrecht, S. and Chretien, S.** 1998 A sequence of the CIS gene promoter interacts preferentially with two associated STAT5A dimers: a distinct biochemical difference between STAT5A and STAT5B. Mol Cell Biol 18: 5852-60.

167. **Matsumoto, A., Masuhara, M., Mitsui, K., Yokouchi, M., Ohtsubo, M., Misawa, H., Miyajima, A. and Yoshimura, A.** 1997 CIS, a cytokine inducible SH2 protein, is a target of the JAK-STAT5 pathway and modulates STAT5 activation. Blood 89: 3148-54.
168. **Matsumoto, A., Seki, Y., Kubo, M., Ohtsuka, S., Suzuki, A., Hayashi, I., Tsuji, K., Nakahata, T., Okabe, M., Yamada, S. and Yoshimura, A.** 1999 Suppression of STAT5 functions in liver, mammary glands, and T cells in cytokine-inducible SH2-containing protein 1 transgenic mice. Mol Cell Biol 19: 6396-407.
169. **Verdier, F., Chretien, S., Muller, O., Varlet, P., Yoshimura, A., Gisselbrecht, S., Lacombe, C. and Mayeux, P.** 1998 Proteasomes regulate erythropoietin receptor and signal transducer and activator of transcription 5 (STAT5) activation. Possible involvement of the ubiquitinated Cis protein. J Biol Chem 273: 28185-90.
170. **Okabe, S., Tauchi, T., Morita, H., Ohashi, H., Yoshimura, A. and Ohyashiki, K.** 1999 Thrombopoietin induces an SH2-containing protein, CIS1, which binds to Mpl: involvement of the ubiquitin proteosome pathway. Exp Hematol 27: 1542-7.
171. **Yasukawa, H., Misawa, H., Sakamoto, H., Masuhara, M., Sasaki, A., Wakioka, T., Ohtsuka, S., Imaizumi, T., Matsuda, T., Ihle, J.N. and Yoshimura, A.** 1999 The JAK-binding protein JAB inhibits Janus tyrosine kinase activity through binding in the activation loop. Embo J 18: 1309-20.
172. **Sasaki, A., Yasukawa, H., Shouda, T., Kitamura, T., Dikic, I. and Yoshimura, A.** 2000 CIS3/SOCS-3 suppresses erythropoietin (EPO) signaling by binding the EPO receptor and JAK2 [In Process Citation]. J Biol Chem 275: 29338-47.
173. **Marine, J.C., McKay, C., Wang, D., Topham, D.J., Parganas, E., Nakajima, H., Pendeville, H., Yasukawa, H., Sasaki, A., Yoshimura, A. and Ihle, J.N.** 1999 SOCS3 is essential in the regulation of fetal liver erythropoiesis. Cell 98: 617-27.
174. **Alexander, W.S., Starr, R., Fenner, J.E., Scott, C.L., Handman, E., Sprigg, N.S., Corbin, J.E., Cornish, A.L., Darwiche, R., Owczarek, C.M., Kay, T.W., Nicola, N.A., Hertzog, P.J., Metcalf, D. and Hilton, D.J.** 1999 SOCS1 is a critical inhibitor of interferon gamma signaling and prevents the potentially fatal neonatal actions of this cytokine. Cell 98: 597-608.
175. **Wang, Q., Miyakawa, Y., Fox, N. and Kaushansky, K.** 2000 Interferon-alpha directly represses megakaryopoiesis by inhibiting thrombopoietin-induced signaling through induction of SOCS-1. Blood 96: 2093-9.
176. **Bjorbaek, C.,Elmquist, J. K., Frantz, J. D., Shoelson, S. E., Flier, J. S.** 1998 Identification of SOCS-3 as a potential mediator of central leptin resistance. Mol. Cellules,1: 619-25
177. **Bjorbaek, C., El-Haschimi, K., Frantz, J.D. and Flier, J.S.** 1999 The role of SOCS-3 in leptin signaling and leptin resistance. J Biol Chem 274: 30059-65.
178. **Bjorbak, C., Lavery, H.J., Bates, S.H., Olson, R.K., Davis, S.M., Flier, J.S. and Myers, M.G., Jr.** 2000 SOCS3 mediates feedback inhibition of the leptin receptor via tyr985 [In Process Citation]. J Biol Chem 275: 40649-57.
179. **Stoffel, R., Ziegler, S., Ghilardi, N., Ledermann, B., de Sauvage, F.J. and Skoda, R.C.** 1999 Permissive role of thrombopoietin and granulocyte colony-stimulating factor receptors in hematopoietic cell fate decisions in vivo. Proc Natl Acad Sci U S A 96: 698-702.
180. **Socolovsky, M., Fallon, A.E. and Lodish, H.F.** 1998 The prolactin receptor rescues EpoR-/- erythroid progenitors and replaces EpoR in a synergistic interaction with c-kit. Blood 92: 1491-6.
181. **Socolovsky, M., Dusanter-Fourt, I. and Lodish, H.F.** 1997 The prolactin receptor and severely truncated erythropoietin receptors support differentiation of erythroid progenitors. J Biol Chem 272: 14009-12.
182. **Chida, D., Miura, O., Yoshimura, A. and Miyajima, A.** 1999 Role of cytokine signaling molecules in erythroid differentiation of mouse fetal liver hematopoietic cells: functional analysis of signaling molecules by retrovirus-mediated expression. Blood 93: 1567-78.

183. **Fichelson, S., Chretien, S., Rokicka-Piotrowicz, M., Bouhanik, S., Gisselbrecht, S., Mayeux, P. and Lacombe, C.** 1999 Tyrosine residues of the erythropoietin receptor are dispensable for erythroid differentiation of human CD34+ progenitors. Biochem Biophys Res Commun 256: 685-91.

Section IV

G Protein-Coupled Receptors

9

G Proteins and G Protein-Coupled Receptors: *Overview*

Michael A. Shetzline* **and Marc G. Caron**

Howard Hughes Medical Institute Laboratories and Department of Cell Biology and Medicine, Duke University Medical Center, Durham, NC 27710, USA

CELL SURFACE RECEPTORS

Cells live in a constantly changing milieu. The structure and biochemical nature of this environment is dynamic and in order for normal cellular function to proceed in a manner appropriate for the benefit of the organism, cells must be able to access this changing information. The biochemical mediators of this information exchange are cell surface receptors. Receptors transduce information from the extracellular space to the intracellular compartment. This process involves the following steps: receptor activation, G protein coupling, and second messenger generation. Equally important in cell signaling is the termination of this signal including desensitization, internalization and resensitization. All these steps represent potential regulatory checkpoints and possible targets for therapeutic intervention. Cell surface receptors include G protein-coupled receptors (GPCRs), as well as, ion channels and enzyme-linked receptors. GPCRs are responsible for physiological responses as diverse as visual perception, neurotransmission, cell growth and cell differentiation.

G protein-coupled receptors are encoded by a significant portion of the genetic code. In the *Caenorhabditis elegans* genome about 5% of genes

---

* shetz001@mc.duke.edu

encode GPCRs [3]. In the human, GPCRs represent 2-3 percent of the genome; or at least 616 receptors, excluding sensory receptors [61].

Table I: Classification of G Protein-Coupled Receptors

Class I
- Rhodopsin, catecholamines, acetylcholine, serotonin, angiotensin, olfactory
- Gastrin/CCK, neuropeptide Y, substance P, endothelin, opioid, chemokine (CCR 1-5, CXCR1-4), histamine, vasopressin

Class II
- Secretin, calcitonin, parathyroid hormone, vasoactive intestinal polypeptide, glucagon, glucagon-like peptide 1, PACAP

Class III
- Glutamate-metabotrophic receptors, calcium receptors, $GABA_B$

G PROTEIN-COUPLED RECEPTORS

G protein-coupled receptors are seven membrane spanning domain proteins that are functionally associated with heterotrimeric G proteins. The membrane regions consist of α-helical domains with a well-conserved structural motif of 22 to 28 residues each [34; 45]. For a variety of GPCRs, the third intracellular loop has been found to provide a binding motif for the G protein. GPCRs are classified by their structure into three distinct classes (Table I). Class I, the largest, contains the receptors for light perception, catecholamines, many peptide hormones, neuropeptides, and glycoproteins. These receptors share 30-65% sequence homology and often the presence of a "D-R-Y" motif in the second intracellular loop. Class II receptors include the secretin/glucagon/vasoactive intestinal peptide receptor family. This group is distinct from Class I, sharing only approximately 12% sequence homology with Class I and the absence of the "D-R-Y" motif in the N-terminal portion of the intracellular loop between membrane spanning domains 3 and 4. Class III GPCRs contain the metabotrophic receptors (calcium-sensing and glutamate receptors) and with the exception of the heptahelical spanning domain organization, are significantly different from the other two receptor classes. However, there are a number of receptors that have not been classified. These receptors have been called Orphan GPCRs.

Receptor discovery has been based on receptor function, bioassay, pharmacology, biochemical purification and the use of receptor specific ligands. With the sequences known for many hundreds of GPCRs, it is now possible to discover novel GPCRs based on sequence similarity with an established receptor. Presently, well over 100 receptors have been found using sequence homology screening but, as of yet, few have known endogenous ligands. This group of receptors is fertile ground for discovery and some orphan GPCRs have found their 'family' and are no longer 'orphaned'. For example, using the growth hormone secretagogue receptor sequence, a GPCR was identified with significant sequence similarity, however no specific ligand could be determined and this construct was named GPR38. Subsequently, this receptor was found to be the motilin receptor [14]. Another previously "orphaned" GPCR originally misclassified due to a weak ligand similarity to the gastrointestinal hormone secretin is the hypocretin/orexin receptor [27]. Absence of this receptor has been demonstrated to produce narcolepsy in targeted orexin receptor gene knockout mice [7]. Recently, Lee et al have reviewed and updated the growing list of orphan receptors, including those whose ligand has been found [31].

G protein-coupled receptors contain an extracellular amino terminus and an intracellular carboxyl terminus (Fig. 1). The N- and C-termini can contain anywhere from less than 10 to greater than 500 residues. When stimulated by the appropriate chemical messenger, the GPCR undergoes a conformational change that provides coupling to a specific G protein. Recently, the crystal structure of rhodopsin has been elucidated to 2.8 angstroms [45]. This three-dimensional structure reveals a highly organized heptahelical transmembrane component with a portion of the C-terminus perpendicular to the seventh membrane-spanning domain. The availability of structural information will likely prove useful in the further study of GPCR regulation.

The structural and regulatory information obtained from studying GPCRs has established the presence of constitutively active receptors, constitutively desensitized receptors, co-receptors involved in viral entry and receptors whose N-terminal segment functions as a native ligand for the receptor subunit [2; 40].

Mutations in GPCR structure can produce a receptor construct that is continuously active [33]. This lack of ligand induced activation causes persistent second messenger generation and loss of receptor regulation. Although the majority of constitutively active receptors have been produced by *in vitro*, mutant receptors that activate G protein subunits in the absence of agonist have been found *in vivo* [1; 23].

Mutation of the highly conserved "DRY" motif in the second intracellular loop of the histamine receptor produces a receptor construct that generated cAMP in the absence of agonist [1]. Also, somatic mutations of the thyrotropin receptor are the pathophysiologic basis for hyperthyroidism and hypothyroidism [46]. Point mutations have been located in the thyrotropin receptor that produce a constitutively active receptor and consequently, increased cAMP and a hyperthyroid state. Similarly, specific point mutations have been found in the luteinizing hormone receptor and result in precocious puberty in male children [52]. Cell proliferation is controlled by a variety of GPCRs. Recently, a mutated cholecystokinin B/gastrin receptor was found in human colon cancer cells and adenomatous colon polyps. This receptor contained a splice variant in the third intracellular loop that produced a cell that generated spontaneous oscillations in intracellular calcium in the absence of ligand application [23]. Receptor mediated cell proliferation as a mechanism for neoplasia may suggest a role for receptor inverse agonists in their treatment.

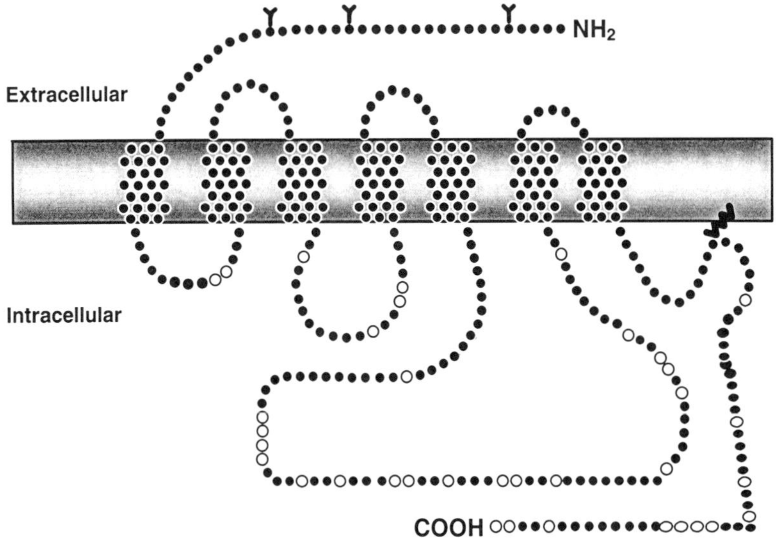

Figure 1. Representative secondary structure of a typical heptahelical GPCR. The N-terminus is of variable length and it often contains N-linked glycoslyation sites (Y) important in ligand binding. There are seven transmembrane domains and intracellular loops that contain the binding sites for G proteins and multiple kinase phosphorylation sites (o). Frequently, the C terminal peptide is palmitoylated at a specific cysteine residue (ᴡᴡ).

Certainly the presence of permanently "turned on" signaling pathways has been apparent for some time. However, with our increased understanding of receptor desensitization, permanently "turned off" receptors have now been identified [2]. Mutations of the arginine vasopressin receptor have long been recognized in nephrogenic diabetes insipidus. However, alteration in receptor structure, that resulted in substitution of arginine by histidine at position 137, was believed to cause loss of function via inability to activate G proteins. It has now been found that this receptor, in the absence of agonist, is phosphorylated and has an enhanced affinity for β-arrestin [2]. Therefore, the receptor appears not to signal, however it is in reality desensitized via its interaction with β-arrestin and constitutively sequestered. Therefore, how a mutation in a receptor will manifest will be the result of the interplay of the mutation's effect on G protein coupling and signaling, as well as the ability of components of the desensitization process to function.

G PROTEINS

G-proteins are molecular intermediaries that mediate the intracellular communication process [42; 60]. (Fig. 2) G proteins contain three subunits, alpha (α), beta (β) and gamma (γ) [25]. When a receptor-ligand interaction occurs the receptor changes conformation and guanosine diphosphate (GDP) is exchanged on the α subunit for guanosine triphosphate (GTP). This GTP activated G protein then proceeds to activate intracellular signaling substrates as outlined in Table II. Receptor activation also initiates G protein subunit dissociation, whereby the α subunit separates from the βγ subunits. The Gα-GTP complex is gradually inactivated by intrinsic GTPase that converts GTP to GDP. The conversion of GTP to GDP no longer allows G protein stimulation of adenylate cyclase and is one way by which the hormone signal is terminated and the basal condition restored. G-proteins are classified according to their alpha subunit [67]. G-proteins that stimulate adenylyl cyclase are classified as the Gs type; those that inhibit adenylyl cyclase are called Gi [42]. The β and γ subunits has been shown to serve a vast array of signaling functions that include activation or inhibition of effector systems, participation in receptor desensitization by translocation of G protein receptor kinases to agonist occupied receptors and stimulation of Ras-mediated MAP kinase regulation [10].

Presently, 16 alpha subunits have been identified, with 4 additional variants, and are organized according to the activity of the second-messenger system [67] (Table II). The α subunit often contains residues that can be myristoylated and/or palmitoylated. The C-terminus of this subunit is important in receptor recognition and the application of antibodies specific to this region can abrogate G protein interaction with the activated receptor. The βγ subunits are divided into several subfamilies. The Gs family, originally identified for its ability to stimulate adenylyl cyclase, contains two splice variants and the olfactory protein, Golf.

Table II: Classification of G Protein α Subunits

Class	Subunit	Signaling
$G_\alpha S$	$G_\alpha S$, $G_\alpha olf$	Adenylyl cyclase Ca^{2+} channels
$G_\alpha i$	$G_\alpha i$, $G_\alpha o$, $G_\alpha t$ $G_\alpha gust$, $G_\alpha z$	Adenylyl cyclase, cGMP Phophodiesterase
$G_\alpha q$	$G_\alpha q$, $G_\alpha 11$, $G_\alpha 14$ $G_\alpha 15$/ $G_\alpha 16$	Phospholipase Cβ
$G_\alpha 12$	$G_\alpha 12$, $G_\alpha 13$	Na^+/H^+ exchange Rho guanine NEF

The Gαi family of subunits includes Gαi1, Gαi2, Gαi3, Gαo, transducin and gustducin (Table II). Gαi activation initiates an inhibition of a subset of adenylyl cyclases within the cell. In this manner, GPCRs can maintain fine control of the cellular cAMP concentration and subsequent intracellular signaling. Members of this family also activate phospholipases, phosphodiesterases [22] and participate in ion channel regulation [10; 21]. It was long known that the inhibitory activity of G protein-coupled receptors on adenylate cyclase could be blocked by pertussis toxin. It is now established that this inhibition results from the ability of Gα to serve as a substrate for pertussis toxin-mediated ADP-ribosylation [49]. Other GPCRs couple with Gq and G11. The Gq family of G protein subunits regulates the production of inositol 1,4,5-trisphosphate (IP3) and diacylglycerol (DAG) [50].

Little is known about Gα12 and Gα13. Targeted deletion of Gα13 is embryonic lethal [43]. Gα12 and Gα13 are activated by lysophosphatidic acid, thrombin and thromboxane A_2 receptors. Recently, evidence has accumulated indicating that Gα12 and Gα13 may be involved in cell migration, proliferation and transformation. Gα12 and Gα13 activate tyrosine phosphorylation of paxcillin and focal adhesion kinase, substrates important in the regulation of cell migration and proliferation [41]. Moreover, both Gα12 and Gα13 regulate the Na+/H+ antiporter [63].

Presently there are 6 known subtypes of the β subunit (β_1, β_2, β_3, β_4, β_5, β_6) and two groups of γ subunits which includes a total of 12 members. Gamma subunits can be modified by farnesylation (including γ_1, γ_{11}, γ_{8cone}) or geranyl-geranylation (γ_2, γ_3, γ_4, γ_5, γ_7, γ_{8olf}, γ_{10}, γ_{12}). Interestingly, neither free β nor free γ subunits are normally found in cells. The β subunit contains a WD40 repeat of β strands. The crystal structure of the βγ subunit has revealed this region to be a β propeller [55]. There is a very tight association between these two subunits and it appears that there is a specificity of interaction among the differing β and γ subunits with specific regions on the Gγ domain providing a nidus for interaction with Gβ [54]. After subunit dissociation, the βγ subunits remain tightly associated and also act as a signal transducer. It is now established that the βγ subunit participates in a vast array of cellular signals, including activation of MAP kinase, phospholipase A_2 and PI3 kinase [19](Table III). When the α subunit is in the GDP-bound form, it re-associates with βγ. With re-establishment of the heterotrimer, αβγ, along with other mechanisms of desensitization, receptor signaling via the separate subunits ceases.

Table III: Gβγ Effectors

- G protein coupled receptor kinases
- Adenylyl cyclase (types I, II, III, IV, VII)
- PI3-Kinase (γ)
- MAPK (Shc, Raf 1, Ras exchange factor)
- Phospholipase A_2
- Yeast pheromone response
- Ion channels (potassium, calcium)
- PLCβ (β1 - β4)
- p21-activated kinases (PAKs)

Adapted from Ref. [10]

Regulators of G Protein Signaling

The GTPase terminating event for GPCR signaling is enhanced by a family of proteins designated RGS (regulator of G-protein signaling) [12]. The RGS family of proteins contains a domain that binds to the Gα subunit (RGS domain) [11]. Presently, over 25 mammalian proteins have been identified that contain this regulatory domain. RGS proteins can turn off G proteins by stimulating their GTPase activity. This mechanism requires the RGS to function as a GAP (GTPase activating protein). Also, RGS proteins can physically prevent the G protein from interacting with, and stimulating its effector, as has been demonstrated by RGS4 [24]. Lastly, RGS proteins may accelerate reformation of the heterotrimer by enhancing the affinity of Gα for

$\beta\gamma$. This mechanism has been shown for overexpressed RGS1, RGS3 and RGS4 [13].

Although phosphorylation by G protein receptor kinases and subsequent arrestin interaction play an essential role in GPCR signal termination, (see below) RGS proteins serve to assist acute disruption of signaling via rapid turnoff of the G protein. The ability to terminate a G protein effect by stimulating GTP hydrolysis provides the cell another mechanism to swiftly abrogate a signaling event at a level in addition to changes in the receptor. While GRK dependent receptor phosphorylation prevents further agonist-receptor activation, RGS proteins act directly on activated G proteins to cease effector stimulation.

Effector Systems

Following receptor activation, GTP-bound G-protein subunits cause activation of specific enzymes or other proteins, ultimately resulting in an intracellular signaling event (Fig. 2). Effector enzymes, such as adenylyl cyclase or phospholipase C, may generate specific intracellular messengers cyclic adenosine monophosphate (cAMP), inositol 1,4,5 trisphosphate (IP3), and diacylglycerol, or may couple directly with specific ion channels, such as potassium or calcium channels, and initiate changes in ion permeability. The effector systems are not fully understood for some receptors such as those involved with cell growth and differentiation.

The adenylyl cyclase system represents one of the most studied receptor activated effector systems. Nine distinct membrane associated adenylyl cyclases have been cloned [53] and recently, a cytosolic form has been purified from the mammalian testis [5]. This cyclase is insensitive to G protein and/or forskolin regulation and may imply the existence of cytoplasmic cyclase activity.

Recently, the activity of adenylyl cyclase has become clearer with the elucidation of its structure. Hurley et al [69] determined the three dimensional crystal structure of the adenylyl cyclase catalytic core while Tesmer determined the resolution of the catalytically active form of adenylyl cyclase in a complex with Gsα [57]. These data have produced a model for the interaction of Gsα and adenylyl cyclase that shows the catalytic domain to be composed of two distinct parts. These regions are brought together by G-protein interaction or forskolin.

When an agonist binds to a Gs coupled receptor, the receptor associates with the Gαs subunit. Subsequent activity is dependent on the type of adenylyl

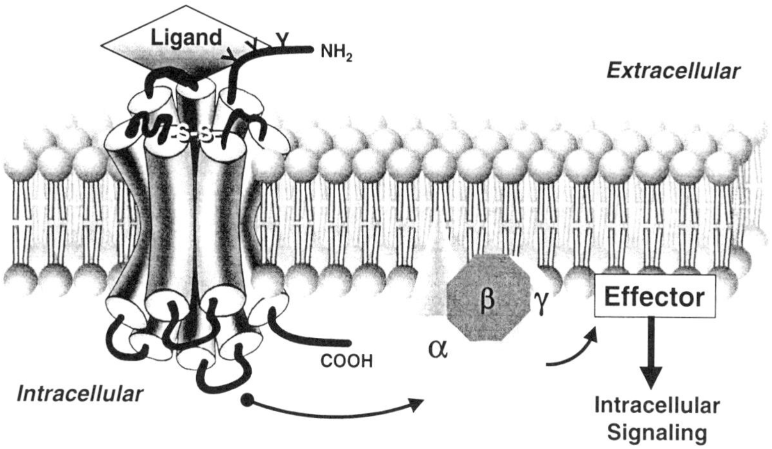

Figure 2. A specific ligand interacts with its cognate GPCR by binding to a unique location within the receptor. Upon binding, the receptor conformation is altered such that GDP bound to the G protein α subunit is dissociated from the G protein and is replaced by GTP. GTP activates the G protein and leads to dissociation of the α subunit from the βγ subunit and activation of effector pathways. These effector substances include stimulation of adenylate cyclase, ion channel opening or closing or an array of other functions (see text).

cyclase activated by the G protein (Table IV). All adenylyl cyclases are stimulated by Gs and can be further classified into 6 subtypes based on sequence and functional similarities [48]. In addition to localization specificity of the various adenylyl cyclases, biochemical analysis of overexpressed adenylyl cyclase isoforms has demonstrated different sensitivity to receptor signal regulators like protein kinase A, protein kinase C, calcium, calmodulin, and the Gβγ subunit. This subtype specificity of signal regulation provides an opportunity for the cell to integrate signals from various receptors and initiate a unique response. With activation of Gs and increased cellular cAMP, cAMP is then capable of producing effects within the cell such as phosphorylation of effector proteins that ultimately lead to responses such as secretion, cell movement or growth. With the crystal structure of Gsα now known [56], including the conformation of adenylyl cyclase with Gsα [57; 58], new insight into the structural determinants important for substrate recognition of the cyclase and catalysis have been identified. This will provide new information into the mechanisms of G protein activation and will likely assist our understanding of the regulation of other members of the Gα family, including Gαi.

Other effector systems include phospholipase C, phospholipase A, cyclic GMP, phosphodiesterase, and potassium and calcium channels.

Table IV: Mammalian Adenylyl Cyclase Isoforms

Type	Tissue location (mRNA)
1	brain, retina, adrenal medulla
2	brain, olfactory bulb, lung
3	brain, retina, aorta, lung, testis
4	kidney, brain, heart, liver, lung
5	heart, brain (striatum), kidney
6	heart, brain, kidney, testis, spleen, liver
7	lung, heart, spleen, kidney, brain
8	brain
9	skeletal muscle, brain, kidney, lung, liver, heart

Adapted from Ref. [53]

Signal Transduction and Desensitization

To insure the sensitivity of hormone signaling, cells adjust their responsiveness based on their current level of stimulation. Shortly after receptor activation, a series of events are initiated which ultimately turn off signaling. The principal events in this process involve receptor desensitization and internalization, which re-establish cell responsiveness.

While eliciting a receptor-mediated signal is important to normal physiologic processes, attenuation of that signal is just as critical for normal cellular function. Phosphorylation and sequestration are involved in turning off the signal after agonist binding to GPCRs [17; 18]. Signal termination by phosphorylation occurs via two distinct pathways [18]. Mechanisms that modulate only stimulated GPCRs, are termed homologous desensitization. Phosphorylation by GRKs is a major component of this initial diminished responsiveness [32; 35]. Studies of the regulation of the rhodopsin and β-adrenergic receptors have demonstrated the importance of these receptor serine/threonine kinases [18; 47]. Other modes of signal attenuation which act on both agonist-occupied as well as agonist-unoccupied receptors, and often involve second messenger-dependent processes, are termed heterologous desensitization [37]. Protein kinase A (PKA) and protein kinase C (PKC) play a role in heterologous desensitization [37]. Protein kinase A, the cyclic AMP dependent kinase, phosphorylates receptors at serine and threonine residues often contained within specific consensus sites.

The molecular substrates of GPCR desensitization have recently become apparent. GPCRs are phosphorylated in an agonist-dependent fashion by a family of receptor kinases [4; 47; 66]. However, biochemical work with the β_2-adrenergic receptor and in the visual system suggested the involvement of an additional protein [4; 66]. Using purified GRKs, Benovic et al were unable to completely quench receptor signaling, but reconstitution with purified visual arrestin could re-establish receptor uncoupling [4]. Shortly, thereafter Lohse and coworkers cloned a protein similar to arrestin (59% sequence homology) and demonstrated its role in β_2-adrenergic receptor desensitization [38]. This protein is called β-arrestin. Both GRK and β-arrestin could fully terminate β_2-adrenergic receptor signaling [37]. Arrestins, as cytoplasmic proteins, do not reside in close proximity to the receptors they modulate. The interaction of β-arrestin with the β_2-adrenergic receptor is facilitated by phosphorylation [15]. β-arrestin contains an amino terminal domain that appears to bind to the third intracellular loop of the agonist-occupied receptor; this region has been called the activation-recognition domain [68]. Arrestin proteins contain another region that appears to interact with the C-terminal serine and threonine residues phosphorylated by G protein receptor kinases, the phosphorylation-recognition domain [26]. It is now clear that GRK-dependent receptor phosphorylation and association of arrestin proteins with the phosphorylated receptor initiate GPCR desensitization [15]. Also, β-arrestins play an important role in the mobilization of GPCR from the cell's surface to clathrin-coated pits for endocytosis, possibly by serving as an intermediary to target the receptor to the pits [70]. β-arrestin contains distinct binding sites for the adaptor protein (AP-2) and clathrin in its carboxyl terminus [30]. In the adrenergic system, phosphorylation of the receptor by receptor specific kinases promotes binding of β-arrestin, which uncouples the receptor from the G-protein [37]. This uncoupling and subsequent receptor internalization (sequestration) continues the process of signal termination and eventually leading to the re-establishment of responsiveness.

Receptor Sequestration and Resensitization

Receptor sequestration occurs within minutes of receptor occupancy. It is best characterized as the beginning of resensitization, as opposed to the end of desensitization since, presumably, only receptors that have been internalized can be dephosphorylated and returned to the cell surface for subsequent activation [28; 36](Fig. 3). It has been demonstrated that sequestration of both the muscarinic M_2 and the β_2-adrenergic receptor is facilitated by G protein-coupled receptor kinase mediated phosphorylation [16; 59]. GPCRs internalize via the clathrin-coated vesicle-mediated endocytic pathway [62]. Receptors for LDL, transferrin, and insulin share this pathway [8; 39; 65]. Dynamin, a 100 kDa GTPase, which is involved in the fission of clathrin-coated vesicles from the plasma membrane, has been shown to participate in endocytosis of the β_2-adrenergic receptor [70]. In contrast, sequestration of certain receptors has been shown not to be affected by dynamin mutants [64; 70]. This has prompted investigators to postulate the presence of alternative pathways of GPCR internalization [6; 9]. These pathways might be distinguished by their use of clathrin-coated pits or caveolae and by their sensitivity to dominant-negative forms of β-arrestin and/or dynamin. However, the breath of evidence supports receptor endocytosis *via* a β-arrestin-mediated, clathrin-dependent internalization of the hormone-receptor complex [30].

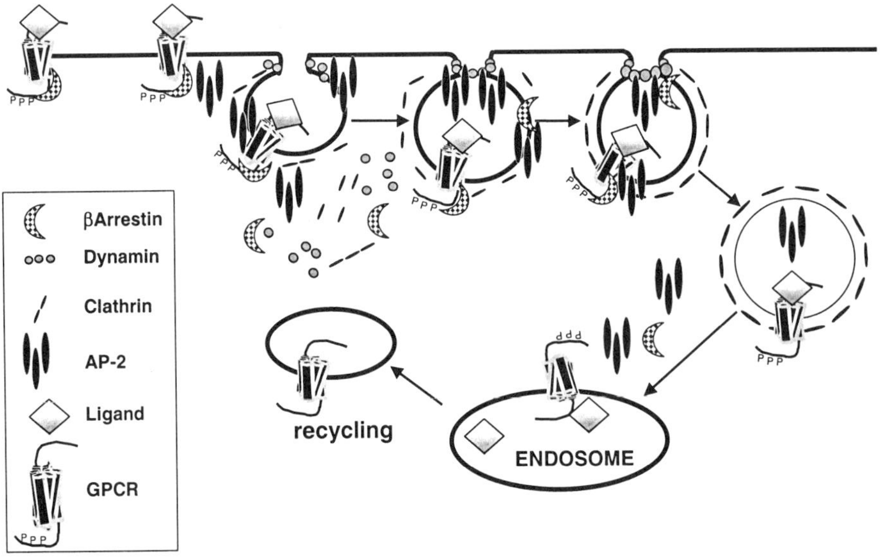

Figure 3. GPCR Internalization - Agonist-activated receptors are phosphorylated by GRKs at specific intracellular residues (P). The phosphorylated receptor is uncoupled from its G protein and β–arrestins translocate from the cytosol to the phosphorylated receptor, initiating receptor endocytosis. β–arrestin associated GPCRs interact with adaptor proteins (AP-2) and β–arrestin also directly interacts with clathrin. Dynamin causes scission of the clathrin-coated vesicle from the plasma membrane. Specific residues in β–arrestin mediate binding of AP-2. GPCR endocytosis is important for receptor dephosphorylation, recycling and resensitization (see text for details).

GPCRs enter the cell through clathrin-coated vesicles via adaptor proteins [51]. β-arrestin plays an integral role in agonist-stimulated receptor association with adaptor proteins and the formation of the clathrin lattice. β-arrestins interact with AP-2 [29] and clathrin [20]. AP-2 contains a clathrin binding domain and a motif capable of recognizing cytoplasmic tyrosine-based signals of membrane proteins [44]. With agonist stimulation, the receptor is phosphorylated and β-arrestin associates with the hormone-receptor complex. βarrestin, via its AP-2 and clathrin interacting domains, initiates a stable complex in punctate areas of the plasma membrane [30]. This coordinated interaction of the GPCR, β-arrestin, clathrin, and AP-2 targets the complex to an endocytic pathway [51]. Receptor endocytosis is important for receptor recycling and dephosphorylation, processes necessary for the re-establishment of receptor responsiveness [28].

SUMMARY

Since the original description of G proteins in receptor signaling over a generation ago, the number of molecular substrates involved in signal transduction has increased dramatically. The crystal structure and basic cell physiology has been determined for some of these transducers, however, many aspects of signal regulation remain unanswered. Advances continue to provide cell biologists, molecular biologists, pharmacologists and physiologists with new targets in their quest to elucidate the basic mechanisms of cellular communication. The past has proved quite fruitful, now with over 10% of the human genome identified as participants in signal transduction [61], the future will likely produce even more exciting discoveries.

Acknowledgments

This work was supported in part by the National Institute of Health Grants DK02544-04 and DK58828-01 and by the Mal Tyor Scholarship Award in Gastroenterology (to M.A.S).

REFERENCES

1. **Alewijnse, A. E., Timmerman, H., Jacobs, E. H., Smit, M. J., Roovers, E., Cotecchia, S. and Leurs, R.** 2000 The effect of mutations in the DRY motif on the constitutive activity and structural instability of the histamine H(2) receptor. Mol Pharmacol 57, 890-8.
2. **Barak, L. S., Oakley, R. H., Laporte, S. A. and Caron, M. G.** 2001 Constitutive arrestin-mediated desensitization of a human vasopressin receptor mutant associated with nephrogenic diabetes insipidus. Proc Natl Acad Sci U S A 98, 93-98.
3. **Bargmann, C. I.** 1998 Neurobiology of the Caenorhabditis elegans genome. Science 282, 2028-33.

4. **Benovic, J. L., Kuhn, H., Weyand, I., Codina, J., Caron, M. G. and Lefkowitz, R. J.** 1987 Functional desensitization of the isolated beta-adrenergic receptor by the beta-adrenergic receptor kinase: potential role of an analog of the retinal protein arrestin (48-kDa protein). Proc Natl Acad Sci U S A 84, 8879-82.
5. **Buck, J., Sinclair, M. L., Schapal, L., Cann, M. J. and Levin, L. R.** 1999 Cytosolic adenylyl cyclase defines a unique signaling molecule in mammals. Proc Natl Acad Sci U S A 96, 79-84.
6. **Bunemann, M., Lee, K. B., Pals-Rylaarsdam, R., Roseberry, A. G. and Hosey, M. M.** 1999 Desensitization of G-protein-coupled receptors in the cardiovascular system. Annu Rev Physiol 61, 169-92.
7. **Chemelli, R. M., Willie, J. T., Sinton, C. M., Elmquist, J. K., Scammell, T., Lee, C., Richardson, J. A., Williams, S. C., Xiong, Y., Kisanuki, Y., Fitch, T. E., Nakazato, M., Hammer, R. E., Saper, C. B. and Yanagisawa, M.** 1999 Narcolepsy in orexin knockout mice: molecular genetics of sleep regulation. Cell 98, 437-51.
8. **Chen, W. J., Goldstein, J. L. and Brown, M. S.** 1990 NPXY, a sequence often found in cytoplasmic tails, is required for coated pit-mediated internalization of the low density lipoprotein receptor.J Biol Chem 265, 31.
9. **Claing, A., Perry, S. J., Achiriloaie, M., Walker, J. K., Albanesi, J. P., Lefkowitz, R. J. and Premont, R. T.** 2000 Multiple endocytic pathways of G protein-coupled receptors delineated by GIT1 sensitivity. Proc Natl Acad Sci U S A 97, 1119-1124.
10. **Clapham, D. E. and Neer, E. J.** 1997 G protein beta gamma subunits. Annu Rev Pharmacol Toxicol 37, 167-203.
11. **De Vries, L., Mousli, M., Wurmser, A. and Farquhar, M. G.** 1995 GAIP, a protein that specifically interacts with the trimeric G protein G alpha i3, is a member of a protein family with a highly conserved core domain. Proc Natl Acad Sci U S A 92, 11916-20.
12. **De Vries, L., Zheng, B., Fischer, T., Elenko, E. and Farquhar, M. G.** 2000 The regulator of G protein signaling family. Annu Rev Pharmacol Toxicol 40, 235-71.
13. **Doupnik, C. A., Davidson, N., Lester, H. A. and Kofuji, P.** 1997 RGS proteins reconstitute the rapid gating kinetics of gbetagamma-activated inwardly rectifying K+ channels. Proc Natl Acad Sci U S A 94, 10461-6.
14. **Feighner, S. D., Tan, C. P., McKee, K. K., Palyha, O. C., Hreniuk, D. L., Pong, S. S., Austin, C. P., Figueroa, D., MacNeil, D., Cascieri, M. A., Nargund, R., Bakshi, R., Abramovitz, M., Stocco, R., Kargman, S., O'Neill, G., Van Der Ploeg, L. H., Evans, J., Patchett, A. A., Smith, R. G. and Howard, A. D.** 1999 Receptor for motilin identified in the human gastrointestinal system. Science 284, 2184-8.
15. **Ferguson, S. S., Downey, W. E., 3rd, Colapietro, A. M., Barak, L. S., Menard, L. and Caron, M. G.** 1996 Role of beta-arrestin in mediating agonist-promoted G protein-coupled receptor internalization. Science 271, 363-6.
16. **Ferguson, S. S., Menard, L., Barak, L. S., Koch, W. J., Colapietro, A. M. and Caron, M. G.** 1995 Role of phosphorylation in agonist-promoted beta 2-adrenergic receptor sequestration. Rescue of a sequestration-defective mutant receptor by beta ARK1. J Biol Chem 270, 24782-9.
17. **Ferguson, S. S., Zhang, J., Barak, L. S. and Caron, M. G.** 1998 Molecular mechanisms of G protein-coupled receptor desensitization and resensitization. Life Sci 62, 1561-5.
18. **Freedman, N. J. and Lefkowitz, R. J.** 1996 Desensitization of G protein-coupled receptors. Recent Prog Horm Res 51, 319-51.
19. **Gautam, N., Downes, G. B., Yan, K. and Kisselev, O.** 1998 The G-protein betagamma complex. Cell Signal 10, 447-55.
20. **Goodman, O. B., Jr., Krupnick, J. G., Santini, F., Gurevich, V. V., Penn, R. B., Gagnon, A. W., Keen, J. H. and Benovic, J. L.** 1996 Beta-arrestin acts as a clathrin adaptor in endocytosis of the beta2- adrenergic receptor. Nature 383, 447-50.

21. **Gutkind, J. S.** 2000 Regulation of Mitogen-Activated Protein Kinase Signaling Networks by G Protein-Coupled Receptors.
Science's STKE: stke.sciencemag.org/cgi/content/full/OC_sigtrans;2000/40/re1, 1-13.
22. **Hamm, H. E.** 1998 The many faces of G protein signaling. J Biol Chem 273, 669-72.
23. **Hellmich, M. R., Rui, X. L., Hellmich, H. L., Fleming, R. Y., Evers, B. M. and Townsend, C. M., Jr.** 2000 Human colorectal cancers express a constitutively active cholecystokinin-B/Gastrin receptor that stimulates cell growth [In Process Citation]. J Biol Chem 275, 32122-8.
24. **Hepler, J. R., Berman, D. M., Gilman, A. G. and Kozasa, T.** 1997 RGS4 and GAIP are GTPase-activating proteins for Gq alpha and block activation of phospholipase C beta by gamma-thio-GTP-Gq alpha. Proc Natl Acad Sci U S A 94, 428-32.
25. **Hildebrandt, J. D.** 1997 Role of subunit diversity in signaling by heterotrimeric G proteins. Biochem Pharmacol 54, 325-39.
26. **Kieselbach, T., Irrgang, K. D. and Ruppel, H.** 1994 A segment corresponding to amino acids Val170-Arg182 of bovine arrestin is capable of binding to phosphorylated rhodopsin. Eur J Biochem 226, 87-97.
27. **Kilduff, T. S. and Peyron, C.** 2000 The hypocretin/orexin ligand-receptor system: implications for sleep and sleep disorders. Trends Neurosci 23, 359-65.
28. **Krueger, K. M., Daaka, Y., Pitcher, J. A. and Lefkowitz, R. J.** 1997 The role of sequestration in G protein-coupled receptor resensitization. Regulation of beta2-adrenergic receptor dephosphorylation by vesicular acidification. J Biol Chem 272, 5-8.
29. **Laporte, S. A., Oakley, R. H., Holt, J. A., Barak, L. S. and Caron, M. G.** 2000 The interaction of beta-arrestin with the AP-2 adaptor is required for the clustering of beta 2-adrenergic receptor into clathrin-coated pits. J Biol Chem 275, 23120-6.
30. **Laporte, S. A., Oakley, R. H., Zhang, J., Holt, J. A., Ferguson, S. S., Caron, M. G. and Barak, L. S.** 1999 The beta2-adrenergic receptor/betaarrestin complex recruits the clathrin adaptor AP-2 during endocytosis. Proc Natl Acad Sci U S A 96, 3712-3717.
31. **Lee, D. K., George, S. R., Evans, J. F., Lynch, K. R. and O'Dowd, B. F.** 2001 Orphan G protein-coupled receptors in the CNS. Curr Op Pharmacol 1, (in press).
32. **Lefkowitz, R. J.** 1993 G protein-coupled receptor kinases. Cell 74, 409-12.
33. **Lefkowitz, R. J.** 1993 G-protein-coupled receptors. Turned on to ill effect. Nature 365, 603-4.
34. **Lefkowitz, R. J.** 2000 The superfamily of heptahelical receptors. Nat Cell Biol 2, E133-6.
35. **Lefkowitz, R. J., Hausdorff, W. P. and Caron, M. G.** 1990 Role of phosphorylation in desensitization of the beta-adrenoceptor. Trends Pharmacol Sci 11, 190-4.
36. **Lefkowitz, R. J., Pitcher, J., Krueger, K. and Daaka, Y.** 1998 Mechanisms of beta-adrenergic receptor desensitization and resensitization. Adv Pharmacol 42, 416-20.
37. **Lohse, M. J., Andexinger, S., Pitcher, J., Trukawinski, S., Codina, J., Faure, J. P., Caron, M. G. and Lefkowitz, R. J.** 1992 Receptor-specific desensitization with purified proteins. Kinase dependence and receptor specificity of beta-arrestin and arrestin in the beta 2-adrenergic receptor and rhodopsin systems. J Biol Chem 267, 8558-64.
38. **Lohse, M. J., Benovic, J. L., Codina, J., Caron, M. G. and Lefkowitz, R. J.** 1990 beta-Arrestin: a protein that regulates beta-adrenergic receptor function. Science 248, 1547-50.
39. **Miller, K., Shipman, M., Trowbridge, I. S. and Hopkins, C. R.** 1991 Transferrin receptors promote the formation of clathrin lattices. Cell 65, 621-32.
40. **Morris, A. J. and Malbon, C. C.** 1999 Physiological regulation of G protein-linked signaling. Physiol Rev 79, 1373-430.
41. **Needham, L. K. and Rozengurt, E.** 1998 Galpha12 and Galpha13 stimulate Rho-dependent tyrosine phosphorylation of focal adhesion kinase, paxillin, and p130 Crk-associated substrate. J Biol Chem 273, 14626-32.

42. **Neer, E. J.** 1995 Heterotrimeric G proteins: organizers of transmembrane signals. Cell 80, 249-57.
43. **Offermanns, S., Mancino, V., Revel, J. P. and Simon, M. I.** 1997 Vascular system defects and impaired cell chemokinesis as a result of Galpha13 deficiency. Science 275, 533-6.
44. **Ohno, H., Stewart, J., Fournier, M. C., Bosshart, H., Rhee, I., Miyatake, S., Saito, T., Gallusser, A., Kirchhausen, T. and Bonifacino, J. S.** 1995 Interaction of tyrosine-based sorting signals with clathrin-associated proteins. Science 269, 1872-5.
45. **Palczewski, K., Kumasaka, T., Hori, T., Behnke, C. A., Motoshima, H., Fox, B. A., Le Trong, I., Teller, D. C., Okada, T., Stenkamp, R. E., Yamamoto, M. and Miyano, M.** 2000 Crystal structure of rhodopsin: A G protein-coupled receptor. Science 289, 739-45.
46. **Paschke, R. and Ludgate, M.** 1997 The thyrotropin receptor in thyroid diseases. N Engl J Med 337, 1675-81.
47. **Premont, R. T., Inglese, J. and Lefkowitz, R. J.** 1995 Protein kinases that phosphorylate activated G protein-coupled receptors. Faseb J 9, 175-82.
48. **Premont, R. T., Matsuoka, I., Mattei, M. G., Pouille, Y., Defer, N. and Hanoune, J.** 1996 Identification and characterization of a widely expressed form of adenylyl cyclase. J Biol Chem 271, 13900-7.
49. **Rens-Domiano, S. and Hamm, H. E.** 1995 Structural and functional relationships of heterotrimeric G-proteins. Faseb J 9, 1059-66.
50. **Rhee, S. G. and Bae, Y. S.** 1997 Regulation of phosphoinositide-specific phospholipase C isozymes. J Biol Chem 272, 15045-8.
51. **Schmid, S. L.** 1997 Clathrin-coated vesicle formation and protein sorting: an integrated process. Annu Rev Biochem 66, 511-48.
52. **Shenker, A., Laue, L., Kosugi, S., Merendino, J. J., Minegishi, T. and Cutler, G. B.** 1993 A constitutively activating mutation of the luteinizing hormone receptor in familial male precocious puberty. Nature 365, 652-4.
53. **Simonds, W. F.** 1999 G protein regulation of adenylate cyclase. Trends Pharmacol Sci 20, 66-73.
54. **Snow, B. E., Krumins, A. M., Brothers, G. M., Lee, S. F., Wall, M. A., Chung, S., Mangion, J., Arya, S., Gilman, A. G. and Siderovski, D. P.** 1998 A G protein gamma subunit-like domain shared between RGS11 and other RGS proteins specifies binding to Gbeta5 subunits. Proc Natl Acad Sci U S A 95, 13307-12.
55. **Sondek, J., Bohm, A., Lambright, D. G., Hamm, H. E. and Sigler, P. B.** 1996 Crystal structure of a G-protein beta gamma dimer at 2.1A resolution. Nature 379, 369-74.
56. **Sunahara, R. K., Tesmer, J. J., Gilman, A. G. and Sprang, S. R.** 1997 Crystal structure of the adenylyl cyclase activator Gsalpha. Science 278, 1943-7.
57. **Tesmer, J. J., Sunahara, R. K., Gilman, A. G. and Sprang, S. R.** 1997 Crystal structure of the catalytic domains of adenylyl cyclase in a complex with Gsalpha.GTPgammaS. Science 278, 1907-16.
58. **Tesmer, J. J., Sunahara, R. K., Johnson, R. A., Gosselin, G., Gilman, A. G. and Sprang, S. R.** 1999 Two-metal-Ion catalysis in adenylyl cyclase. Science 285, 756-60.
59. **Tsuga, H., Kameyama, K., Haga, T., Kurose, H. and Nagao, T.** 1994 Sequestration of muscarinic acetylcholine receptor m2 subtypes. Facilitation by G protein-coupled receptor kinase (GRK2) and attenuation by a dominant-negative mutant of GRK2. J Biol Chem 269, 32522-7.
60. **Vaughan, M.** 1998 Signaling by heterotrimeric G proteins minireview series. J Biol Chem 273, 667-8.
61. **Venter, J. C., et al.** 2001 The Sequence of the Human Genome. Science 291, 1304-1351.
62. **von Zastrow, M. and Kobilka, B. K.** 1994 Antagonist-dependent and -independent steps in the mechanism of adrenergic receptor internalization. Biol Chem 269, 18448-52.

63. **Voyno-Yasenetskaya, T., Conklin, B. R., Gilbert, R. L., Hooley, R., Bourne, H. R. and Barber, D. L.** 1994 G alpha 13 stimulates Na-H exchange. J Biol Chem 269, 4721-4.
64. **Walker, J. K., Premont, R. T., Barak, L. S., Caron, M. G. and Shetzline, M. A.** 1999 Properties of secretin receptor internalization differ from those of the beta(2)-adrenergic receptor. J Biol Chem 274, 31515-23.
65. **Watts, C. and Marsh, M.** 1992 Endocytosis: what goes in and how? J Cell Sci 103, 1-8.
66. **Wilden, U., Hall, S. W. and Kuhn, H.** 1986 Phosphodiesterase activation by photoexcited rhodopsin is quenched when rhodopsin is phosphorylated and binds the intrinsic 48-kDa protein of rod outer segments. Proc Natl Acad Sci U S A 83, 1174-8.
67. **Wilkie, T. M., Gilbert, D. J., Olsen, A. S., Chen, X. N., Amatruda, T. T., Korenberg, J. R., Trask, B. J., de Jong, P., Reed, R. R., Simon, M. I. and et al.** 1992 Evolution of the mammalian G protein alpha subunit multigene family. Nat Genet 1, 85-91.
68. **Wu, G., Krupnick, J. G., Benovic, J. L. and Lanier, S. M.** 1997 Interaction of arrestins with intracellular domains of muscarinic and alpha2-adrenergic receptors. J Biol Chem 272, 17836-42.
69. **Zhang, G., Liu, Y., Ruoho, A. E. and Hurley, J. H.** 1997 Structure of the adenylyl cyclase catalytic core. Nature 386, 247-53.
70. **Zhang, J., Ferguson, S. S. G., Barak, L. S., Menard, L. and Caron, M. G.** 1996 Dynamin and beta-arrestin reveal distinct mechanisms for G protein- coupled receptor internalization. J Biol Chem 271, 18302-5.

10

Gonadotropin and TSH Receptors

Nicolas de Roux*, Beate Doeker, Edwin Milgrom

INSERM Unit135, Laboratoire d'Hormonologie et de Biologie Moleculaire, Hôpital Bicêtre, 78 rue du Général Leclerc, 94270 Le Kremlin Bicêtre, France

INTRODUCTION

Luteinizing hormone (LH), follicule-stimulating hormone (FSH) and thyroid-stimulating hormone (TSH) are glycoproteins synthesized by the pituitary. In females, LH stimulates androgen synthesis by ovarian thecal cells and triggers ovulation. In males, LH stimulates testosterone production by Leydig cells. In the ovary, FSH stimulates granulosa cell estrogen production from androgens synthesized in thecal cells. In males, FSH regulates Sertoli cell function in the testis. TSH stimulates thyroid hormone synthesis by thyroid follicular cells. All three are heterodimeric hormones with a common α subunit and a specific β subunit. Chorionic gonadotropin (CG) is the fourth member of this family found only in human (hCG) and equine (eCG) species where it is synthesized by the placenta. It is also a heterodimeric hormone composed of the common α subunit and a specific β subunit highly homologous to the β LH subunit.

All glycoprotein hormones act on target cells through G-protein coupled receptors (GPCR; see Chapter 9). LH and hCG bind to the same receptor (LH/CG receptor). FSH and TSH act through their own specific receptors. These receptors share some homology with other G-protein coupled receptors such as rhodopsin. Natural mutations of the genes of gonadotropin and TSH receptors were recently characterized in several endocrine diseases. Gene defects leading to constitutive activation of the receptor were linked to gain of

* nicolas.deroux@bct.ap-hop-paris.fr

function phenotypes such as precocious puberty or non-autoimmune hyperthyroidism. Conversely, loss of function phenotypes (hypogonadism and hypothyroidism) were linked to gene defects located in other parts of the coding sequence.

STRUCTURE OF TSH, LH/CG AND FSH RECEPTORS

The LH receptor cDNA was cloned by expression and screening with an antibody directed against the purified receptor [1] or by use of degenerated oligonucleotides [2]. Cross hybridization led to the cloning of the TSH receptor [3] and finally of the FSH receptor [4]. The TSH receptor is the largest, comprising 764 amino acids as compared to the 699 amino acids for the LH receptor and 695 for the FSH receptor. These receptors are composed of two functional domains: An extracellular hormone binding domain (ECD) and a transmembrane/intracellular domain (TMD/ICD) which bears the seven transmembrane helix specific features of G-protein coupled receptors. The overall homology is very high between these receptors (see Fig. 1). The highest homology is located within the transmembrane region. Several sequence stretches of the extracellular domain show 100% homology. The

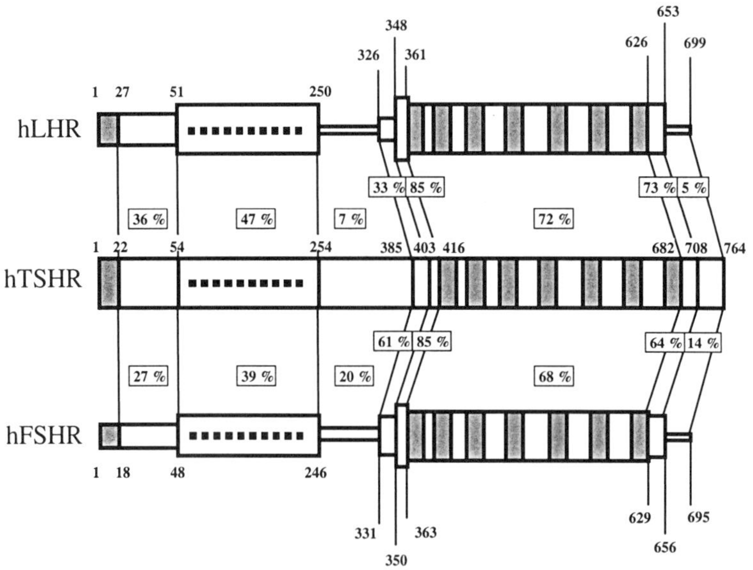

Figure 1. Comparison of the human TSH, LH/CG and FSH receptor protein sequences.
■ *leucine-rich repeat* ▯ *transmembrane span*

TSH receptor has a specific insertion of 50 amino acids located before this high homology region.

The extracellular domains are composed of several leucine-rich repeats (see Fig. 1). These motifs are known to be involved in protein-protein and cell-matrix interactions. The glycoprotein hormone receptors were the first G-protein coupled receptors to display such repeats. Other G-protein coupled receptors with leucine-rich repeats within the extracellular domain were cloned subsequently in mammals [5-7]. The high amino acid homology between these receptors and rhodopsin led to the speculation that they all arose from a common G-protein coupled ancestor receptor.

Consensus sequence analysis predicted a signal peptide cleavage site and several glycosylation sites. A recent mass spectrometric analysis showed variable cleavage of the signal peptide of the pig LHR expressed in L cells [8]. Western blot analysis showed that deglycosylation of receptors either by endoglycosidase F or endoglycosidase H yielded a polypeptide identical in size to that predicted by primary amino acid sequence [9-11]. Deglycosylated receptors still bind the hormone showing that glycosylation is not involved in hormone binding. *In vitro* mutagenesis studies have shown that glycosylation seems to be essential for receptor folding and intracellular trafficking to the cell surface [12-13]. A mass spectrometry analysis showed that all glycosylation sites but one (residue 299) of the precursor and the mature LH receptor are glycosylated [8]. The LH receptor is palmitoylated at a cysteine located in the intracellular domain [14]. For TSH and FSH receptors, cysteine residues located in the intracellular domain might also be palmitoylated

The structure of the mature TSH receptor has been the object of divergent reports (for review see [15-16]). It is now accepted that the TSH receptor is cleaved within the extracellular domain [17]. The mature receptor is composed of two subunits held by disulfide bridges [17]. In the thyroid, the cleavage is almost complete [10]. Monomeric uncleaved precursors may also be observed in transfected cells. Two different cleavage sites were described [18-19]. The cleavage was shown to occur first after amino acid 314 [19]. Then a sequential cleavage was found up to amino acid 378 leading to excision of this part of the receptor. This region shows no homology with the LH and the FSH receptors. The TSHR cleavage enzyme remains unknown but may be related to the adamalysin family of metzincin metalloproteases [19-20]. The role of this cleavage in receptor activation remains unknown. The uncleaved receptor is able to bind the TSH and transduces the signal similarly to the wild type receptor [21]. In human thyroid cell membrane extracts, the concentration of TMD is 2.5-3 fold higher than the concentration of the ECD [17]. This difference may be explained by the shedding of the ECD [22].

GENES

The FSH and LH receptor genes were localized in human species to chromosome 2p21 by *in situ* hybridization. In contrast, the TSH receptor has been assigned to chromosome 14q31. FSH and TSH receptor have 10 exons [23]. The LH/CG receptor gene has 11 exons [24] except for the marmoset monkey gene in which exon 10 is spliced out as an intronic sequence [25]. The adenylate cyclase stimulation and cell surface expression in COS cells is similar for this exon 10-deleted marmoset receptor as compared to the wild type human LHR [25]. However, cell surface expression of the human LH receptor deleted of exon 10 was decreased [26].

Exon sizes are highly homogenous between the three receptors. Several polymorphisms are now well described in these genes. One polymorphism consisting of a 2 amino acid insertion at position 19 of the human LH receptor (LQ allele) has been described [27]. These two residues are located within the signal peptide sequence.

RECEPTOR VARIANTS

Cloning of the pig LH receptor showed that several transcripts could be detected in testicular tissues [1]. These transcripts code for truncated proteins, but in all cases, the N-terminal parts of the proteins are identical. The divergence occurs at residue 316. Two variants are deleted of the transmembrane domain and intracellular domain. A third variant lacks a transmembrane domain but still contains the intracellular domain. In the rat, other variants of 81, 116 and 294 amino acids were cloned. They also arose by alternative splicing [28].

A similar pattern of variants was also observed for the FSH and the TSH receptors [29-30]. For the TSH receptor, several variants lacking the transmembrane domain were isolated during the initial cloning of the full length cDNA. These variants contain the first 8 exons. The added sequences downstream of the 8 exons are not found within the full length receptor. A variant lacking exon 3 was found in dog only [31].

Truncated LH receptor isoforms were found by immunoblotting in testicular extracts [32]. In LH receptor transfected cells, the variant forms lacking the transmembrane domain were found in the cell culture media and within the cell or were trapped within the endoplasmic reticulum [32]. The physiological role of these variants remains unknown. Inhibitory binding proteins were found for the TSH receptor within the retro-orbital tissue and for the FSH receptor within testicular extracts. Such inhibitory proteins may be encoded by mRNA variants lacking the transmembrane domain.

Immunoreactivity with monoclonal antibodies against the TSH receptor was found in human plasma [22-33]. This soluble protein may be due to the secretion of soluble receptor or to the shedding of the ECD from the cleaved receptor.

CELLULAR TRAFFICKING OF GLYCOPROTEIN RECEPTORS

Cloning of these receptors led to production of highly specific monoclonal antibodies. Immunocytochemical studies were performed to localize the receptors within the cells.

Internalization of Glycoprotein Receptors

Internalization is one of the mechanisms involved in the regulation of signal transduction by GPCR. Following ligand binding to the receptor, the cells rapidly internalize the ligand-receptor complex *via* clathrin-coated pits and caveolae. Then two situations may occur. In some cases, the ligand-receptor complex is dissociated and the receptor recycles to the cell surface, while in other cases, the receptor is directed into the lysosomes where it is degraded. Internalization is usually measured as the radioactive ligand remaining bound after an acid wash of the cell surface. The use of monoclonal antibodies against the extracellular domain of the glycoprotein receptors has greatly helped in the description of this process [34-35]. Antibodies were conjugated to gold particles and the intracellular routes were followed by electronic microscopy.

The internalization process of the LH receptor was studied in stably transfected L cell and in Leydig cells [34]. In both cell lines, LH receptor internalization occurred through the same pathway. First, LH receptor complexes were present in clathrin coated pits, internalized by clathrin coated vesicles and finally degraded in the lysosomes. No recycling occurred for this receptor. A constitutive internalization was observed in transfected L cells which increased about 11 fold under LH stimulation [34] Recently, Baratti and coworkers have performed similar experiments with the TSH receptor in L cells and thyroid cells [35]. The internalization process of the TSH receptor followed a pathway similar to the LHR through the clathrin coated pits and vesicles. However, only a three-fold increase of internalization was observed under TSH stimulation. Moreover, the internalized TSH receptor was almost totally recycled to the cell surface and the hormone was degraded in the lysosomes. These results are surprising as both receptors present a very high homology. Studies were performed with LHR/TSHR chimeric constructs to find peptide sequences responsible for this difference in intracellular trafficking [35]. A chimera with the extracellular domain of the LH receptor fused to the transmembrane and intracellular domains of the TSH receptor was completely recycled. The information leading to the choice between the two pathways is thus borne by the transmembrane or the intracellular domains.

Biochemical methods were also used to study the internalization process of the glycoprotein receptors. The rate of the internalization was closely related to the activation process. The LH receptor was phosphorylated upon hCG stimulation (for a review see Ref. [36]). Mutations in the rat LHR which impaired the agonist induced phosphorylation also impaired the internalization process. The internalization rate of the human LH receptor was 7 time faster than that of the rat LH receptor. This difference was ascribed to 7 discrete residues located in the second intracellular loop (4 residues), the third intracellular loop (2 residues) and one in the intracellular domain.

Cellular Polarization of Receptors in Target Cells

Antibodies directed against the extracellular part of the TSH receptor showed the localization of the receptor at the basolateral pole of the thyrocytes [17]. An identical polarized expression was found for the FSH receptor within Sertoli cells [37].

The LH receptor was found all over the surface of thecal, granulosa and luteal cells and Leydig cells in the ovary and testis [38]. Developmental regulation of LH receptor expression was studied in the ovaries during follicle maturation. LH receptor immunoreactivity appeared in thecal cells of secondary follicles. Granulosa cells were also labeled in large antral follicles and preovulatory follicles whereas two zones were observed in the theca interna [39]. The most external cells of the cyclic corporea lutea were stained and no staining was found for the most abundant cells of granulosa origin.

The difference observed in the cellular distribution between these three receptors could be due to a difference in their molecular structure or to the fact that thyroid and Sertoli cells are polarized. Polarized Madin-Darby canine kidney (MDCK) cells were used as a model system to study the cellular polarization of the receptors. In these cells, the three receptors were found to be addressed directly to the basolateral pole [40] This result suggests that all three receptors possess a basolateral targeting signal. A specific signal for the basolateral localization was delineated at the C-terminal part of the intracellular domain between amino acids 678-691 of the FSH receptor [41]. When grafted into the neurotrophin receptor, this signal was able to localize this apical receptor to the basolateral pole of the MDCK cells [41].

Transcytosis of the LH Receptor in Endothelial Cells

The mechanism by which LH reaches its target from the blood stream was poorly understood until recently. In most organs, the vascular endothelium prevents the free passage of macromolecules. LH as well as FSH receptors are expressed in testicular endothelial cells [37-38]. Ghinea and coworkers [42] showed a LH/hCG receptor specific passage of the hCG through endothelial cells in rat testicular microvasculature. hCG transcytosis occurred in several steps. The hormone-receptor complex was first internalized at the luminal endothelial side *via* clathrin coated pits and transferred *via* clathrin coated vesicles to the endosomal compartment. The hCG-receptor complexes were then transferred to the abluminal side of the cells *via* smooth vesicles that apparently pinched off from tubular extensions of the endosomes. The hormone was then removed from the receptor and found in the interstitial fluid. This intracellular traffic was completely different that observed in target organ cells or in transfected L cells (see above). Limited transcytosis was also observed in polarized MDCK cells expressing the FSH and the LH receptors [40].

ECTOPIC EXPRESSION OF GLYCOPROTEIN RECEPTORS

LH regulates gonadal steroidogenesis, FSH regulates Sertoli cell function and TSH regulates thyroid hormone synthesis. However, recent evidence has shown the presence of these receptors in other tissues. LH receptor expression was found in normal breast [43], in breast cancer [43], in epidermis [44] and in pig umbilical cord [45]. The expression of LHR in breast cancer is an interesting finding. The protective effect of an early pregnancy on breast cancer development is well known and is attributed to the differentiation elicited by hCG which renders the cells less susceptible to neoplastic transformation.

TSH receptor expression was found in lymphocytes, heart, adipocytes and retro-orbital fat. The physiological significance of such expression remains unknown. However, the TSH receptor expressed in the eye may explain the exophtalmia observed in Graves' disease. In this auto-immune thyroid disease, the patients develop stimulating auto-antibodies against the TSH receptor.

SIGNAL TRANSDUCTION

Glycoprotein hormone receptors mainly activate the adenylate cyclase pathway. A weak activation of the phospholipase C pathway was shown for the TSH and the LH receptors. Thyroid growth and cell proliferation were clearly linked to adenylate cyclase stimulation by the TSH receptor whereas the phospholipase C pathway was involved in thyroid hormone synthesis. cAMP stimulates steroid hormone synthesis in the testis and in the ovary. A physiological role of the phopholipase C pathway remains less clear in gonads than in the thyroid.

Ligand Binding and Signal Transduction Domains

The ligand binding domain is now well defined for various GPCRs. Several subfamilies of GPCR have been defined on the basis of sequence homology. The glycoprotein hormone receptors belong to the subfamily showing homology with key residues of rhodopsin and of β2 adrenergic receptor. This subfamily termed "family A" displays several subgroups based on sequence homologies and on the nature of the ligand. The large extracellular domain of the glycoprotein hormone receptors defines one of these subgroup.

As shown by *in vitro* mutagenesis, the high affinity binding of FSH, LH and TSH occurs in the N-terminal part of the extracellular domain of the respective receptors. Construction of chimeras between various extracellular and transmembrane domains led the definition of the high-affinity ligand binding domain of each receptor. Several mutagenesis studies tried to pinpoint the specific amino acids involved in this binding. However, such *in vitro* mutagenesis strategies have been criticized since they may change the overall conformation of the protein (for reviews, see Ref. [16,30,36]).

Sequence analogy with other leucine-rich repeat proteins suggests that the structure of the ECD is similar to the horseshoe-shaped structure defined for the ribonuclease inhibitor (RI) [46]. In this protein, the inner circumference of the horseshoe is formed by β sheets and the outer circumference by α helices. For the TSH receptor, the ligand may interact with the concave surface of the ECD.

Involvement of cysteine residues in ligand binding was shown by *in vitro* mutagenesis. Twelve cysteines were located within the extracellular domain of the LH/CG receptor and 11 in the TSH receptor. In the latter, the receptor is cleaved in two subunits held by disulfide bridges. Three groups of cysteines are located in the N-terminal part (around amino acids 280-300) and in the C-terminal part (around amino acids 30-50) of the extracellular domain. Mutation of the N-terminal cysteines of the LH receptor to serines led to decreased ligand binding [47]. A similar mutagenesis analysis of the cluster of cysteines located at the C-terminal part of the ECD of the LH receptor showed a decrease in cell surface expression [47]. For the TSH receptor, mutagenesis of cysteine 41 to serine led to a decrease of TSH binding [16]. Substitution of cysteine residues located in the middle part of the ECD dramatically decreased TSH binding [48]. However, it is unclear if these cysteines are involved in the overall conformation of the ECD or in the disulfide bridges which attach the ECD to the transmembrane and intracellular domains. The deletion of exon 10 found in marmoset monkey receptor removes cysteine 282. It is interesting to note that at position 314, the invariant cysteine in all LH, FSH and TSH receptors is replaced by a histidine in the marmoset receptor. The human LH receptor deleted of exon 10 is poorly expressed at the cell surface [26]. Cysteine 314 is thus probably involved in a disulfide bridge with cysteine 282 located in exon 10 of LH receptor. Leaving the cysteine 314 free in human LH receptor may interfere with receptor normal cellular trafficking. As found in other members of GPCR sub family A, two cysteines located in the first and the second extracellular loops form a disulfide bridge. A substitution of these cysteines to serines impairs signal transduction.

LHR, TSHR and FSHR interact primarily with Gs. TSHR interacts directly with Gq/11 α to activate PLC [49] whereas LHR may stimulate PLC pathway by the release of $\beta\gamma$ subunits from Gs or Gi [50]. The TSH receptor may also interact directly with Gi and G12 proteins [49]. The physiological role of these interactions remains unclear. Many studies have tried to dissect the molecular mechanisms of signal transduction by these receptors. Several common features with other G-protein coupled receptors have been described. The C-terminal part of the third intracellular loop contributes to adenylate cyclase stimulation [16,30,36]. The conserved triplet including an acidic amino acid (Asp or Gln), an arginine, and an aromatic amino acid (Tyr or Trp) at the end of the third transmembrane domain and the beginning of the second intracellular loop is involved in adenylate cyclase stimulation. It is present in all rhodopsin like GPCRs. The role of the sixth transmembrane domain of the LH and TSH receptors has been clearly showed by the description of many activating mutations (see below).

Receptor Activation

Ligand binding is known to take place within the ECD of glycoprotein receptors but further involvement of this domain in signal transduction was unexpected. Natural mutations of the extracellular domain of TSHR were characterized in hypothyroid and hyperthyroid patients (see below). One residue located in this domain, Ser281 mutated in isoleucine, threonine and asparagine led to constitutive activation of the TSH receptor [51]. The homologous Ser277 of the LH receptor replaced by all natural amino acids also led to constitutive activation [52]. This part of the receptor is not involved in hormone binding. Activating mutations were also characterized in the first and the second extracellular loops of THSR [53]. These mutations constitutively activate the phospholipase C pathway [53]. Conversely mutations of the extracellular domain of the TSH receptor leading to normal binding of the ligand but to a decreased signal transduction were described in hypothyroid patients [54] Similar results were obtained by *in vitro* mutagenesis in the LH receptor [55]. These mutations are located in a highly homologous region located before the first transmembrane domain. These results strongly suggest a close interaction between the extracellular and the transmembrane domains probably mediated by the extracellular loops.

Two models have been proposed to explain glycoprotein hormone receptor activation. First, the extracellular domain may exert an inhibitory effect on the transmembrane domain. The ligand may decrease this interaction leading to a stabilization of the active state. Constitutive activation of the TSH receptor mimics the ligand and favors the equilibrium towards the active state [56]. The second model was established by studying the LH receptor [57]. It suggests that agonist binding brings the hormone or the C-terminal part of the extracellular domain close to the extracellular loops leading to a stabilization of the transmembrane domain in the active form. This model may explain the sequential interaction between the hormone and the receptor. However, these hypotheses are difficult to test experimentally since expression of the TSH or LH receptors at the cell surface requires the presence of their extracellular domain. Recently, an ECD truncated TSH receptor was shown to have increased constitutive activity when compared to the wild type receptor at similar concentration on the cell surface [58]. This result favors the hypothesis of an inhibitory effect of the ECD on transmembrane domain mediated signal transduction.

ACTIVATING AND LOSS OF FUNCTION MUTATIONS

The cloning of receptor genes led to the definition of sequencing strategies directly from genomic DNA. Descriptions of natural activating and inactivating mutations in several endocrine diseases were thus made possible. Somatic or germinal mutations were described. Germinal activating mutations led to a dominant transmission model. Inactivating mutations were found at homozygote or compound heterozygote states with a full penetrance. These studies have highlighted important residues for ligand binding or signal transduction *in vivo*.

Activating Mutations

TSH Receptor

Several observations argued for the existence of activating mutations of the TSH receptor. Growth and differentiation of the thyroid gland are controlled at least in part by TSH. GTPase inactivating mutations were found in thyroid tumors tissues as well as in McCune Albright syndrome. *In vitro* mutagenesis studies showed that substitution of a residue located in the third intracelllular loop (Ala293) by any other amino acid may spontaneously activate the α1 adrenergic receptor.

Solitary autonomous adenoma, an encapsulated benign tumor with hyperthyroidism and slow growth, was a good candidate for activating mutations of the TSH receptor. The first natural mutations of the TSH receptor were described in the C-terminal part of the third intracellular loop [59]. One of these residues is homologous to the Ala293 of the α1 adrenergic receptor. Heterologous expression of these mutated receptors in COS-7 cells showed an increase of the basal level of cAMP. The mutation was somatic since it was absent in the DNA of normal thyroid tissue surrounding the adenoma. Several other somatic mutations were described [56] dispersed throughout the transmembrane domain of the TSH receptor but mainly localized within the third intracellular loop and the cytoplasmic part of the sixth transmembrane α helix (Fig. 2). Unexpected activating mutations were also localized within the extracellular domain or the extracellular loops (see above).

Germinal activating mutations of the TSHR were described in familial non-autoimmune hyperthyroidism [60]. This phenotype is inherited as an autosomal dominant trait and the clinical expression is usually delayed. The expressivity of the phenotype may be variable within the same family [60]. Activating neomutations of the TSH receptor were found in sporadic congenital hyperthyroidism [60]. In these cases, hyperthyroidism occurred during fetal life [61].

Genotype-phenotype correlations showed that most of the mutations described in sporadic congenital hyperthyroidism were also found in toxic adenomas. Mutations found in familial hyperthyroidism are private except in one case. Mutations leading to toxic adenomas or congenital hyperthyroidism are probably more prone to strong receptor activation than the mutations found in delayed hyperthyroidism. Genetic or environmental factors leading to severe or mild non-autoimmune familial hyperthyroidism remain unknown.

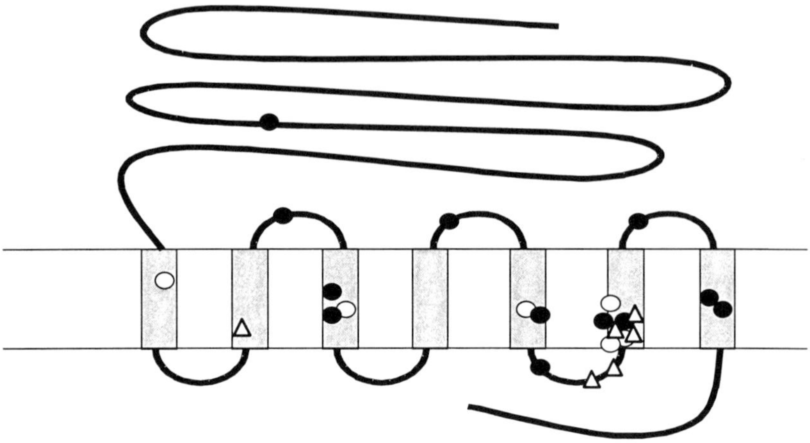

Figure 2 : Activating mutations found in the TSH receptor (●), the LH receptor (○), or both receptors (Δ).

LH Receptor

Gonadotropin independent precocious puberty was shown to be linked to LH receptor activating mutations [62]. This male limited phenotype is familial and inherited as a dominant trait. The patients have development of the secondary sexual characteristics, increased secretion of testosterone, prepubertal levels of serum gonadotropins unresponsive to GnRH stimulation test. Most of the activating mutations of the LH receptor are localized within the C-terminal part of the third intracellular loop and in the sixth transmembrane domain. In the USA, a founder effect was discussed to explain the high frequency of mutation Asp578Gly. This mutation accounts for 80% of all reported familial male precocious puberty cases. Other mutations are localized in the first, second, third and fifth transmembrane domains. No activating mutation of the ECD has been described for the LH receptor. Activating mutations of the TSH receptor are more diverse than LH receptor mutations. The TSH receptor shows a basal constitutive activity which is absent in wild type LH receptor. This constitutive activation was explained by a less constraint conformation. Mutations in the TSH receptor thus appear more prone to induce a conformational change from the inactive to the active form than mutations in the LH receptor. Individuals harboring mutations of the ECD of the LH receptor may thus be clinically normal.

The penetrance of LH receptor activating mutations is not complete. Three familial cases with normal carriers have been described. As for the TSH receptor, modifier genes or environmental factors may explain this observation.

A single somatic mutation was found in Leydig cell tumors from three boys with precocious puberty [63]. This mutation substitutes a histidine for the glycine 578. This mutation induces a very high constitutive activation of the adenylate cyclase pathway. Moreover, the basal production of IP3 was seven times that observed with the wild type receptor. Leydig cell tumors may necessitate stimulation of both pathways.

Loss of Function Mutations

TSH Receptor

The TSH receptor is expressed during fetal life after migration of the thyroid precursor cells in front of the trachea. TSH is needed for the growth of the thyroid but not for the migration process. Loss of function mutations of the TSH receptor lead to a variable extent of hypothyroidism presenting as compensated hypothyroidism or thyroid hypoplasia. Inactivating TSHR mutations were initially described in three siblings with increased levels of TSH and normal levels of thyroid hormones [64]. All affected siblings were compound heterozygotes for mutations located within the extracellular domain. Transmission of the inactivating mutations as a recessive trait was confirmed by several other descriptions of familial cases of TSH resistance [54-65]. A major defect of the TSH receptor should result in thyroid hypoplasia with a normally situated gland. Indeed, Abramowicz and coworkers described an inactivating mutation in a familial case of thyroid hypoplasia [66]. This mutation Ala553Thr, located within the fourth transmembrane domain, severely decreased the expression of the TSH receptor at the cell surface.

Other mutations leading to thyroid hypoplasia have been described [67-70]. In most cases, they severely impaired adenylate cyclase stimulation by TSH. Homozygous mutations were more prevalent in thyroid hypoplasia patient as compared to the milder TSH resistance phenotype. The expressivity of the phenotype was strictly the same between patients having the same mutation within one family. An identical phenotype was also observed between patients from different families bearing the same mutation.

The first case of TSH resistance was reported by Stanbury and coworkers in 1968 [70a]. Other clinical cases were then described in the literature before the cloning of TSH receptor cDNA. Codaccioni *et al.* published a documented case in 1982 [71]. They showed that TSH could bind to follicular cell membranes of this patient but was unable to stimulate the adenylate cyclase.

However, no mutation of the coding sequence of the TSH receptor was found in DNA from blood lymphocytes of this patient (de Roux et al., unpublished result) suggesting that other genes might be involved in TSH resistance [72].

LH Receptor

Male sex differentiation during fetal life depends on androgen synthesized by Leydig cells. This synthesis is regulated by hCG at the beginning of the pregnancy and then by LH. Leydig cell hypoplasia with high LH levels responding normally to GnRH stimulation has been previously reported. Such male patients had female phenotype with two abdominal testes and absent müllerian structures. *In vitro* studies showed the loss of hCG binding on membranes from removed testes whereas *in vivo* hCG stimulation failed in these patients .

The first description of inactivating mutations was made in two siblings with a complete form of pseudohermaphrodism [73]. An homozygous Ala593Pro mutation was found in both patients. This mutation decreased hCG binding to cell membranes without changing apparent affinity. hCG stimulation did not elicit adenylate cyclase activation. Proline substitution to an alanine probably resulted in a dramatic change of LH receptor conformation leading to a decreased expression on the cell surface [73]. A partial case of male pseudohermaphrodism was found in a patient with a homozygous mutation Cys133Arg [74]. This residue is located in the extracellular domain of the receptor. A complete lack of hCG binding was found for the mutated Cys133Arg receptor. A conventional microscopic study of the testis did not reveal the presence of Leydig cells. However, immunohistochemical staining with LH receptor and anti-P450c17 antibodies showed dual labeling in about one third of the interstitial cells.

Although LH and hCG, the LH receptor ligands, are highly homologous in amino acid sequence, their physiological functions may be different. hCG induces masculinization in XY fetuses during the first trimester of pregnancy whereas LH stimulates the growth of gonad and penis during the second and the third trimester. After birth, LH synthesis becomes significant at the onset of puberty. This physiological difference was recently confirmed by the description of a homozygous LH receptor exon 10 deletion in a patient presenting a delayed puberty associated with a normal masculinization [75]. Stimulation with hCG of the deleted exon 10 human LH receptor induced a normal cAMP production whereas the expression of this receptor at the cell surface was markedly decreased [26]. It is thus probable that the normal masculinization of this patient was related to the normal response to hCG of the mutated receptor during fetal life. Conversely, delayed puberty probably relates to the lack of LH stimulation at the beginning of the puberty. A comparable situation may be involved in another case [76].

In females, the phenotype of LH receptor inactivating mutations becomes apparent at puberty. Four cases of 46 XX females have been described [77]. Three of them had primary amenorrhea, one showed irregular menses. All of these patients were sisters of XY patients presenting with pseudohermaphrodism.

FSH Receptor

Ovarian dysgenesis is characterized by a block of follicular maturation at the stage of primary follicles. The patients present a primary amenorrhea with streak or hypoplastic ovaries [78]. In the Finnish population presenting ovarian dysgenesis, a homozygous mutation (Ala189Val) of the FSH receptor was described [78]. This mutation, located within the extracellular domain impaired receptor trafficking and FSH binding. Loss of function of the FSH receptor in men of the same families resulted in reduced testicular volume and variable degree of impaired spermatogenesis [62].

A different inactivating phenotype was described in a female patient with secondary amenorrhea. She had a normal puberty but developed secondary amenorrhea at the age of 16 [79]. She had high levels of FSH and LH and low estradiol leading to the diagnosis of premature menopause. Pelvic ultrasonography showed two ovaries slightly enlarged with numerous follicles up to 5 mm. Two compound heterozygote mutations were found in exon 6 (Ile160Thr) and exon 10 (Arg573Cys) of the FSH receptor gene. The Ile160Thr mutation dramatically decreased FSH binding on the cell surface whereas the Arg573Cys mutation decreased signal transduction. The lack of FSH binding in cells transfected with the Ile160Thr mutated receptor was shown to be related to the lack of expression of the protein in the cell surface. Histological studies of the ovaries of this patient showed a normal follicular development up to the small antral stage. Another case of partial inactivating FSH receptor mutation was recently described [80]. Two compound heterozygote mutations were found in this patient: Asp224Val in the ECD and Leu601Val in the third extracellular loop. Asp224Val mutation led to a lack of cell surface expression of the mutated receptor. FSH binding on the Leu601Val mutated receptor showed a normal affinity while signal transduction was markedly decreased.

As for the LH and the TSH receptors, the partial phenotype is probably related to a milder defect in receptor function as compared to the complete phenotype [80]. *In vivo* stimulation with recombinant FSH showed an absence of response in the patient with the most severe phenotype. Conversely, FSH stimulation elicited a progressive increase in follicular growth and plasma concentration of inhibin B in the patient with the milder phenotype [80].

CONCLUSIONS

Glycoprotein hormone LH/CG, FSH and TSH receptors compose a sub-family of the very large G protein coupled receptor family. This sub-family is defined by the combination of a large extracellular domain with leucine rich repeats and a transmembrane domain bearing extensive homology with rhodopsin-like receptors. The ECD determines the high affinity binding domain of this receptor but also plays a role in signal transduction. The transmembrane domain activates adenylate cyclase and to a less extent phospholipase C. The TSH receptor differs from the gonadotropin receptors as it is cleaved at two sites within the extracellular domain. A basolateral polarization signal has been identified in the intracellular domain of the FSH receptor.

Activating as well as loss of function mutations have been described in several reproduction and thyroid diseases.

Acknowledgments

We thank JF Savouret for critical reading of the manuscript.

REFERENCES

1. **Loosfelt H, Misrahi M, Atger M, Salesse R, Vu Hai-Luu Thi MT, Jolivet A, Guiochon-Mantel A, Sar S, Jallal B, Garnier J, Milgrom E** 1989 Cloning and sequencing of porcine LH-hCG receptor cDNA: variants lacking transmembrane domain. Science 245:525-528.
2. **McFarland KC, Sprengel R, Phillips HS, Kohler M, Rosemblit N, Nikolics K, Segaloff DL, Seeburg PH** 1989 Lutropin-choriogonadotropin receptor: an unusual member of the G protein-coupled receptor family. Science 245:494-499.
3. **Parmentier M, Libert F, Maenhaut C, Lefort A, Gerard C, Perret J, Van Sande J, Dumont JE, Vassart G** 1989 Molecular cloning of the thyrotropin receptor. Science 246:1620-1622.
4. **Minegishi T, Nakamura K, Takakura Y, Ibuki Y, Igarashi M, Minegish T** 1991 Cloning and sequencing of human FSH receptor cDNA [published erratum appears in Biochem Biophys Res Commun 1994 Jun 15;201(2):1057]. Biochem Biophys Res Commun 175:1125-1130.
5. **Hsu SY, Liang SG, Hsueh AJ** 1998 Characterization of two LGR genes homologous to gonadotropin and thyrotropin receptors with extracellular leucine-rich repeats and a G protein-coupled, seven-transmembrane region. Mol Endocrinol 12:1830-1845.
6. **Hsu SY, Kudo M, Chen T, Nakabayashi K, Bhalla A, van der Spek PJ, van Duin M, Hsueh AJ** 2000 The three subfamilies of leucine-rich repeat-containing G protein-

coupled receptors (LGR): identification of LGR6 and LGR7 and the signaling mechanism for LGR7 [In Process Citation]. Mol Endocrinol 14:1257-1271.

7. **Hermey G, Methner A, Schaller HC, Hermans-Borgmeyer I** 1999 Identification of a novel seven-transmembrane receptor with homology to glycoprotein receptors and its expression in the adult and developing mouse. Biochem Biophys Res Commun 254:273-279.
8. **Vu-Hai MT, Huet JC, Echasserieau K, Bidart JM, Floiras C, Pernollet JC, Milgrom E** 2000 Posttranslational modifications of the lutropin receptor: mass spectrometric analysis. Biochemistry 39:5509-5517.
9. **Hipkin RW, Sanchez-Yague J, Ascoli M** 1992 Identification and characterization of a luteinizing hormone/chorionic gonadotropin (LH/CG) receptor precursor in a human kidney cell line stably transfected with the rat luteal LH/CG receptor complementary DNA. Mol Endocrinol 6:2210-2218.
10. **Misrahi M, Ghinea N, Sar S, Saunier B, Jolivet A, Loosfelt H, Cerutti M, Devauchelle G, Milgrom E** 1994 Processing of the precursors of the human thyroid-stimulating hormone receptor in various eukaryotic cells (human thyrocytes, transfected L cells and baculovirus-infected insect cells). Eur J Biochem 222:711-719.
11. **Vuhai-Luuthi MT, Jolivet A, Jallal B, Salesse R, Bidart JM, Houllier A, Guiochon-Mantel A, Garnier J, Milgrom E** 1990 Monoclonal antibodies against luteinizing hormone receptor. Immunochemical characterization of the receptor. Endocrinology 127:2090-2098.
12. **Davis D, Liu X, Segaloff DL** 1995 Identification of the sites of N-linked glycosylation on the follicle- stimulating hormone (FSH) receptor and assessment of their role in FSH receptor function. Mol Endocrinol 9:159-170.
13. **Davis DP, Rozell TG, Liu X, Segaloff DL** 1997 The six N-linked carbohydrates of the lutropin/choriogonadotropin receptor are not absolutely required for correct folding, cell surface expression, hormone binding, or signal transduction. Mol Endocrinol 11:550-562.
14. **Zhu H, Wang H, Ascoli M** 1995 The lutropin/choriogonadotropin receptor is palmitoylated at intracellular cysteine residues. Mol Endocrinol 9:141-150.
15. **Misrahi M, Couet J, Milgrom E** 1997 Mechanisms of shedding of a soluble form of the TSH receptor. Ann Endocrinol 58:365-369.
16. **Rapoport B, Chazenbalk GD, Jaume JC, McLachlan SM** 1998 The thyrotropin (TSH) receptor: interaction with TSH and autoantibodies [published erratum appears in Endocr Rev 1999 Feb;20(1):100]. Endocr Rev 19:673-716.
17. **Loosfelt H, Pichon C, Jolivet A, Misrahi M, Caillou B, Jamous M, Vannier B, Milgrom E** 1992 Two-subunit structure of the human thyrotropin receptor. Proc Natl Acad Sci U S A 89:3765-3769.
18. **Chazenbalk GD, Tanaka K, Nagayama Y, Kakinuma A, Jaume JC, McLachlan SM, Rapoport B** 1997 Evidence that the thyrotropin receptor ectodomain contains not one, but two, cleavage sites. Endocrinology 138:2893-2899.
19. **de Bernard S, Misrahi M, Huet JC, Beau I, Desroches A, Loosfelt H, Pichon C, Pernollet JC, Milgrom E** 1999 Sequential cleavage and excision of a segment of the thyrotropin receptor ectodomain. J Biol Chem 274:101-107.
20. **Couet J, de Bernard S, Loosfelt H, Saunier B, Milgrom E, Misrahi M** 1996 Cell surface protein disulfide-isomerase is involved in the shedding of human thyrotropin receptor ectodomain. Biochemistry 35:14800-14805.
21. **Chazenbalk GD, McLachlan SM, Nagayama Y, Rapoport B** 1996 Is receptor cleavage into two subunits necessary for thyrotropin action? Biochem Biophys Res Commun 225:479-484.

22. **Couet J, Sar S, Jolivet A, Hai MT, Milgrom E, Misrahi M** 1996 Shedding of human thyrotropin receptor ectodomain. Involvement of a matrix metalloprotease. J Biol Chem 271:4545-4552.
23. **Gross B, Misrahi M, Sar S, Milgrom E** 1991 Composite structure of the human thyrotropin receptor gene. Biochem Biophys Res Commun 177:679-687.
24. **Atger M, Misrahi M, Sar S, Le Flem L, Dessen P, Milgrom E** 1995 Structure of the human luteinizing hormone-choriogonadotropin receptor gene: unusual promoter and 5' non-coding regions. Mol Cell Endocrinol 111:113-123.
25. **Zhang FP, Rannikko AS, Manna PR, Fraser HM, Huhtaniemi IT** 1997 Cloning and functional expression of the luteinizing hormone receptor complementary deoxyribonucleic acid from the marmoset monkey testis: absence of sequences encoding exon 10 in other species. Endocrinology 138:2481-2490.
26. **Zhang FP, Kero J, Huhtaniemi I** 1998 The unique exon 10 of the human luteinizing hormone receptor is necessary for expression of the receptor protein at the plasma membrane in the human luteinizing hormone receptor, but deleterious when inserted into the human follicle-stimulating hormone receptor. Mol Cell Endocrinol 142:165-174.
27. **Rodien P, Cetani F, Costagliola S, Tonacchera M, Duprez L, Minegishi T, Govaerts C, Vassart G** 1998 Evidences for an allelic variant of the human LC/CG receptor rather than a gene duplication: functional comparison of wild-type and variant receptors. J Clin Endocrinol Metab 83:4431-4434.
28. **Tsai-Morris CH, Buczko E, Wang W, Dufau ML** 1990 Intronic nature of the rat luteinizing hormone receptor gene defines a soluble receptor subspecies with hormone binding activity. J Biol Chem 265:19385-19388.
29. **Graves PN, Tomer Y, Davies TF** 1992 Cloning and sequencing of a 1.3 KB variant of human thyrotropin receptor mRNA lacking the transmembrane domain. Biochem Biophys Res Commun 187:1135-1143.
30. **Simoni M, Gromoll J, Nieschlag E** 1997 The follicle-stimulating hormone receptor: biochemistry, molecular biology, physiology, and pathophysiology. Endocr Rev 18:739-773.
31. **Libert F, Lefort A, Gerard C, Parmentier M, Perret J, Ludgate M, Dumont JE, Vassart G** 1989 Cloning, sequencing and expression of the human thyrotropin (TSH) receptor: evidence for binding of autoantibodies. Biochem Biophys Res Commun 165:1250-1255.
32. **VuHai-LuuThi MT, Misrahi M, Houllier A, Jolivet A, Milgrom E** 1992 Variant forms of the pig lutropin/choriogonadotropin receptor. Biochemistry 31:8377-8383.
33. **Murakami M, Miyashita K, Yamada M, Iriuchijima T, Mori M** 1992 Characterization of human thyrotropin receptor-related peptide-like immunoreactivity in peripheral blood of Graves' disease. Biochem Biophys Res Commun 186:1074-1080.
34. **Ghinea N, Vu Hai MT, Groyer-Picard MT, Houllier A, Schoevaert D, Milgrom E** 1992 Pathways of internalization of the hCG/LH receptor: immunoelectron microscopic studies in Leydig cells and transfected L-cells. J Cell Biol 118:1347-1358.
35. **Baratti-Elbaz C, Ghinea N, Lahuna O, Loosfelt H, Pichon C, Milgrom E** 1999 Internalization and recycling pathways of the thyrotropin receptor. Mol Endocrinol 13:1751-1765.
36. **Dufau ML** 1998 The luteinizing hormone receptor. Annu Rev Physiol 60:461-496.
37. **Vannier B, Loosfelt H, Meduri G, Pichon C, Milgrom E** 1996 Anti-human FSH receptor monoclonal antibodies: immunochemical and immunocytochemical characterization of the receptor. Biochemistry 35:1358-1366.

38. **Meduri G, Vuhai-Luuthi MT, Jolivet A, Milgrom E** 1992 New functional zonation in the ovary as shown by immunohistochemistry of luteinizing hormone receptor. Endocrinology 131:366-373.
39. **Meduri G, Vu Hai MT, Jolivet A, Takemori S, Kominami S, Driancourt MA, Milgrom E** 1996 Comparison of cellular distribution of LH receptors and steroidogenic enzymes in the porcine ovary. J Endocrinol 148:435-446.
40. **Beau I, Misrahi M, Gross B, Vannier B, Loosfelt H, Hai MT, Pichon C, Milgrom E** 1997 Basolateral localization and transcytosis of gonadotropin and thyrotropin receptors expressed in Madin-Darby canine kidney cells. J Biol Chem 272:5241-5248.
41. **Beau I, Groyer-Picard MT, Le Bivic A, Vannier B, Loosfelt H, Milgrom E, Misrahi M** 1998 The basolateral localization signal of the follicle-stimulating hormone receptor. J Biol Chem 273:18610-18616.
42. **Ghinea N, Mai TV, Groyer-Picard MT, Milgrom E** 1994 How protein hormones reach their target cells. Receptor-mediated transcytosis of hCG through endothelial cells. J Cell Biol 125:87-97.
43. **Meduri G, Charnaux N, Loosfelt H, Jolivet A, Spyratos F, Brailly S, Milgrom E** 1997 Luteinizing hormone/human chorionic gonadotropin receptors in breast cancer. Cancer Res 57:857-864.
44. **Venencie PY, Meduri G, Pissard S, Jolivet A, Loosfelt H, Milgrom E, Misrahi M** 1999 Luteinizing hormone/human chorionic gonadotrophin receptors in various epidermal structures. Br J Dermatol 141:438-446.
45. **Wasowicz G, Derecka K, Stepien A, Pelliniemi L, Doboszynska T, Gawronska B, Ziecik AJ** 1999 Evidence for the presence of luteinizing hormone-chorionic gonadotrophin receptors in the pig umbilical cord. J Reprod Fertil 117:1-9.
46. **Kajava AV, Vassart G, Wodak SJ** 1995 Modeling of the three-dimensional structure of proteins with the typical leucine-rich repeats. Structure 3:867-877.
47. **Zhang R, Buczko E, Dufau ML** 1996 Requirement of cysteine residues in exons 1-6 of the extracellular domain of the luteinizing hormone receptor for gonadotropin binding. J Biol Chem 271:5755-5760.
48. **Kosugi S, Ban T, Akamizu T, Kohn LD** 1992 Role of cysteine residues in the extracellular domain and exoplasmic loops of the transmembrane domain of the TSH receptor: effect of mutation to serine on TSH receptor activity and response to thyroid stimulating autoantibodies. Biochem Biophys Res Commun 189:1754-1762.
49. **Laugwitz KL, Allgeier A, Offermanns S, Spicher K, Van Sande J, Dumont JE, Schultz G** 1996 The human thyrotropin receptor: a heptahelical receptor capable of stimulating members of all four G protein families. Proc Natl Acad Sci U S A 93:116-120.
50. **Herrlich A, Kuhn B, Grosse R, Schmid A, Schultz G, Gudermann T** 1996 Involvement of Gs and Gi proteins in dual coupling of the luteinizing hormone receptor to adenylyl cyclase and phospholipase C. J Biol Chem 271:16764-16772.
51. **Kopp P, Muirhead S, Jourdain N, Gu WX, Jameson JL, Rodd C** 1997 Congenital hyperthyroidism caused by a solitary toxic adenoma harboring a novel somatic mutation in the extracellular domain of the thyrotropin receptor. J Clin Invest 100:1634-1639.
52. **Nakabayashi K, Kudo M, Kobilka B, Hsueh AJ** 2000 Activation of the luteinizing hormone receptor following substitution of ser-277 with selective hydrophobic residues in the ectodomain hinge region. J Biol Chem 275:30264-30271.
53. **Parma J, Van Sande J, Swillens S, Tonacchera M, Dumont J, Vassart G** 1995 Somatic mutations causing constitutive activity of the thyrotropin receptor are the major cause of hyperfunctioning thyroid adenomas: identification of additional mutations

activating both the cyclic adenosine 3',5'-monophosphate and inositol phosphate-Ca2+ cascades. Mol Endocrinol 9:725-733.

54. **de Roux N, Misrahi M, Brauner R, Houang M, Carel JC, Granier M, Le Bouc Y, Ghinea N, Boumedienne A, Toublanc JE, Milgrom E** 1996 Four families with loss of function mutations of the thyrotropin receptor. J Clin Endocrinol Metab 81:4229-4235.
55. **Alvarez CA, Narayan P, Huang J, Puett D** 1999 Characterization of a region of the lutropin receptor extracellular domain near transmembrane helix 1 that is important in ligand-mediated signaling. Endocrinology 140:1775-1782.
56. **Duprez L, Parma J, Van Sande J, Rodien P, Sabine C, Abramowicz M, Dumont JE, Vassart G** 1999 Pathology of the TSH receptor. J Pediatr Endocrinol Metab 12 Suppl 1:295-302.
57. **Ji TH, Ryu KS, Gilchrist R, Ji I** 1997 Interaction, signal generation, signal divergence, and signal transduction of LH/CG and the receptor. Recent Prog Horm Res 52:431-453.
58. **Zhang M, Tong KP, Fremont V, Chen J, Narayan P, Puett D, Weintraub BD, Szkudlinski MW** 2000 The extracellular domain suppresses constitutive activity of the transmembrane domain of the human TSH receptor: implications for hormone-receptor interaction and antagonist design. Endocrinology 141:3514-3517.
59. **Parma J, Duprez L, Van Sande J, Cochaux P, Gervy C, Mockel J, Dumont J, Vassart G** 1993 Somatic mutations in the thyrotropin receptor gene cause hyperfunctioning thyroid adenomas. Nature 365:649-651.
60. **Duprez L, Parma J, Van Sande J, Allgeier A, Leclere J, Schvartz C, Delisle MJ, Decoulx M, Orgiazzi J, Dumont J, et al.** 1994 Germline mutations in the thyrotropin receptor gene cause non-autoimmune autosomal dominant hyperthyroidism. Nat Genet 7:396-401.
61. **de Roux N, Polak M, Couet J, Leger J, Czernichow P, Milgrom E, Misrahi M** 1996 A neomutation of the thyroid-stimulating hormone receptor in a severe neonatal hyperthyroidism. J Clin Endocrinol Metab 81:2023-2026.
62. **Themmen APN, Huhtaniemi IT** 2000 Mutations of gonadotropins and gonadotropin receptors: elucidating the physiology and pathophysiology of pituitary-gonadal function . Endocr Rev 21:551-583.
63. **Liu G, Duranteau L, Carel JC, Monroe J, Doyle DA, Shenker A** 1999 Leydig-cell tumors caused by an activating mutation of the gene encoding the luteinizing hormone receptor. N Engl J Med 341:1731-1736.
64. **Sunthornthepvarakul T, Gottschalk M-E, Hayashi Y, Refetoff S** 1995 Resistance to thyrotropin caused by mutations in the thyrotropin receptor gene. N Engl J Med 332:155-160.
65. **Clifton-Bligh RJ, Gregory JW, Ludgate M, John R, Persani L, Asteria C, Beck-Peccoz P, Chatterjee VK** 1997 Two novel mutations in the thyrotropin (TSH) receptor gene in a child with resistance to TSH. J Clin Endocrinol Metab 82:1094-1100.
66. **Abramowicz MJ, Duprez L, Parma J, Vassart G, Heinrichs C** 1997 Familial congenital hypothyroidism due to inactivating mutation of the thyrotropin receptor causing profound hypoplasia of the thyroid gland. J Clin Invest 99:3018-3024.
67. **Tiosano D, Pannain S, Vassart G, Parma J, Gershoni-Baruch R, Mandel H, Lotan R, Zaharan Y, Pery M, Weiss RE, Refetoff S, Hochberg Z** 1999 The hypothyroidism in an inbred kindred with congenital thyroid hormone and glucocorticoid deficiency is due to a mutation producing a truncated thyrotropin receptor. Thyroid 9:887-894.
68. **Tonacchera M, Agretti P, Pinchera A, Rosellini V, Perri A, Collecchi P, Vitti P, Chiovato L** 2000 Congenital hypothyroidism with impaired thyroid response to

thyrotropin (TSH) and absent circulating thyroglobulin: evidence for a new inactivating mutation of the TSH receptor gene. J Clin Endocrinol Metab 85:1001-1008.
69. **Gagne N, Parma J, Deal C, Vassart G, Van Vliet G** 1998 Apparent congenital athyreosis contrasting with normal plasma thyroglobulin levels and associated with inactivating mutations in the thyrotropin receptor gene: are athyreosis and ectopic thyroid distinct entities? J Clin Endocrinol Metab 83:1771-1775.
70. **Biebermann H, Schoneberg T, Krude H, Schultz G, Gudermann T, Gruters A** 1997 Mutations of the human thyrotropin receptor gene causing thyroid hypoplasia and persistent congenital hypothyroidism. J Clin Endocrinol Metab 82:3471-3480.
70a. **Stanbury JB, Rocmans P, Buhler UK, Ochi Y** 1968 Congenital hypothyroidism with impaired thyroid response to thyrotropin. N Engl J Med 279:1132-1136
71. **Codaccioni JL, Carayon P, Michel-Bechet M, Foucault F, Lefort G, Pierron H** 1980 Congenital hypothyroidism associated with thyrotropin unresponsiveness and thyroid cell membrane alterations. J Clin Endocrinol Metab 50:932-937.
72. **Xie J, Pannain S, Pohlenz J, Weiss RE, Moltz K, Morlot M, Asteria C, Persani L, Beck-Peccoz P, Parma J, Vassart G, Refetoff S** 1997 Resistance to thyrotropin (TSH) in three families is not associated with mutations in the TSH receptor or TSH [see comments]. J Clin Endocrinol Metab 82:3933-3940.
73. **Kremer H, Kraaij R, Toledo SP, Post M, Fridman JB, Hayashida CY, van Reen M, Milgrom E, Ropers HH, Mariman E, et al.** 1995 Male pseudohermaphroditism due to a homozygous missense mutation of the luteinizing hormone receptor gene. Nat Genet 9:160-164.
74. **Misrahi M, Meduri G, Pissard S, Bouvattier C, Beau I, Loosfelt H, Jolivet A, Rappaport R, Milgrom E, Bougneres P** 1997 Comparison of immunocytochemical and molecular features with the phenotype in a case of incomplete male pseudohermaphroditism associated with a mutation of the luteinizing hormone receptor. J Clin Endocrinol Metab 82:2159-2165.
75. **Gromoll J, Eiholzer U, Nieschlag E, Simoni M** 2000 Male hypogonadism caused by homozygous deletion of exon 10 of the luteinizing hormone (LH) receptor: differential action of human chorionic gonadotropin and LH. J Clin Endocrinol Metab 85:2281-2286.
76. **Laue LL, Wu SM, Kudo M, Bourdony CJ, Cutler GB, Hsueh AJ, Chan WY** 1996 Compound heterozygous mutations of the luteinizing hormone receptor gene in Leydig cell hypoplasia. Mol Endocrinol 10:987-997.
77. **Arnhold IJ, Latronico AC, Batista MC, Mendonca BB** 1999 Menstrual disorders and infertility caused by inactivating mutations of the luteinizing hormone receptor gene. Fertil Steril 71:597-601.
78. **Aittomaki K, Lucena JL, Pakarinen P, Sistonen P, Tapanainen J, Gromoll J, Kaskikari R, Sankila EM, Lehvaslaiho H, Engel AR, et al.** 1995 Mutation in the follicle-stimulating hormone receptor gene causes hereditary hypergonadotropic ovarian failure. Cell 82:959-968.
79. **Beau I, Touraine P, Meduri G, Gougeon A, Desroches A, Matuchansky C, Milgrom E, Kuttenn F, Misrahi M** 1998 A novel phenotype related to partial loss of function mutations of the follicle stimulating hormone receptor. J Clin Invest 102:1352-1359.
80. **Touraine P, Beau I, Gougeon A, Meduri G, Desroches A, Pichard C, Detoeuf M, Paniel B, Prieur M, Zorn JR, Milgrom E, Kuttenn F, Misrahi M** 1999 New natural inactivating mutations of the follicle-stimulating hormone receptor: correlations between receptor function and phenotype. Mol Endocrinol 13:1844-1854.

11

Endothelin G Protein-Coupled Receptors

Gastone G. Nussdorfer[1*] and Gian Paolo Rossi[2]

Departments of [1]Human Anatomy and Physiology, and, [2]Clinical and Experimental Medicine, School of Medicine, University of Padua, Via Gabelli 65, I-35121, Padua, Italy.

INTRODUCTION

Endothelins (ETs) are a family of endogenous peptides, mainly secreted by endothelial cells, which exert a potent vasoconstrictor and pressor activity, acting through two classes of G protein-coupled receptors, called ETA and ETB. ETs were originally isolated from porcine aortic endothelium [1], but subsequent investigations revealed that these peptides are synthesized and their receptors are expressed in a great variety of tissues and organs, where they play important physiological and pathophysiological roles [2-6]. At present, ETs are known to regulate the cardiovascular system and blood pressure (BP), kidney excretion, pulmonary function, endocrine-gland secretion, digestive functions, and the development of tissues derived from embryonic neural-crest lineage and of the central and peripheral nervous systems. The ability of all these tissues to synthesize ETs and the fact that the blood concentrations of ETs are below those able to evoke sizeable biological effects led to the contention that these peptides almost exclusively act through autocrine-paracrine mechanisms.

Before surveying the currently known signaling mechanisms of ET receptors in the various tissues and organs, the general biological features of ETs and their receptor subtypes will be briefly outlined.

---

* gastone.nusdorfer@unipd.it

Biological Features of ETs

ETs are 21-amino acid peptides (ET[1-21]), with four cysteine residues forming two intramolecular disulfide bonds. The ET family includes three distinct isoforms, referred to as ET-1, ET-2 and ET-3, which are synthesized as prepro(pp) proteins of about 200 amino acids (ppET-1, ppET-2 and ppET-3) encoded by three distinct genes [7], whose expression in cells and tissues is depicted in Table I [2-6, 8-11].

ET[1-21] biologically active isopeptides are generated *via* a two-step processing, which first involves the sequential cleavage by signal peptide peptidases and furin-like convertases of ppETs to form the 37/41-amino acid inactive big ETs. Then, big ETs are cleaved at the Trp^{21}-Val/Ile^{22} bond to form physiologically active ET[1-21] by specific endopeptidases, called endothelin-converting enzymes (ECE-1, ECE-2 and ECE-3) [2-6].

More recently, evidence has been provided that big ET-1 may be also cleaved by a chymase at the Tyr^{31}-Gly^{32} bond, thereby generating ET-1[1-31] [12]. Moreover, the cleavage of big ET-1 by matrix metalloproteinase-2 at the Gly^{32}-Leu^{33} bond can generate ET-1[1-32] [13].

Biological Features of ET Receptors

Within 2 years from the discovery of ET, the genes encoding ETA and ETB receptors were identified and cloned [14-16]. They are structured into 8 and 7 exons and have been mapped on chromosomes 4 and 13, respectively. The two genes are thought to derive from a common ancestral precursor, because they display marked similarities in their structural organization [17].

ETA and ETB receptors belong to the superfamily of the G protein-coupled receptors (see Chapter 9), and entail 7 transmembrane domains with an extracellular N-terminus and an intracytoplasmic C-terminus. Depending on the tissue, they range in size from 45 to 50 kDa. The two receptors share a high degree of homology in the transmembrane domains, but differ in their N-terminus and C-terminus, which are involved in the ligand binding and G protein-coupling, respectively. Their amino acid structure is highly conserved (about 90%) among the mammalian species [2]. The expression of ETA and ETB receptors in cells and tissues is summarized in Table I [2-6,8-11].

ETA and ETB receptors can be differentiated pharmacologically based on their affinity for the isopeptides of the et family. The ETA receptor displays very high (subnanomolar) affinity for Et-1[1-21] and Et-2[1-21], and a 70- to 100-fold lower affinity for ET-3 [1-21]. The ETB receptor is non-isopeptide selective, as it binds with high and similar affinity all the isopeptides. The existence of different subtypes of ETA and ETB receptors (provisionally termed ETA1, ETA2 etc., and ETB1, ETB2 etc.) has been postulated in light of their different affinity for selective non-ET ligands [18]. Recent findings suggest that ET-1[1-31] behaves as a selective agonist of the ETA receptor [19,20].

Table I. Expression of ETs and Their Receptors in Cells and Tissues.

Cells and tissues	**ETs**	**ET receptors**
Breast epithelial cells	ET-1	?
Blood vessels		
endothelial cells	ET-1	ETB>>>ETA
vascular SMC	ET-1	ETA>ETB
Connective tissue cells		
mast cells	ET-1	?
macrophages	ET-1, ET-3	?
Digestive system		
small intestine	ET-1, ET-2, ET-3	ETA, ETB
large intestine (colon)		ETA
liver		ETA, ETB
exocrine pancreas		ETA>ETB
Endocrine system		
adrenal cortex	ET-1>ET-3	ETA, ETB
adrenal medulla	ET-1	ETA>>ETB
endocrine pancreas	?	ETB ?
pituitary gland	ET-1	ETA, ETB
thyroid gland	ET-1	?
parathyroid gland	ET-1	ETA, ETB
Immune system		
thymus	?	ETA
Heart		
endocardium	ET-1	ETB>>ETA
myocardium	ET-1	ETA, ETB
Nervous system		
neurons	ET-1	ETA, ETB
astrocytes	ET-1	ETA, ETB
Respiratory system		
trachea	?	ETA, ETB
lungs	ET-1, ET-3	ETA, ETB
Reproductive system		
testis	ET-1	ETA>>ETB
prostate	ET-1	ETA (stroma), ETB (epithelium)
ovary	ET-1	ETA, ETB
uterus (endometrium)	ET-1, ET-2, ET-3	ETA, ETB
placenta	ET-1, ET-3>ET-2	
Urinary system		
kidney	ET-1, ET-3	ETA, ETB
bladder/urethra	ET-1	?

SIGNALING MECHANISMS OF ET RECEPTORS

Structure-Function Arrangement of ET Receptors

The two receptor subtypes share the following common structural features (Fig. 1): (i) an extracellular NH_2-terminal region, that possesses two potential N-linked glycosylation sites and is involved in ligand binding; (ii) seven transmembrane helices (I-VII), separated by three extracellular and three intracellular loops; (iii) a cytoplasmic COOH-terminal region, that is involved in the G protein-coupling and possesses a conserved six-amino acid sequence (Cys-Leu-Cys-Cys-Cys-Cys) important in the stabilization of the extracellular ligand-binding site; and (iv) a three-amino acid sequence (Asp-Arg-Tyr) at the end of the transmembrane helix III, which is believed to be crucial for G protein-coupling [2,5].

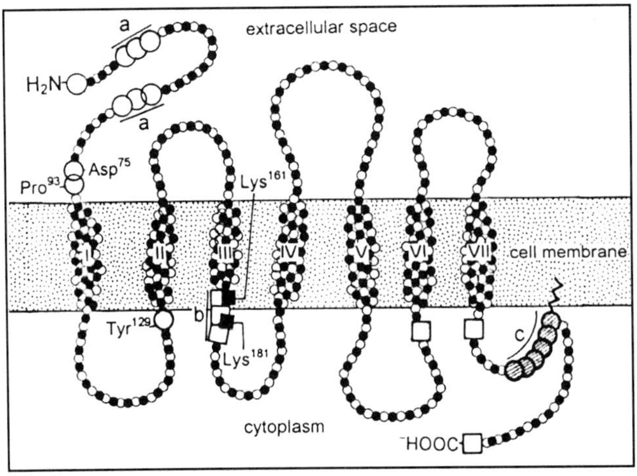

Figure 1. Schematic representation of the primary structure of ET receptors a, Glycosylation sites in the NH_2-terminal region; b, three-amino acid sequence (Asp-Arg-Tyr) at the cytoplasmic end of helix III; c, six-amino acid sequence (Cys-Leu-Cys-Cys-Cys-Cys) in the COOH-terminal region; I-VII, transmembrane helices. See text for details.

The ETA receptor possesses a Tyr residue (position 129) at the cytoplasmic end of helix II, which plays a role in the ligand binding, and two sequences (□) at the cytoplasmic ends of helices VI and VII involved in G protein-coupling. The ETB receptor has two amino acid residues (Asp^{75} and Pro^{93}) at the extracellular end of helix I (positions 161 and 181) which are essential for ligand binding, and two lysine residues at the cytoplasmic end of helix III, that are thought to be involved in the nitric oxide (NO) synthase (NOS)-coupling and ET-3 binding [3] (Fig. 1).

G Protein-Coupling of ET Receptors

Binding of ET isopeptides to ETA and ETB receptors results in the activation of at least three classes of Gα proteins: Gsα, Giα and Gqα [2-6]. Interestingly, different domains of the ET-receptor structure are required for the C-terminus coupling to different Gα proteins. Gα protein-coupling leads to the activation of the 2nd messenger system (Fig. 2). Briefly, (i) Gsα- and Giα-coupling results in the activation and inhibition of adenylate cyclase, respectively; (ii) Giα-coupling also activates voltage-gated Ca^{2+} channels; and (iii) Gqα-coupling is most likely linked to the activation of phospholipase C (PLC) and non-selective cation channels. ETA receptors are coupled with Gsα and Gqα proteins, and ETB receptors with Giα and Gqα proteins. Coupling of ET receptors with different G proteins may explain their divergent effects in the various cell types (see below).

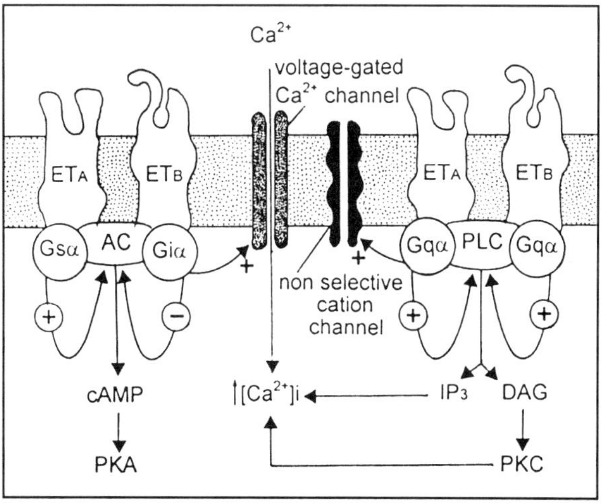

Figure 2. Coupling of the ET receptors to G proteins. AC, adenylate cyclase. Other abbreviations are as indicated in the text.

Signaling Pathways of ET Receptors and Their Functional Relevance

Adenylate Cyclase Pathway

ETA and ETB receptors activate and inhibit adenylate cyclase, respectively, in some cell systems, leading to the rise and fall in intracellular cAMP concentration, consequently modulating protein kinase (PK) A activity [2-6] (Fig. 3). Under normal conditions, because ET-1, which binds ETA and ETB receptors with equal affinity, is the main endogenous ET, the two effects on adenylate cyclase may conceivably annul each other. This observation might explain why investigations aimed at linking the ET-induced adenylate-cyclase modulation with a clear-cut physiological response have so far been inconclusive.

However, few data suggest that ET-1 may use the adenylate cyclase-PKA pathway to modulate secretory activity of some endocrine cells (Table II) [2-6,9,21]. A selective PKA inhibitor was found to partially prevent the ETA receptor-mediated aldosterone and corticosterone response to ET-1 of amphibian interrenal cells [22]. ET-1 was reported to raise cAMP concentration and progesterone secretion in the rat ovary [23]. In contrast, ETs have been shown to lower LH- and FSH-stimulated cAMP production by cultured pig and rat ovary granulosa cells [24-26], as well as to suppress forskolin- and dibutyryl-cAMP-induced progesterone release, thereby suggesting that ETs may also interfere with the PKA-dependent cascade [25]. There is also indication that ETA receptor-evoked cAMP production is involved in the catecholamine secretion in response to ETs [3].

Table II. Main Effects of ET-Receptor Activation

Target cells, tissues and organs	**Effects**	**ET-receptors**
Blood vessels	Constriction (increase in BP)	ETA
	Transient dilation	ETB
	Mitogenic effect on endothelial and smooth muscle cells and inhibition of apoptosis	ETA
Breast epithelial cells	Mitogenic effect	?
Connective-tissue cells		
fibroblasts	Mitogenic effect	ETA (?)
macrophages	Chemotactic effect	?
Digestive system		
small intestine (ileum)	Contraction/relaxation	ETA, ETB
large intestine (colon)	Stimulation of ion secretion	ETA
liver	Cholestasis, glycogenolysis	ETA, ETB
Endocrine system		
adrenal cortex	Stimulation of steroid synthesis	ETA, ETB (humans), ETB (rats)
	Mitogenic effect (zona glomerulosa)	ETA
adrenal medulla	Stimulation of catecholamine release	ETA, ETB
pancreatic islets	Stimulation of insulin and glucagon release	ETB (?)
thyroid gland	Inhibition of thyroglobulin synthesis	?
	Mitogenic effect on follicular cells	?
parathyroid gland	Stimulation (?) of PTH secretion	?
pituitary gland	Stimulation of ACTH and gonadotropin secretion	ETA, ETB
	Inhibition of PRL secretion	ETA, ETB
	Stimulation of glucose metabolism	ETA, ETB

Table II (continued)

Target cells, tissues and organs	Effects	ET-receptors
Heart		
ventricle	Inotropic and chronotropic effects	ETA, ETB
	Hypertrophy	ETA
atria	Chronotropic effect	ETA, ETB
	Stimulation of ANP release	ETA, ETB
Immune system		
thymus	Mitogenic effect on thymocytes	ETA
Nervous system	Increase/decrease of BP, stimulation of pyramidal neurons	ETA, ETB
hypothalamus	Stimulation of AVP, oxytocin, LH-RH and (?) CRH release	ETA, ETB
	Inhibition of PRH release	ETA, ETB
	Inhibition of water intake	ETA, ETB
brainstem	Respiratory depression	ETA, ETB
Respiratory system		
trachea and bronchi	Contraction	ETA, ETB
	Mitogenic effect on epithelial cells, stimulation of mucus secretion and Cl^- secretion	ETA
Reproductive system		
testis (Leydig cells)	Stimulation of testosterone secretion	ETA
vas deferens	Contraction	ETA
ovary (granulosa cells)	Inhibition of progesterone secretion	ETA, ETB ?
corpus luteum	Luteolysis	ETA ?
myiometrium	Contraction	ETA, ETB
Urinary system		
kidney	Decrease of GFR and kf	ETA, ETB
	Contraction and mitogenic effect on mesangial cells	ETA
	Inhibition of renin release by juxtaglomerular cells	ETA
	Decrease of Na^+ reabsorption by tubules and collecting ducts	?

PLC Pathway and Ca^{2+} Signaling

Both ETA and ETB receptors by coupling Gqα proteins are known to activate PLC, causing phosphatidylinositol (PIP_2) hydrolysis, and rapid and sustained accumulation of inositol triphosphate (IP3) and diacylglycerol (DAG), respectively. IP3 is an important intracellular signaling molecule, that triggers the release of Ca^{2+} from intracellular stores, and DAG activates PKC (Fig. 3). ET receptors also increase extracellular Ca^{2+} influx, through the activation of either dihydropiridine-sensitive voltage-dependent L-type Ca^{2+} channels or receptor-operated dihydropiridine-insensitive non-selective cation channels [2-6]. The IP3-induced Ca^{2+} release from intracellular stores accounts for the initial transient rise in $[Ca^{2+}]i$, which is independent of extracellular Ca^{2+}, while the stimulation of Ca^{2+} influx causes the sustained increase in $[Ca^{2+}]i$, which is dependent on extracellular Ca^{2+} [4]. The ET-evoked rise in $[Ca^{2+}]i$, in addition to activating calmodulin-dependent PKs, contributes to the stimulation of both PKC and PKA [2-6] (Fig. 3).

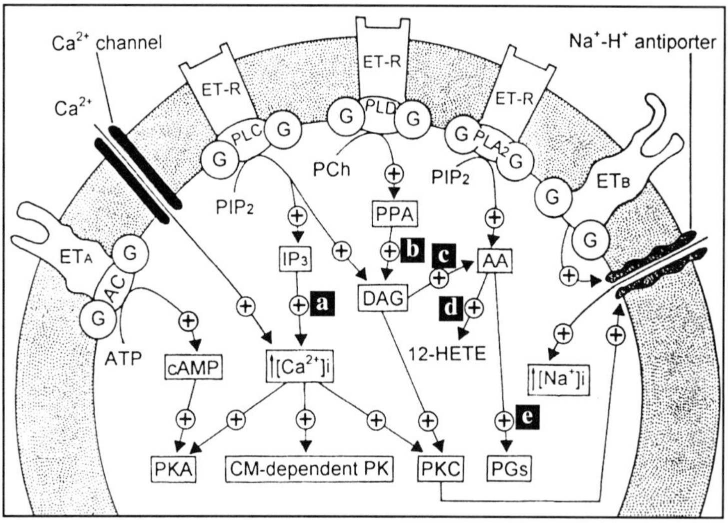

Figure 3. Signaling pathways of ET receptors. a, Ca^{2+} mobilization from intracellular stores; b, phosphohydroxylase; c, diglyceride lipase; d, lipoxigenase pathway; e, COX pathway. AA, arachidonic acid; AC, adenylate cyclase; CM, calmodulin; ET-R, ET receptors A and B; PCh, phosphatidylcholine; PPA, phosphatidic acid. Other abbreviations are as indicated in the text.

The above depicted pathways mediate several of the well-defined effects of ETs summarized in Table II [2-6,23,24], among which: (i) the contractory action on smooth-muscle cells (SMC), leading to vasoconstriction and the ensuing rise in BP, and to the increased motility of the intestine, excretory passages of the reproductive and urinary systems and conducting passages of the respiratory system [2]. Since these effects are mainly ETA receptor-mediated and these target tissues are provided with both ET-receptor subtypes (Table I), it may be conceived that ETB-receptor activating effects on PLC/Ca^{2+} pathways are of minor relevance and/or are overcome by the stimulation of other cascades eliciting antagonistic actions (e.g. NO-cGMP pathway) [2-6]. However, there is evidence that the pure activation of the ETB receptors located in vascular SMC (exposure of endothelium-deprived blood-vessel rings to non-ET ETB-receptor agonists, e.g. BQ-3020 or IRL-1620) may evoke a mild contraction [27]; (ii) the ETA- and ETB-mediated and the exclusive ETB receptor-mediated stimulatory action on corticosteroid secretion of human and rat adrenocortical cells, respectively [28,29]; (iii) the stimulatory effect on catecholamine secretion from medullary chromaffin cells [30]; (iv) the inhibitory effect on agonist-stimulated progesterone release by ovarian granulosa cells [24,31]; (v) the stimulatory action on PTH secretion from rat and bovine parathyroid cells [32,33], and (vi) the stimulating effect on the secretory activity of pituitary gonadotropes and hypothalamic neurosecreting neurons [3].

PLA2 Pathway

Evidence indicates that ETs activates PLA2, leading to the production of arachidonic acid from PIP2. It is not yet settled whether the effect of ETs is directly mediated by G protein-coupled ET receptors or indirectly caused by the rise in $[Ca^{2+}]i$ [2-6]. Arachidonic acid, which may be also derived from DAG *via* the action of a diglyceride lipase, may be metabolized through the cyclooxygenase (COX) or the lypoxygenase pathway, leading to the formation of prostaglandins (PGs) and 12-hydroxyeicosatetraneoic acid, respectively (Fig. 3).

PGs are involved in many of the ET receptor-mediated effects (Table II), among which: (i) the ETA receptor-mediated contractory action on SMC (PGI2 and PGF2α) [2-6,27]; (ii) the ETB receptor-mediated secretagogue action on human adrenocortical cells (PGE2?) [29]; (iii) the ETA receptor-mediated stimulating effect on basal and LH-enhanced testosterone secretion from rat Leydig cells (PGE2) [34]; (iv) the ETA receptor-mediated inhibitory action on progesterone secretion from granulosa cells and corpus luteum slices (PGE2, PGF2α) [35,36], although Calogero *et al.* [26] denied this contention based on the lack of effect of the COX inhibitor indomethacin; (v) the ETA receptor-mediated (?) luteolytic effect [37,38]; (vi) the ETB receptor-mediated effects on kidney [2]; and (vii) the stimulation of ion secretion from intestine [39].

PLD Pathway

The agonist-induced PLD activation is commonly associated with PLC stimulation operated by tyrosine kinase (TK) and G protein-coupled receptors [3]. PLD hydrolyzes phosphatidylcholine to phosphatidic acid, that in turn is converted to DAG by a phosphohydroxylase, thereby contributing to the activation of both PKC- and arachidonate-dependent cascades (Fig. 3). Evidence is available that PLD may be activated not only directly, but also in a PKC-dependent manner [40], which stresses the strict cross-talk between PLC and PLD pathways.

Findings indicate that ETs, acting mainly *via* ETA receptors, activate PLD in SMC [41,42], cardiomyocytes [43], rat fibroblasts [3], neuroblastoma cells

[3], cerebellar granules [3], astrocytes [44,45], and kidney mesangial and medullary interstitial cells [46].

Stimulation of Ion Exchange

There is evidence that ETs, through Giα protein-coupled ETB receptors activate the Na^+-H^+ antiporter, leading to intracellular alcalinization due to H^+ excretion and Na^+ influx [3]. The stimulation of Na^+-H^+ exchange may be also dependent on the ET-induced activation of PLC-PKC cascade [2,3] (Fig. 3). The activation of Na^+-H^+ antiporter, by increasing $[Na^+]$i, may stimulate Na^+/K^+-ATPase activity, which in turn by decreasing $[Na^+]$i and increasing the Na^+ gradient across the plasma membrane may enhance Na^+-H^+ exchange [2,3,6].

ETs have been found to activate Na^+-H^+ exchange in vascular SMC and human skin fibroblasts [2]. Moreover, the activation of Na^+/K^+-ATPase has been reported to be involved in the ETB receptor-mediated secretagogue effect of ETs on rat zona glomerulosa cells [47].

Mitogen-Activated Protein Kinase (MAPK) Pathway

p42/p44 MAPKs are ubiquitous members of a family of serine/threonine kinases which are known to play a crucial role in cell proliferation [48,49](see Chapter 3). MAPK cascade involves a series of cytoplasmic phosphorylations, where p21-activated kinase (PAK or MAPK kinase kinase kinase) activates Raf (MAPK kinase kinase), which in turn phosphorylates MEK1/2 (MAPK kinase), which eventually activates extracellular signal-regulated kinase-1/2 (ERK1/2). ERK1/2 translocate from the cytosol to the nucleus, where they phosphorylate transcriptional factors, that induce expression of the growth-associated nuclear proto-oncogenes, including *c-fos*, *c-jun* and *c-myc*. Increased transcription of these immediate-early gene results in the G0 to G1 and G2 to M transition of the cell cycle. Receptor TK plays a pivotal role in the MAPK cascade, because by binding to its agonists it activates Ras, a peptide belonging to the family of low-molecular weight GTP-binding proteins, which activates PAK and Raf. Compelling evidence also indicates that, in addition to RTK, G protein-coupled receptors can stimulate MAPK cascade: PKC and PKA activate PAK and Raf through a Ras independent mechanism or *via* the transactivation of the TK-dependent cascade and thereby *via* a Ras-dependent mechanism [3-5,50-52] (Fig. 4).

ETs, mainly *via* the ETA receptor, are potent mitogen in cultured vascular SMC, cardiomyocytes, kidney mesangial cells and adrenal zona glomerulosa cells (Table II), by enhancing the MAPK-dependent transcription of *c-fos* and *c-jun* proto-oncogenes [3-5]. ETA receptor activation may also stimulate *c-myc* and other immediate-early genes in the heart, leading to increased protein production and hypertrophy of cardiomyocytes [3]. Fig. 4 clearly shows that not only ETA, but also ETB receptors could activate MAPK cascade, either by coupling with a Gqα protein and activating PLC/PKC pathway or by coupling with a Giα or a Giβ/γ pertussis toxin-sensitive protein and activating Ras [4,5,53]. There is also evidence that the coupling with Giα protein may lead to the transactivation of the receptor TK. Findings are available that ETs may stimulate receptor TK or activate cytosolic TK also through an ET receptor-dependent mechanism. Why the currently known growth-promoting effects of ETs are almost exclusively ETA receptor-mediated remains an open problem. The ETA receptor-induced activation of the adenylate cyclase-PKA pathway could be crucial, but evidence suggests that this is not true for the ETA-receptor mediated zona glomerulosa proliferogenic effect of ETs, which is not blunted by the PKA inhibitor H-89 [54].

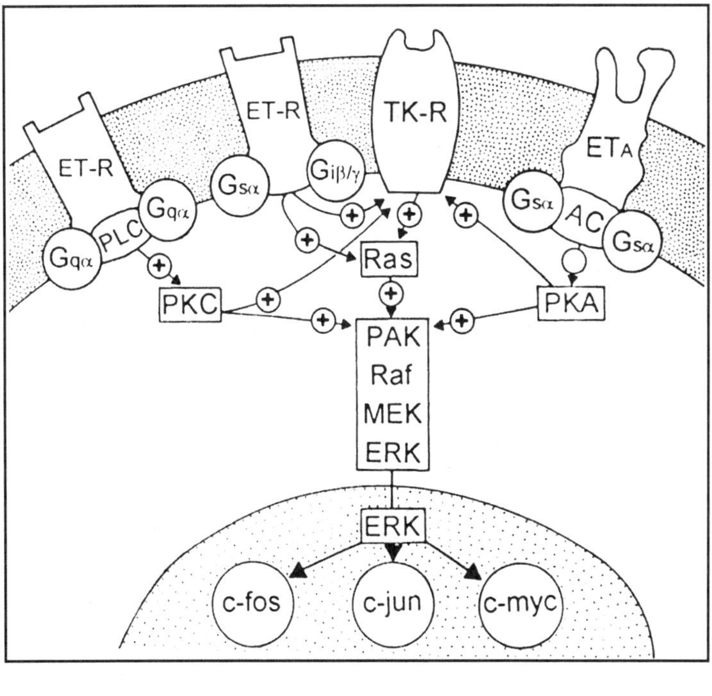

Figure 4. Signaling pathways involved in the ET-induced activation of MAPK cascade. AC, adenylate cyclase; ET-R, ET receptors A and B. Other abbreviations are as indicated in the text.

NO-cGMP Pathway

Three different NOS have been identified: the neural NOS (nNOS) and the Ca^{2+}-regulated endothelial form (eNOS), which are considered to be constitutively expressed, and the inducible form (iNOS), that is expressed in cells upon immunological and inflammatory stimulation [55,56]. Evidence suggests that the ETB receptor is coupled with eNOS in endothelial cells (Fig. 1). NOS in the presence of O_2 oxidizes L-arginine to NO and L-citrullin. The full activity of the enzyme is displayed only in the presence of Ca^{2+}-activated calmodulin, which also may result from ETB-receptor activation (Fig. 3). The principal physiological actions of NO are associated with the activation of a soluble (cytosolic) guanylate cyclase leading to cGMP production [55,56]. NO is also known to inhibit ET-1 synthesis and release, and to accelerate the dissociation of ET-1 binding from its receptors, thereby terminating ET signaling [57] (Fig. 5).

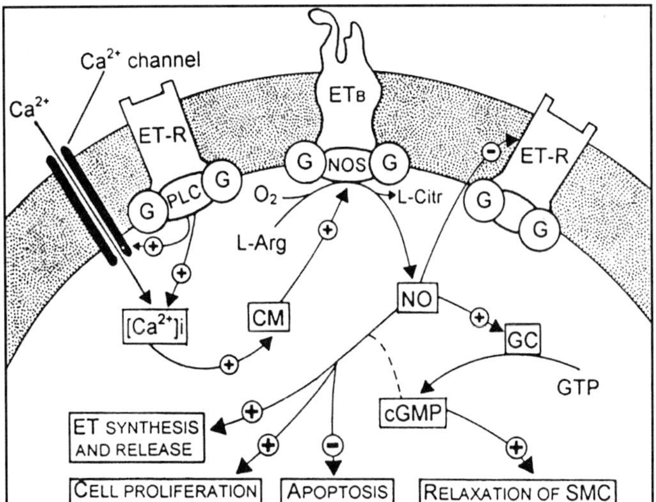

Figure 5. Signaling mechanisms involved in the ET-induced activation of the NO-cGMP pathway. CM, calmodulin; ET-R, ET receptors A and B; GC, guanylate cyclase; L-Arg, L-arginine; L-Citr, L-citrulline. Other abbreviations are as indicated in the text.

NO-cGMP pathway plays a role in the following ETB receptor-mediated effects of ETs: (i) transient endothelium-dependent vasodilation in most vascular beds (including brain, coronary, kidney and adrenal beds), thereby decreasing vascular resistances and raising local blood flow [56,58]; (ii) inhibition of vascular SMC and fibroblast proliferation and growth and stimulation of apoptosis, thereby being involved in cardiovascular remodeling [59,60]; and (iii) inhibitory effect on basal and agonist-stimulated secretory activity of adrenocortical, Leydig and lutein cells [6]. Collectively, these three effects suggest that NO exerts a buffering action on ETA and ETB receptor-mediated vasoconstrictor, vascular growth-promoting and adrenocortical and testis secretagogue actions of ETs. In contrast, ETB receptor-induced NO production might agree with the anti-secretagogue and luteolytic effects of ETs.

CONCLUSIONS

The preceding sections of this brief survey have clearly shown that in the 12 years since the discovery of ETs an impressive mass of data has been accumulated, which has elucidated, at least in part, the molecular biology of ETs and their receptors. The wide distribution of ET receptors in tissues and organs has been demonstrated, and many biological effects ensuing from their ET-induced activation have been discovered. However, the investigations dealing with the relevance of the endogenous paracrine-autocrine ET system in the physiological regulation of the body function gave rather disappointing results.

In fact, the short-term blockade of ET receptors by selective antagonists failed to evoke sizeable functional alterations of tissues and organs provided with an active ET system [2-6]. Aganglionic megacolon has been observed in humans bearing a mutation of the ETB-receptor gene and in ETB-receptor "knockout" mice [61,62], which stresses the possible critical role of this ET-receptor subtype in the normal development and differentiation of cells derived from neural crest (see Introduction). It is even conceivable that the effects of acute experimentally-induced alterations of ET-receptor signaling may be obscured by the activation of compensatory mechanisms mediated by other local systems with which the ET system is known to interact (e.g. tissue renin-angiotensin and adrenomedullin systems).

Conversely, alterations of the ET system are implicated in the pathogenesis of many diseases especially of the cardiovascular system [3,5,27]. Several lines of evidence strongly suggest that enhanced expression or exceedingly increased activation of ETA receptors (and/or perhaps reduced activity of ETB receptors coupled with NO-cGMP pathway) may play a role in vascular-wall thickening and in the genesis of artheriosclerotic lesions, cardiac hypertrophy, chronic heart failure, and many forms of systemic and pulmonary hypertension, where beneficial effects of ETA-receptor blockade has been demonstrated. A role of the ET system in the development of kidney diseases (e.g., acute renal failure) and brain injury and cerebral vasospasm has been also postulated. Finally, because of their ETA receptor-mediated proliferogenic actions, ETs have been implicated in the pathogenesis of some kinds of benign and malignant tumors [2-6], among which breast and prostate tumors (and prostatic hyperplasia), aldosterone producing adenomas, ovarian carcinomas, parathyroid adenomas, pheochromocytomas, endometral and renal adenocarcinomas, and hepatocellular carcinomas.

Further studies on the ET system will not only increase our knowledge of the basic regulatory mechanisms involved in the maintenance of body homeostasis, but also will open novel and important perspectives in the therapy of several human diseases.

Acknowledgements

This work was supported by Ministero dell'Università e Ricerca Scientifica Grant COFIN 98-06197882 to Gastone G. Nussdorfer. We thank Giuseppe Gottardo and Alberta Coi for outstanding secretarial support.

REFERENCES

1. **Yanagisawa M, Kurihara H, Kimura S, Tomobe Y, Kobayashi M, Yazaki Y, Goto K, Masaki T.** 1988 A novel potent vasoconstrictor peptide produced by vascular endothelial cells. Nature (London) 332:411-415.
2. **Rubanyi GM, Polokoff MA.** 1994 Endothelins: molecular biology, biochemistry, pharmacology, physiology, and pathophysiology. Pharmacol Rev 46:325-415.
3. **Levin ER.** 1996 The endothelin peptides. Role in physiology and pathophisiology. In: Sower JR (ed), Endocrinology of the Vasculature. Humana Press, Totowa (NJ), 49-66.
4. **Stojilkovic SS, Catt KJ.** 1996 Expression and signal transduction pathways of endothelin receptors in neuroendocrine cells. Front Neuroendocrinol 17:327-369.
5. **Miyauchi T, Masaki T.** 1999 Pathophysiology of endothelin in the cardiovascular system. Annu Rev Physiol 61:391-415.
6. **Nussdorfer GG, Rossi GP, Malendowicz LK, Mazzocchi G.** 1999 Autocrine-paracrine endothelin system in the physiology and pathology of steroid-secreting tissues. Pharmacol Rev 51:403-437.
7. **Inoue A, Yanagisawa M, Kimura S, Kasuya Y, Miyauchi T, Goto K, Masaki T.** 1989 The human endothelin family: three structurally and pharmacologically distinct isopeptides predicted by three separate genes. Proc Natl Acad Sci USA 86:286-287.
8. **Goto K, Hama H, Kasuya Y.** 1996 Molecular pharmacology and pathophysiological significance of endothelin. Jpn J Pharmacol 72:261-290.
9. **Malendowicz LK, Macchi V, Brelinska R, Trejter M, Gottardo G, Nussdorfer GG.** 1998 Endothelin-1 enhances thymocyte proliferation in monolaterally adrenalectomized rats with controlateral adrenocortical regeneration. Histol Histopathol 13:721-725.
10. **Brock B, Gregersen S, Kristensen K, Thomsen JL, Buschard K, Kofod H, Hermansen K.** 1999 The insulinotropic effect of endothelin-1 is mediated by glucagon release from the islet alpha cells. Diabetologia 42:1302-1307.
11. **De Carlo E, Milanesi A, Martini C, Maffei P, Sicolo N, Scandellari C.** 2000 Endothelin-1 and endothelin-3 stimulate insulin release by isolated rat pancreatic islets. J Endocrinol Invest 23:240-245.

12. **Nakano A, Kishi F, Minami K, Wakabayashi H, Nakaya Y, Kido H.** 1997 Selective conversion of big endothelins to tracheal smooth muscle-constricting 31-amino acid-lenght endothelins by chymase from human mastcells. J Immunol 159:1987-1992.
13. **Fernandez-Patron C, Radomski MW, Davidge ST.** 1999 Vascular matrix metalloproteinase-2 cleaves big endothelin-1 yielding a novel vasoconstrictor. Circ Res 85:906-911.
14. **Arai H, Hori S, Aramori I, Ohkubo H, Nakanishi S.** 1990 Cloning and expression of a cDNA encoding an endothelin receptor. Nature (Lond) 348:730-732.
15. **Sakurai T, Yanagisawa M, Takuwa Y, Miyazaki H, Kimura S, Goto K, Masaki T.** 1990 Cloning of a cDNA encoding a non-isopeptide selective subtype of the endothelin receptor. Nature (London) 348:732-735.
16. **Lin HY, Kaji EH, Winkel GK, Ives HE, Lodish HF.** 1991 Cloning and functional expression of vascular smooth muscle endothelin-1 receptor. Proc Natl Acad Sci USA 88:3185-3189.
17. **Elshourbagy NA, Adamou JE, Cagnon AW, Wu HL, Pullen M, Nambi P.** 1996 Molecular characterization of a novel human endothelin receptor splice variant. J Biol Chem 271:25300-25307.
18. **Ohlstein EH, Eliot JD, Feuerstein GZ, Ruffolo RR Jr.** 1996 Endothelin receptors: receptor classification, novel receptor antagonists, and potential therapeutic targets. Med Res Rev 16:365-390.
19. **Yoshizumi M, Kim S, Kagami S, Hamaguchi A, Tsuchiya K, Houchi H, Iwao H, Kido H, Tamaki T.** 1998 Effect of endothelin-1(1-31) on extracellular signal-regulated kinase and proliferation of human coronary artery smooth muscle cells. Br J Pharmacol 125:1019-1027.
20. **Yoshizumi M, Inui D, Kirima K, Tsuchiya K, Houchi H, Azuma M, Yasuoka H, Kido H, Tamaki T.** 1999 Comparison of the effects of endothelin-1, -2 and –3(1-31) on changes in [Ca^{2+}]i in human coronary artery smooth muscle cells. Jpn J Pharmacol 81:298-304.
21. **Galindo-Pacheco Y, Belloni AS, Malendowicz LK, Meneghelli V, Gottardo G, Nussdorfer GG.** 1996 Endothelin-1 inhibits renin release by rat kidney acting via the ETA receptor subtype. Biomed Res 17:429-434.
22. **Cartier F, Delarue C, Remy-Jouet I, Kodjo MK, Fournier A, Vaudry H.** 1999 The stimulatory effect of endothelin-1 on frog adrenocortical cells is mediated through both the phospholipase C and the adenylate cyclase transduction pathways. Mol Cell Endocrinol 147:27-36.
23. **Usuki S, Otani S, Goto K, Sizuki N, Yanagisawa M, Masaki T.** 1991 Endothelin induces DNA polymerase-α activity in ovaries of hypophysectomized estrogen-treated immature rats. Horm Metab Res 26:621-622.
24. **Flores JA, Quyyumi S, Leong DA, Veldhuis JD.** 1992 Actions of endothelin-1 on swine ovarian (granulosa) cells. Endocrinology 131:1350-1358.
25. **Tedeschi C, Lohman C, Hazum E, Ittoop O, Ben-Shlomo I, Resnick CE, Payne DW, Adashi EY.** 1994 Rat ovarian granulosa cell as a site of endothelin reception and action: attenuation of gonadotropin-stimulated steroidogenesis via perturbation of the A-kinase signaling pathway. Biol Reprod 51:1058-1065.
26. **Calogero AE, Burrello N, Ossino AM.** 1998 Endothelin (ET)-1 and ET-3 inhibit estrogen and cAMP production by rat granulosa cells in vitro. J Endocrinol 157:209-215.
27. **Rossi GP, Belloni AS, Colonna S, Nussdorfer GG, Pessina AC.** 1998 Endothelin receptor subtypes in human arterial hypertension. J Hypertens 16:S81-S84.
28. **Rossi GP, Andreis PG, Neri G, Tortorella C, Pelizzo MR, Sacchetto A, Nussdorfer GG.** 2000 Endothelin-1 stimulates aldosterone synthesis in Conn's adenomas via both A and B receptors coupled with the protein kinase C- and cyclooxygenase-dependent signaling pathways. J Investig Med 48:335-342.

29. **Andreis PG, Tortorella C, Malendowicz LK, Nussdorfer GG.** 2001 Endothelins stimulate aldosterone secretion from dispersed rat adrenal zona glomerulosa cells, acting through ETB receptors coupled with the phospholipase C-dependent signaling pathway. Peptides, 22:117-122.
30. **Garrido MR, Israel A.** 1997 Endothelin ETA receptor subtype mediates phosphoinositide hydrolysis in adrenal medulla. Arch Physiol Biochem 105:467-472.
31. **Kamada S, Blackmore PF, Kubota T, Oehninger S, Asada Y, Gordon K, Hodgen GD, Aso T.** 1995 The role of endothelin-1 in regulating human granulosa cell proliferation and steroidogenesis in vitro. J Clin Endocrinol Metab 80:3708-3714.
32. **Tanini A, Failli P, Maggi M, Franceschelli F, Frediani U, Becherini L, Giotti A, Ruocco C, Brandi ML.** 1993 Effects of endothelin-1 on bovine parathyroid cells. Biochem Biophys Res Commun 193:59-66.
33. **Fujii Y, Tomic M, Stojilkovic SS, Iida T, Brandi ML, Ogino Y, Sakaguchi K.** 1995 Effects of endothelin-1 on Ca^{2+} signaling and secretion in parathyroid cells. J Bone Mineral Res 10:716-725.
34. **Romanelli F, Fillo S, Isidori A, Conte D.** 1997 Stimulatory action of endothelin-1 on rat Leydig cells: involvement of endothelin-A subtype receptor and phospholipase A2-arachidonate metabolism system. Life Sci 61:557-566.
35. **Girsh E, Milvae RA, Wang W, Meidan R.** 1996 Effect of endothelin-1 on bovine luteal cell function: role of prostaglandin F2α-induced antisteroidogenic action. Endocrinology 137:1306-1312.
36. **Acosta TJ, Miyamoto A, Ozawa T, Wijayagunawardane MP, Sato K.** 1998 Local release of steroid hormones, prostaglandin E2 and endothelin-1 from bovine mature follicles in vitro: effects of luteinizing hormone, endothelin-1, and cytokines. Biol Reprod 59:437-443.
37. **Meidan R, Girsch E.** 1997 Role of endothelial cells in the steroidogenic activity of the bovine corpus luteum. Semin Reprod Endocrinol 15:371-382.
38. **Ohtani M, Kobayashi SI, Miyamoto A, Hayashi K, Fukui Y**. 1998 Real-time relationships between intraluteal and plasma concentrations of endothelin, oxytocin, and progesterone during prostaglandin F2α-induced luteolysis in the cow. Biol Reprod 58:103-108.
39. **Brown MA, Smith PL.** 1991 Endothelin: a potent stimulator of intestinal ion secretion in vitro. Regul Pept 36:1-19.
40. **Exton JH.** 1997 Phospholipase D: enzymology, mechanisms of regulation, and function. Physiol Rev 77:303-320.
41. **Zhang Y, Abdel-Latif AA.** 1992 Activation of phospholipase D by endothelin-1 and other pharmacological agents in rabbit iris sphincter smooth muscle. Cell Signal 4:777-786.
42. **Naze S, Le Stunff H, Dokhac L, Thomas G, Harbon S.** 1997 Activation of phospholipase D by endothelin-1 in rat myometrium. Role of calcium and protein kinase C. J Pharmacol Exp Ther 281:15-23.
43. **Clerk A, Sugden PH**. 1997 Regulation of phospholipases C and D in rat ventricular myocytes: stimulation by endothelin-1, bradykinin and phenylephrine. J Mol Cell Cardiol 29:1593-1604.
44. **Desagher S, Cordier J, Glowinski J, Tence M.** 1997 Endothelin stimulates phospholipase D in striatal astrocytes. J Neurochem 68:78-87.
45. **Servitja JM, Masgrau R, Sarri E, Picatoste F.** 1998 Involvement of ETA and ETB receptors in the activation of phospholipase D by endothelins in cultured rat cortical astrocytes. Br J Pharmacol 124:1728-1734.
46. **Kester M, Simonson MS, McDermott RG, Baldi E, Dunn MJ.** 1992 Endothelin stimulates phosphatidic acid formation in cultured rat mesangial cells: role of protein kinase C-regulated phospholipase D. J Cell Physiol 150:578-585.

47. **Pecci A, Romero DG, Lantos C, Cozza EN.** 1998 Mechanisms of action of endothelin-1 in rat adrenal. Life Sci 63:1315-1328.
48. **Campbell JS, Seger R, Graves JD, Jensen AM, Krebs EG.** 1995 The MAP kinase cascade. Recent Prog Horm Res 50:131-159.
49. **Fanger GR.** 1999 Regulation of the MAPK family members: role of subcellular localization and architectural organization. Histol Histopathol 14:887-894.
50. **Blumer KJ, Johnson GL.** 1994 Diversity in function and regulation of MAP kinase pathway. Trends Biochem Sci 19:236-240.
51. **Wang Y, Rose PM, Webb ML, Dunn MJ.** 1994 Endothelins stimulate mitogen-activated protein kinase cascade through either ETA or ETB. Am J Physiol 267:C1130-C1135.
52. **Biesen TV, Luttrell LM, Hawes BE, Lefkowitz RJ.** 1996 Mitogenic signaling via G protein-coupled receptors. Endocr Rev 17:698-714.
53. **Takigawa M, Sakurai T, Kasuya Y, Abe Y, Masaki T, Goto K.** 1995 Molecular characterization of guanine-nucleotide-binding regulatory proteins which couple to endothelin receptors. Eur J Biochem 228:102-108.
54. **Mazzocchi G, Rossi GP, Rebuffat P, Malendowicz LK, Markowska A, Nussdorfer GG.** 1997 Endothelins stimulate deoxyribonucleic acid synthesis and cell proliferation in rat adrenal zona glomerulosa, acting through an endothelin A receptor coupled with protein kinase C- and tyrosine kinase-dependent signaling pathways. Endocrinology 138:2333-2337.
55. **Andrew PG, Mayer B.** 1999 Enzymatic functions of nitric oxide synthases. Cardiovasc Res 43:521-531.
56. **Li H, Föstermann U.** 2000 Nitric oxide in the pathogenesis of vascular disease. J Pathol 190:244-254.
57. **Goligorsky MS, Tsukahara H, Magazine H, Andersen TT, Malik AB, Bahou WF.** 1994 Termination of endothelin signaling: role of NO. J Cell Physiol 158:485-494.
58. **Mazzocchi G, Malendowicz LK, Musajo FG, Gottardo G, Markowska A, Nussdorfer GG.** 1998 The role of endothelins in the regulation of vascular tone in the in situ perfused rat adrenals. Am J Physiol 274:E1-E5.
59. **Rudic RD, Shesely EG, Maeda N, Smithies O, Segal SS, Sessa WC.** 1998 Direct evidence for the importance of endothelium-derived nitric oxide in vascular remodeling. J Clin Invest 101:731-736.
60. **Wever RM, Luscher TF, Cosentino F, Rabelink TJ.** 1998 Atherosclerosis and the two faces of endothelial nitric oxide synthase. Circulation 97:108-112.
61. **Puffenberger EG, Hosoda K, Washington SS, Nakao K, de Witt D, Yanagisawa M, Chakravarti A.** 1994 A missense mutation of the endothelin-B receptor gene in multigenic Hirschsprung's disease. Cell 79:1257-1266.
62. **Hosoda K, Hammer RE, Richardson JA, Baynash AG, Cheung JC, Giaid A, Yanagisawa M.** 1994 Targeted and natural (piebald-lethal) mutations of endothelin-B receptor gene produce megacolon associated with spotted coat color in mice. Cell 79:1267-1276.

Section V

Nuclear Receptors

12

Signal Transduction and Structure of Nuclear Receptors

William Bourguet, Pierre Germain and Hinrich Gronemeyer*

Institut de Génétique et de Biologie Moléculaire et Cellulaire (IGBMC), BP 163, 67404 Illkirch Cedex, C. U. de Strasbourg, France

INTRODUCTION

Multicellular organisms require specific intercellular communication to properly organize the complex body plan during embryogenesis and maintain its properties and functions throughout life. While growth factors, neurotransmitters and peptide hormones bind to membrane receptors thereby inducing the activity of intracellular kinase cascades or the JAK-STAT/Smad signaling pathways, other small, hydrophobic signaling molecules such as steroid hormones, certain vitamins and metabolic intermediates enter, or are generated, within the target cells and bind to cognate members of a large family of nuclear receptors. Nuclear receptors (NRs) are of major importance for metazoan intercellular signaling, as they converge different intra- and extracellular signals to initiate and regulate gene expression programs. They act as transcription factors that i) respond directly through physical association with a large variety of hormonal and other regulatory, as well as metabolic signals, ii) integrate diverse signaling pathways as they correspond themselves to targets of post-translational modifications, and iii) regulate the activities of other signaling cascades (commonly referred to as "signal transduction crosstalk"). The genetic programs that they modulate affect virtually all aspects of the life of a multicellular organism, covering such diverse aspects

* hg@titus.u-strasbg.fr

as, for example, embryogenesis, homeostasis and reproduction, or cell growth and death. Their gene regulatory power and selectivity has prompted intense research on these key factors, which is now starting to decipher a complex network of molecular events accounting for their transcription regulatory capacity. The study of these molecular processes has also shed light on multiple general principles underlying transcription regulation, and it will be a future challenge to uncover the molecular rules that define selective NR-dependent spatial and temporal control of gene expression.

The Family and Its Ligands

To date, about 70 different nuclear receptors have been identified throughout the animal kingdom ranking from nematodes to man. They constitute a family of transcription factors that share a modular structure of six conserved domains encoding specific functions (Fig. 1) [1]. The most prominent distinction from other transcription factors is their capacity to specifically bind small hydrophobic molecules. These ligands constitute regulatory signals that after binding change the transcriptional activity of the corresponding nuclear receptor. For some time a distinction was made between classical nuclear receptors with known ligands and so-called "orphan" receptors, hence receptors without or with unknown ligand.

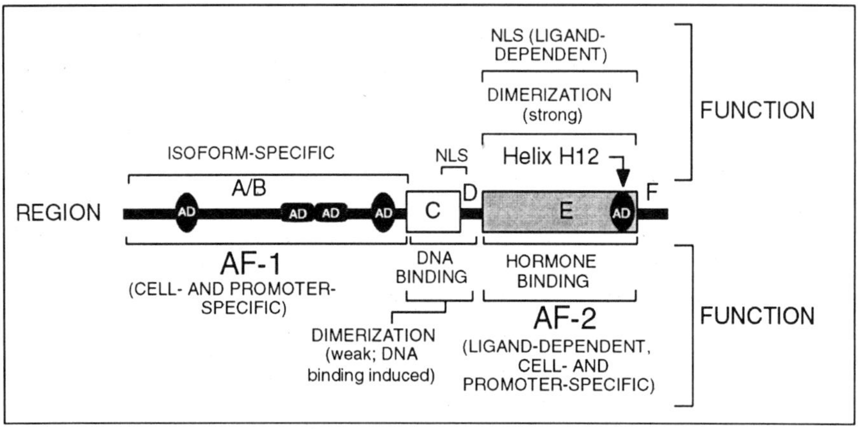

Figure 1. Schematic illustration of the structural and functional organization of nuclear receptors. The evolutionary conserved regions C and E are indicated as boxes and a black bar represents the divergent regions A/B, D, and F. Note that region F may be absent in some receptors. Domain functions are depicted below and above the scheme. Two transcription activation functions (AFs) have been described in several nuclear receptors, a constitutively active (if taken out of the context of the receptor) AF-1 in region A/B and a ligand-inducible AF-2 in region E. Within these activation functions, autonomous transactivation domains (ADs) have been defined.

However, recent years have seen the identification of ligands for many of these orphan receptors, making this distinction rather superficial [2]. Moreover, the classification of nuclear receptors into six-seven phylogenetic subfamilies with groups that comprise both orphan and non-orphan receptors further dismisses such discrimination [3, 4]. The classification of nuclear receptors is done by virtue of the homology to other family members, with the DNA binding (region C; Fig. 1) and ligand binding (region E) domains having the highest evolutionary conservation. It has been proposed that nuclear receptors have evolved from an ancestral orphan "receptor" through early diversification, and only later acquired ligand-binding [3]. However, as long as the identification of ligands for previous orphan receptors continues it can not be formally excluded that all nuclear receptors may have cognate ligands. On the other hand, it is possible that a number of orphan receptors function exclusively as constitutive repressors or activators of transcription. Rev-Erb is an example of such a constitutive repressor that recruits corepressors but lacks a functional AF-2.

MODULAR STRUCTURE AND FUNCTIONS

The N-Terminal Region A/B Harbors Cell-Specific Activation Function(s) AF-1 of Unknown Structure

As depicted in Fig. 1, nuclear receptors are composed of six regions (A to F) that have a modular character. The N-terminal A/B region harbors one (or more) autonomous transcriptional activation function (AF-1) which when linked to a heterologous DNA-binding domain can activate transcription in a constitutive manner. Note, however, that in the context of the full-length receptor AF-1 is silent in the absence of agonist. When comparing nuclear receptors from different sub-families and groups, the A/B region displays the weakest evolutionary conservation, and the distinction between A and B regions is not always evident. A/B regions differ significantly in their length, ranging from 23 (vitamin D receptor) to 550 (androgen receptor) amino acids. No 3D structure of an A/B region has been solved up to now and structure prediction is not straightforward. A/B regions are subject to alternative splicing and differential promoter usage and the majority of known nuclear receptor isoforms differ in their N-terminal region. Through alternative splicing and differential promoter usage (PR forms A + B) the absence or presence of different activation functions found in the AB regions can be regulated (for review see [1]). Moreover, the N-terminus of nuclear receptors is often subject to post-translational modification, such as phosphorylation, which can affect the transactivation potential and AF-1/AF-2 synergy in some receptors [5, 6]. Finally, the activation function(s) AF-1 display cell, DNA-binding domain and promoter specificity [7-9], the origin of which may be related to the cell-specific action and/or expression of AF-1 coactivators [10-13].

DNA Recognition by Nuclear Receptors

Response Elements of Nuclear Receptors

All nuclear receptors recognize derivatives of the same hexameric DNA core motif, 5'-PuGGTCA (Pu = A or G). However, mutation, extension, and duplication, and, moreover, distinct relative orientations of repeats of this

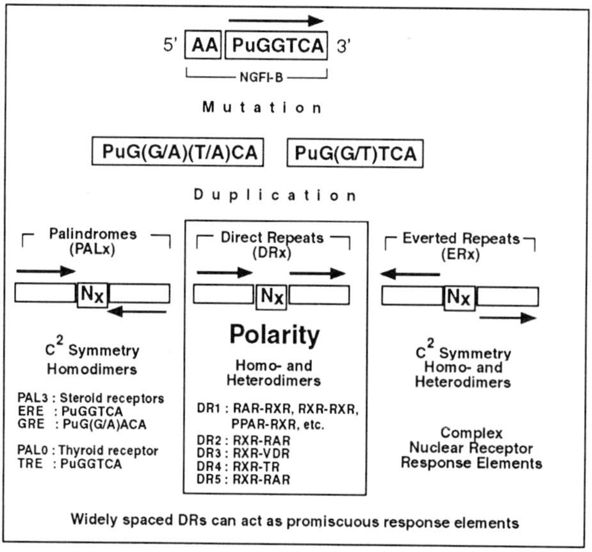

Figure 2. Response elements of nuclear receptors. The canonical core recognition sequence is 5'-PuGGTCA (arrows indicate the 5' to 3' direction) which, together with two 5' As, is a response element of the orphan receptor NGFIB. Duplication of the core sequence generates symmetrical palindromes (PALx), everted repeats (ERx), and polar direct repeats (DRx), with x bp separating the two half-sites. PAL3 is an estrogen (ERE), while PAL0 corresponds to a thyroid hormone (TRE) response element. A single mutation at position 4 of the core sequence from T to A leads to PAL3 response elements (GREs) recognized by the glucocorticoid receptor (and also androgen, progesterone and mineralocorticoid receptors). Note that most response elements are far from ideal; often one of the half-sites contains one or more mutations. Whereas PALs bind homodimers, DRs can bind homo- or heterodimers with the specificities given in the middle box (polarity of the receptors on their cognate DRs: left, 5'; right, 3'). Note that the 3'-positioned receptor makes minor grove DNA contacts in the spacer. Some ERs are response elements for homo- or heterodimers of the thyroid or retinoid receptors. Response elements are known which are composed of complex arrangements of the core motif.

motifs have generated response elements that are selective for a given (class of) receptor(s) (Fig. 2).

The DNA-Binding Domain Encompasses Region C

The highly conserved domain C (Fig.1) harbors the DNA-binding domain (DBD) of nuclear receptors, which confers sequence-specific DNA recognition. This domain has been extensively investigated, especially with respect to its selective response element recognition and dimerization properties. Several X-ray and NMR data sets are available for different nuclear receptor C domains in their DNA complexed and uncomplexed forms. The DNA-binding domain is mainly composed of two zinc-finger motifs, the N-terminal motif Cys-X_2-Cys-X_{13}-Cys-X_2-Cys (CI) and the C-terminal motif Cys-X_5-Cys-X_9-Cys-X_2-Cys (CII); in each motif two cysteine residues chelate one Zn^{2+} ion (Fig.3).

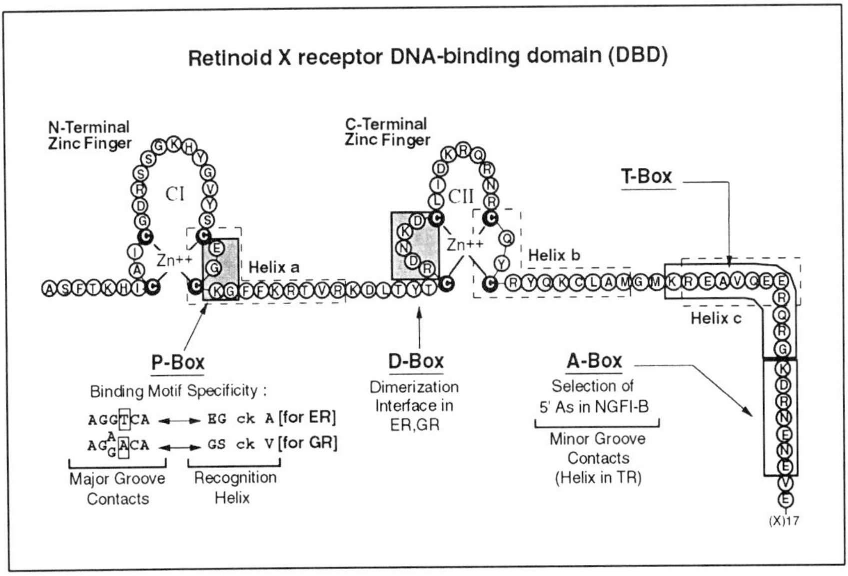

Figure 3. Schematic illustration of the RXR DNA-binding domain (DBD). Boxes illustrate regions involved in response element selection. The P-box is part of the DNA recognition helix (helix a; helices are framed by broken lines) and swapping of the EGA residues of the ER P-box with the corresponding GSV residues of GR switches ERE/GRE recognition as is outlined below the illustration. The D-box is responsible for PAL3/PAL0 selection by estrogen or thyroid hormone receptors and contributes to the homodimerization interfaces of ER and GR DBDs. The T-box region forms a helix (helix c) and corresponds to a dimerization surface in RXR homodimers. CI and CII are the two zinc-finger motifs. For further details see [1] and references therein.

Within the C domain several sequence elements (termed P-, D-, T- and A-boxes) have been characterized that define or contribute to i) response element specificity, ii) a dimerization interface within the DBD, and iii) contacts with the DNA backbone and residues flanking the DNA core recognition sequence. Fig. 3 displays a schematic illustration of the retinoid X receptor DNA-binding domain with an illustration of the different Zn^{2+} fingers and the various boxes (for details see the Figure legend). Fig. 4 illustrates the three prototypic DNA binding modes of nuclear receptors: the estrogen receptor DBD as example of a homodimer that binds to a palindromic response element (Fig. 4, top), the RXR-TR as an example for a anisotropic (5'-RXR-TR-3') heterodimeric complex on a DR4 direct repeat (Fig. 4, middle), and NGFI-B as example for a monomer that binds to an extended hexameric motif, the so-called NBRE (Fig. 4, bottom). The 3D structures of these DBDs all comprise a pair of amphipathic α-helices packed at right angles and crossing near their midpoints. A zinc-binding pocket lies near the N-terminus of each of the two helices. Hydrophobic side-chains form an extensive hydrophobic core between the two helices. The residues N-terminal to the first helix are folded to form two loops. Hydrophobic residues at the tips of the two loops pack with hydrophobic residues in the core between the two helices. Extensive contacts to the phosphate backbone on one side orient the DBD such that the DNA recognition helix enters the major groove, allowing surface side-chains to make sequence-specific contacts to the base pairs.

Homo- and Heterodimerization

Nuclear receptors can bind their cognate response elements as monomers, homodimers, or heterodimers with another family member (Fig. 4; for a review see Ref. [14]). Dimerization is a general mechanism to increase binding site affinity, specificity and diversity due to i) cooperative DNA binding, ii) the lower frequency of two hexamer binding motifs separated by a defined spacer compared to that of single hexamers, and iii) heterodimers may have recognition sites distinct from those of homodimers. Steroid hormone receptors bind generally as homodimers to their response elements, while RAR, RXR, TR and VDR can homo- and heterodimerize. RXRs play a central role in these various signal transduction pathways, since they can both homodimerize and act as promiscuous heterodimerization partner for RAR, TR, VDR and orphan receptors. Heterodimerization has a three-fold effect: it leads to a novel response element repertoire, increases the efficiency of DNA binding relative to the corresponding homodimers, and allows two signaling inputs, that of the ligands of RXR and its partner. Initially, biochemical studies have shown that two dimerization interfaces can be distinguished in nuclear receptors, a weak one by the DBDs and a strong one by the LBDs. LBDs dimerize in solution; the DBD interface is apparently only seen when bound to

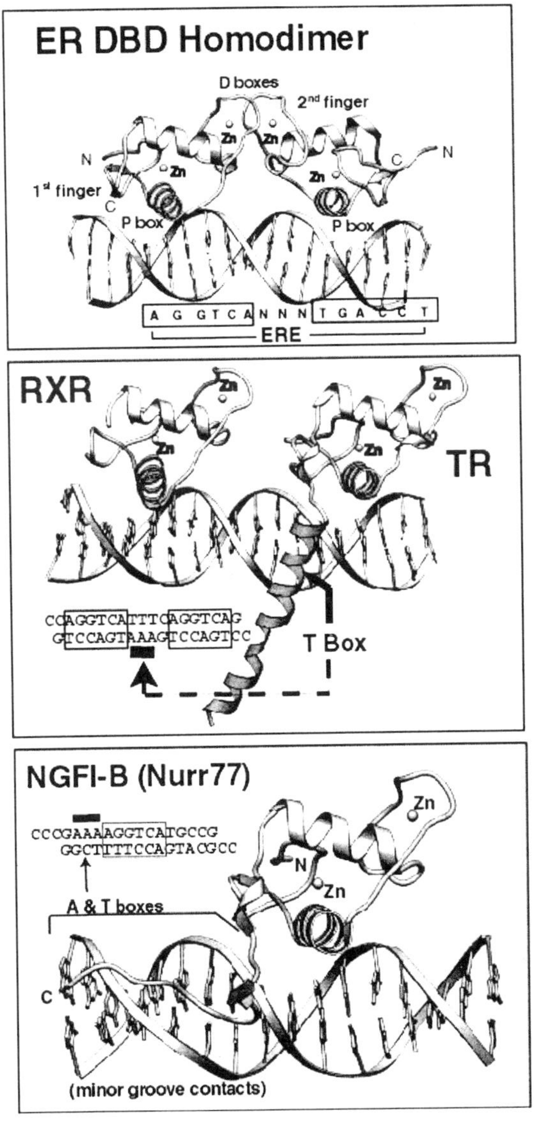

(Top) Structure of the estrogen receptor α (ER) DBD on an estrogen response element (ERE); one strand of the canonical ERE DNA sequence is given at the bottom. The various structural elements (Zn^{++} fingers, D and P boxes) are indicated. Note that the D-boxes form a DNA-induced DBD dimerization interface, while the P-box α-helices establish the selective base contacts in the major groove.

(Middle) Crystal structure of the 5'-RXR-TR-3' heterodimer on a cognate direct repeat response element spaced by 4 base pairs (DR4), depicted as a double-stranded DNA sequence at the bottom left. Note that the T-box makes minor groove contacts thus specifying to some degree the DR4 spacer nucleotide sequence (arrow).

(Bottom) Crystal structure of the monomeric NGFI-B on its response element (NBRE). The double-stranded NBRE sequence is given at the top left. Note that the A and T-boxes define the 5' AAA sequence that contacts the minor groove. This Figure is constructed from the PDB files 1HCQ (ER DBD homodimer), 2NLL (RXR-TR DBD heterodimer) and 1CIT (NGFI-B monomer).

Figure 4. Three-dimensional structures of the three prototype DNA-binding domains (DBDs)-DNA response element co-crystals.

DNA. Crystal structures of DBD homo- and heterodimers have defined the surfaces involved in dimerization [15]. It is important to point out that the response element repertoire described above for receptor homo- and heterodimers (Fig. 2) is dictated by the DBD, while the interface formed by the LBDs stabilizes the dimers, but does not play any role in response element selection.

Region D, a Hinge with Compartmentalization Functions

The D region is less conserved than the surrounding regions C and E. This domain appears to correspond to a "hinge" between the highly structured C and E domains. It might allow the DNA- and ligand-binding domains to adopt several different conformations without creating steric hindrance problems. Note in this respect that the C and the E regions contribute to dimerization interfaces allowing some receptors to accommodate different heterodimerization partners and different types of response elements. Region D contains a nuclear localization signal (NLS), or at least some elements of a functional NLS (reviewed in [1]). The intracellular localization of nuclear receptors is a result of a dynamic equilibrium between nuclear-cytoplasmatic and cytoplasmatic-nuclear shuttling [16]. At equilibrium the large majority of receptors is nuclear, although there is some controversy in the case of corticoid and androgen receptors, which have been reported to reside at cytoplasmic locations in the absence of their cognate ligands and translocate to the nucleus in a ligand-induced fashion [17, 18].

Region E Encompasses the Ligand-Binding Domain and Harbors Activation Function 2 (AF-2)

The hallmark of a nuclear receptor is its ligand-binding domain (LBD) in the E region. This domain is highly structured, and encodes a wealth of distinct functions most of which operate in a ligand-dependent manner. The LBD harbors the ligand-dependant activation function AF-2 and a major dimerization interface. In addition, some nuclear receptors possess a repression function in their LBD which can negatively regulate transcription of target genes in the absence of ligands or in the presence of a certain type of antagonists [19, 20].

The Canonical Fold of Nuclear Receptor Ligand-Binding Domains

Detailed molecular insights into the structure-function relation of signal integration by nuclear receptors have been obtained by the elucidation of the crystal structures of the E region alone or in presence of agonists, antagonists and coregulator peptides. The first 3D structures reported for nuclear receptor LBDs were those of the unliganded RXRα (Fig. 5a), the all-trans retinoic acid bound RARγ (Fig. 5b) and the agonist-bound thyroid receptor TRβ [21-23]. Unliganded receptors are frequently referred to as apo-receptor forms, while the liganded receptor corresponds to the holo-form. These structures demonstrated that apo and holo ligand-binding domains have a common fold. Moreover, a structure-based sequence alignment of all known nuclear receptor strongly supported a common fold for all nuclear receptor LBDs [24]. This hypothesis was fully confirmed when the crystal structures of multiple other NR LBDs were solved. The general fold of nuclear receptors consists of a three-layered α-helical sandwich (Fig. 5). Further structural features are one β-turn and connecting loops of variable lengths. The 12 helices, H1 (N-terminal) to H12 (C-terminal) form a hydrophobic cavity that accommodates hydrophobic ligands. Note that some variability exists; for example, no helix H2 was found in RARγ [22] while an additional short helix H2' is present in PPARγ [25].

The Regulation of the Activation Function AF-2 and the Mouse Trap Model

The different crystal structures of the apo- and holo- forms of receptor ligand-binding domains, as well as extensive biochemical studies demonstrated that the E domain undergoes a major conformational change upon ligand binding. A comparison of the apo- and holo-LBD structures (Fig. 5) suggested a common mechanism by which the activation function AF-2 becomes transcriptionally competent. The structural transition upon ligand binding has been described as the "mouse-trap" mechanism [19, 22, 24, 26]. Pushed by the ligand, helix H11 is repositioned in the continuity of helix H10, and the concomitant swinging of helix H12 unleashes the ω-loop which flips over underneath helix H6, carrying along the N-terminal part of helix H3. In its final position helix H12 seals the "lid" in the ligand-binding pocket and further stabilizes ligand binding by contributing to the hydrophobic environment and, in some cases by making additional contacts with the ligand itself. The repositioning of helix H12 back onto the core of the ligand-binding domain is apparent from the structures of the apo-RXRα and holo-RARγ LBDs shown in Fig. 5. The critical implication in transcription activation of helix H12, also referred to as the activating domain (AD) of the AF-2 function (the features of this AF-2 AD are conserved in all transcriptionally active receptors) was previously deduced from multiple mutagenesis studies (for a review see [1]).

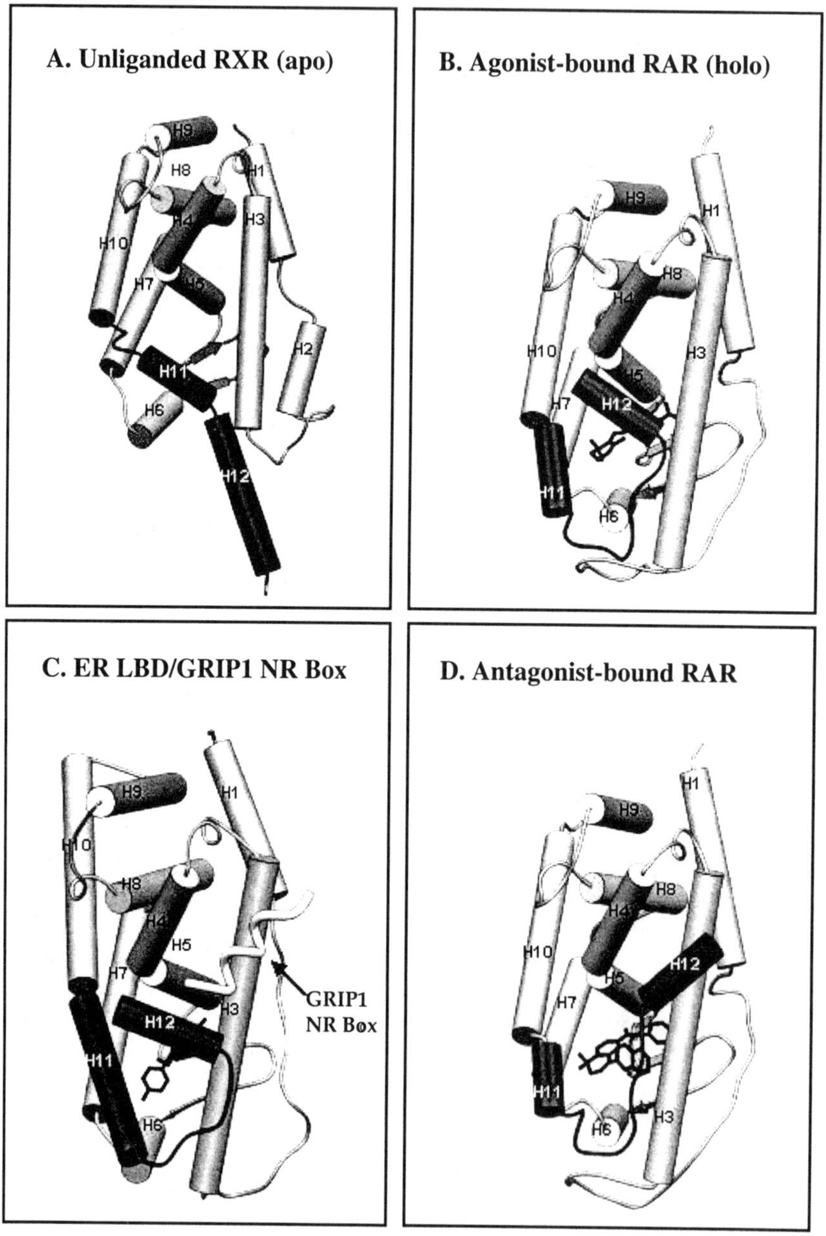
A. Unliganded RXR (apo)
H9
H8
H1
H4
H3
H10
H7
H5
H2
H11
H6
H12
B. Agonist-bound RAR (holo)
H9
H1
H8
H4
H10
H3
H5
H7
H12
H11
H6
C. ER LBD/GRIP1 NR Box
H9
H1
H10
H8
H4
H5
H7
H3
GRIP1
NR Box
H12
H11
H6
D. Antagonist-bound RAR
H1
H9
H8
H4
H10
H12
H5
H7
H11
H6
H3

←
Figure 5. Schematic illustration of the canonical 3D structures of nuclear receptor ligand-binding domains (LBDs). The canonical apo and holo structures of nuclear receptor LBDs are shown in (A) and (B), respectively. The holo structure allows interaction with the NR box of nuclear receptor coactivators, such as TIF2/GRIP1, SRC1, or RAC3/AIB1/ACTR (see text and reviews cited), as depicted in (C). This NR box binding site is occupied by helix H12 in the antagonist-bound LBD illustrated in (D). See also Fig. 6 for a close-up view of this part of the LBD.

Moreover, helix H12 not only contains residues critically required for transcriptional activation but also for coactivator recruitment (coactivators and their interaction with the LBD are discussed below). The structural data reveals that helix H12 when folded back onto the core of the ligand-binding domain forms a hydrophobic cleft together with other surface-exposed amino-acids that accommodates the "NR box" present in the nuclear receptor-interaction domain (NID) of coactivators [27, 28]. Crystal structure data have revealed the details of the interaction of the LxxLL NR box motif with the cognate surface on the ligand-binding domain (originally described by [25, 29, 30]). As an example the interaction of the TIF2/GRIP1 NR box 2 peptide with the agonist (diethylstilbestrol, DES)-bound estrogen receptor ligand-binding domain is illustrated in Fig. 6 (left). Note that helix H12 is crucial in stabilizing this interaction (for details see [30]). The corresponding crystal structure of the antagonist (tamoxifen)-bound estrogen receptor ligand-binding domain was particularly illuminating, as it demonstrated that this (type of) antagonist induces a positioning of H12 which is incompatible with the binding of coactivator NR boxes; indeed H12 binds to the identical cleft to which the NR box binds (compare the two structures in Fig. 6; see also [19, 30-32]). Hence, the information generated by the synthesis or secretion of a nuclear receptor ligand is converted first into a molecular recognition (ligand-receptor) process and subsequently into an allosteric event that allows the secondary recognition of a "downstream" mediator that is more closely connected to the cellular machineries involved in transcription activation.

Some nuclear receptors can also function as DNA-bound repressors of transcription. This phenomenon occurs in the absence of agonists, may be enhanced by certain antagonists, and is attributable to the recruitment of corepressors. Corepressor binding also occurs on the surface of the non-liganded LBD and is similar to coactivator binding in that it also is mediated by a short signature motif, the so-called "CoRNR box" which binds to a surface topologically related to that involved in coactivator interaction [33-35]. Upon ligand binding the transconformation of the α-helical sandwich displaces the corepressors by shifting the equilibrium of the E domain from the apo- to the holo-form.

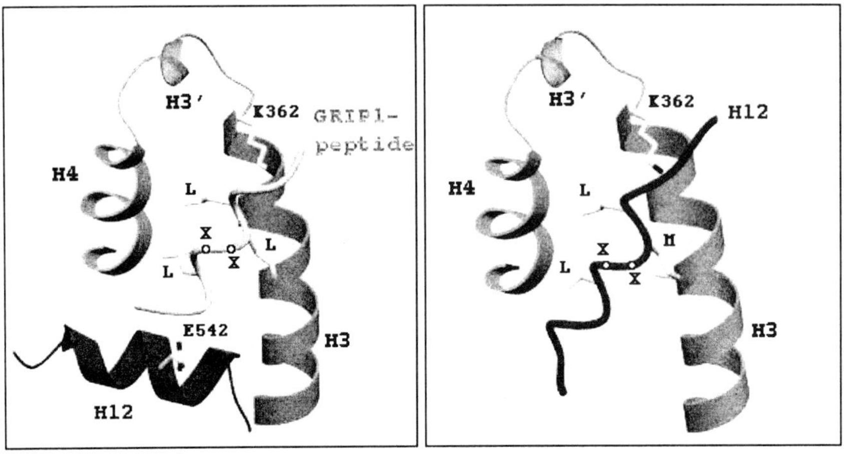

Figure 6. The peptide comprising the NR box of the TIF2/GRIP1 coactivator binds to a groove that is occupied by helix H12 in the antagonist structure. This Figure shows the interfaces between the ERα LBD and both the NR box peptide and helix H12. (Left) Close-up view of the coactivator (GRIP1/TIF2) peptide bound to the ERα/DES LBD complex. The NR box peptide is represented as a Cα "worm" with the leucine side chains of the LxxLL motif shown. For clarity, the hydrophobic residues of the groove interacting with the coactivator peptide have been removed. The two ERα residues K362 and E542, which define the "charge clamp", are depicted. (Right) Close-up view of the ERα/OHT antagonist complex showing helix H12 bound to the static part of the coactivator binding site. Helix H12 is represented as a Cα "worm" depicting the residue side chains that mimic the LxxLL motif of coactivators. This figure was constructed from PDB files 3ERD and 3ERT; see also [30].

The Dimerization Function

Some nuclear receptors form only homodimers (e.g., steroid hormone receptors) while others can form both homodimers (e.g. RXR, TR) and heterodimers with the promiscuous heterodimerization partner RXR (e.g. RAR-RXR, TR-RXR, VDR-RXR, PPAR-RXR, etc.). The ligand-binding domain is the major domain that contributes to dimerization. The recent crystallization of the RAR-RXR [32] and PPAR-RXR [36] ligand-binding domain heterodimers allowed for the first time to compare the homo- and heterodimerization interfaces of several nuclear receptor ligand-binding domains. In principle, the overall heterodimeric arrangement closely resembles that of homodimers of RXRα, ERα and PPARγ [21, 25, 31]. The interfaces involves residues from helices H7, H9, H10 and H11, as well as loops L8-9 and L9-10, with the helices H10 being parallel to the 2-fold axis relating the subunits. The observations that NR LBDs exhibit a common fold and that the dimerization interfaces of all crystallized LBD dimers are topologically conserved suggest that all NRs can establish similar dimer interfaces. Therefore, amino acid variations at the dimerization surface of the various NRs most probably determine their specific dimerization characteristics. An extensive analysis of the dimerization interfaces of nuclear receptor ligand-binding domains, based on crystal structure data and sequence alignment has been reported [32]. The reader is referred to this publication for details and an interpretation of the structural basis that accounts for the homo- and heterodimerization pattern of distinct members of this family. Based on *in vitro* studies it has also been reported that ligand binding can affect the dimerization properties of nuclear receptors. For the thyroid hormone and vitamin D receptors, which can both form homodimers and RXR heterodimers, ligand binding appears to favor heterodimerization [37, 38], but the structural basis of this ligand effect has not been fully elucidated.

The Role of the C-Terminal Region F Is Unknown

Some receptors possess, C-terminal of the ligand-binding domain, a region F that displays little evolutionary conservation. Note that the LBD is structurally defined as the domain generated by the elements between the beginning of helix H1 and the end of helix H12; this sequence is not necessarily identical to what is commonly referred to as region E from sequence alignments and also receptors like the progesterone receptor possess some kind of E region. This region is, however, much longer in the cases of, for example, estrogen and retinoic acid receptors. There are no clues as to the function of the C-terminal sequence. Recent literature suggests that the F region might play a role in coactivator recruitment to the E domain and in determining the specificity of the ligand-binding domain coactivator interface [39, 40]. It is clear that this domain also inherits little structural features. It is tempting to speculate that it sort of fine-tunes the molecular events associated with the transcriptional properties of the E domain, or the entire receptor, or as well it may affect antagonist action [41, 42].

As the A/B domains, also the E and F domains are targets of post-translational modifications, which adds another level of complexity to nuclear receptor signal integration since such events might influence the properties of the encoded functions.

MOLECULAR MECHANISMS OF TRANSCRIPTIONAL REGULATION

The reported effects of nuclear receptors on transcription are restricted to the initiation of transcription by RNA polymerase II. Below we will summarize our current knowledge of the molecular mechanisms/ interactions through which nuclear receptors can positively or negatively regulate the expression of cognate genes (For further details and references, see recent reviews [20,43]).

Chromatin Modifying Nuclear Receptor Coregulators

So-called squelching experiments paved the way to predict the existence of factors that would transmit the signal generated by the holo-receptor to the transcription machineries. Squelching occurs if a receptor inhibits the activity "off the DNA" of the same ("autosquelching", [7]) or a different ("heterosquelching", [44, 45]) receptor in an agonist- and AF-2 integrity-dependent manner. According to the squelching studies, *bona fide* coactivators were predicted to fulfill the following criteria: i) interact directly with nuclear receptor ligand binding domains in an agonist- and activation function-dependent manner, ii) enhance nuclear receptor- dependent reporter gene activity when transiently expressed in mammalian cells, iii) activate transcription autonomously when tethered to DNA *via* a heterologous DNA-binding domain, and iv) relieve squelching. The development of yeast "two hybrid" and direct cDNA expression library screening approaches has allowed identification of a great number of putative and *bona fide* of coactivators that are believed to transmit the nuclear receptor signal to its molecular targets. A summary of these coregulators, their characteristics and their putative mode of action is given in the reviews [20, 46, 47]. Below, we will discuss the general features of different classes of coregulators.

The cloning of coregulators was followed by the definition of the coactivator signature LxxLL NR box motifs (where x is any amino acid) embedded in a short α-helical peptide [27, 28, 48]. These NR boxes are necessary and sufficient for ligand-dependent direct interaction with a cognate surface in the nuclear receptor ligand-binding domain that constitutes the transcriptional activation function AF-2. This surface corresponds to a hydrophobic cleft with "charge clamps" (to which helix H12 contributes when repositioned on the surface of the ligand-binding domain upon agonist binding) which, together with surrounding amino acids, accommodates the amphipathic LxxLL NR box helix as has been revealed by X-ray crystallography (Fig. 6; for a review see [19]). The fact that some coactivators contain multiple LxxLL motifs (up to nine in RIP140), most of which appear to be functional in terms of nuclear receptor binding *in vitro*, brought up the still unsolved question of whether this multiplicity reflects redundancy or may confer some specificity to the interface. Indeed different coactivators, even highly related coactivators from the TIF2/SRC1/RAC3 family, display some degree of nuclear receptor selectivity [29, 49, 50].

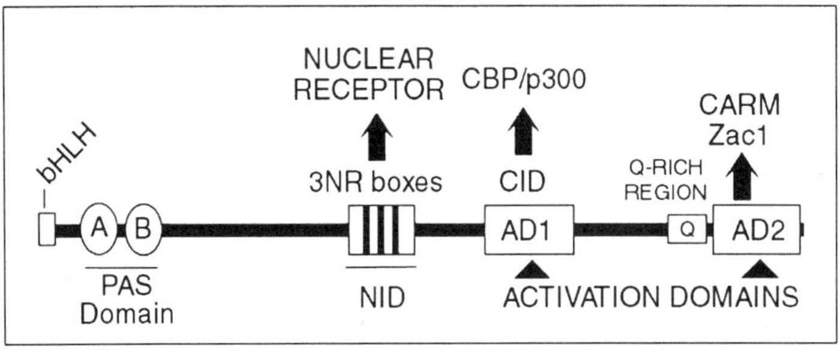

Figure 7. Scheme illustrating the functions associated with the TIF2/GRIP1 nuclear receptor coactivator. Three nuclear receptor boxes (NR boxes encompassing the canonical LxxLL motif) are crucial to the function of the nuclear receptor-interacting domain (NID); for a structural illustration of the NR box peptide-NR interaction see Fig. 6. The activation domain 1 (AD1) acts by recruiting the transcriptional co-modulators CBP/p300, while AD2 was shown to interact with the CARM methylase and/or the coactivator Zac1. The roles of a putative bHLH and PAS domain in the N-terminal region have remained elusive. The two other p160 NR coactivators, SRC1 and RAC3/AIB1/ACTR display similar structural organization.

However, these selectivities have still to be confirmed by *in vivo* studies. In this respect the analysis of different coactivator knockout mice will be particularly instructive. Initial analysis of the TIF2/GRIP1 (M. Gehin, P. Chambon and H.Gronemeyer, unpublished data), RAC3/AIB1/ACTR [51] and SCR1 [52] knockout mice do actually support the concept that *bona fide* coactivators are not fully redundant *in vivo*. The association of nuclear receptors and coactivators is generally believed to be the initiating step for the formation of large activator complexes which are formed due to the ability of coactivators recruit additional transcription factors such as CBP or p300 [20], CARM [53] or ZAC1 [54], through their activation domain AD1 and AD2 (Fig. 7). This complex is believed to mediate all [55] or at least some necessary events leading to transcription activation.

Recently, a second contact site for coactivators has been identified in nuclear receptors. Proteins from the TIF2 family are able to interact specifically with the A/B domains of estrogen and androgen receptors [10-13, 56, 57]. These interactions result in a stimulation of the transcriptional activity originating from AF-1. Moreover, it appears that simultaneous interaction of coactivators with both the AF-1 and the AF-2 of a nuclear receptor accounts for the synergy between both transcriptional activation functions [13, 58]. It is thus possible that the activation functions AF-1 and AF-2 may not in all cases be independent from one another in the context of the entire receptor, but rather constitute two separate regions of a single coactivator-recruiting surface. Hence, the coactivator specificity will have to be reanalyzed, particularly for receptors with proven AF-1 activities, taking into account the effect of the AF-1 binding surface(s). In this respect it should also be noted that for some nuclear receptors the N- and C-terminal domains interact with each other even in the absence of coactivators [10, 11, 59]. Motifs similar to the LxxLL of coactivators appear to mediate, at least in the case of AR, this N-C interaction [60]. Notably, it has been reported for ERβ, that specific phosphorylation of the A/B domain by MAPK pathways (see Chapter 3) results in enhanced recruitment of the coactivator SRC-1 [6].

It is now generally accepted that nuclear receptor coactivators possess or recruit enzymatic activities, and that they form large coactivator complexes. CBP, p300, P/CAF, and, more weakly some p160 coactivators themselves are reported to act as histone acetyltransferases (for recent reviews see [20, 43]). They are capable of acetylating specific residues in the N-terminal tails of different histones, a process that is believed to play an important role in the opening of chromatin during transcription activation, and also to act on non-histone targets [61, 62].

Recruitment of The RNA Polymerase II Holoenzyme: the Second Putative Step of Coactivation

The initial chromatin-modifying step carried out by nuclear receptor coactivators is believed to be followed be the actual recruitment of the RNA polymerase II holoenzyme and initiation of transcription (illustrated in Fig. 8 as the "transcription activation" step). Note, however, that for ER, it has been recently proposed that p160 coactivator interaction is sufficient for target gene activation [55]. While there may be receptor-specific differences in the use of coactivator complexes, comprehension of the recruitment of the polymerase II holoenzyme has been enhanced through the identification and cloning of the mammalian mediator complex as a thyroid hormone and vitamin D receptor-associated complex [63, 64]. The mammalian mediator was identified as the so-called SMCC, the Srb and mediator protein containing complex [65], the TRAP complex, a thyroid hormone receptor associated protein complex [63], or the DRIP complex, a vitamin D receptor interacting protein complex [64].

Figure 8. Hypothetical 3-step mechanism leading to transcriptional activation by nuclear receptors. Some nuclear receptors act as silencers of target gene transcription in the absence of ligand (or in the presence of certain antagonists). This "repression" step is due to the recruitment by the apo-NR of a corepressor complex that exerts histone deacetylase (HDAC) activity. Ligand binding dissociates this complex and recruits the coactivator complex that displays histone acetyltransferase (HAT) activity. The subsequent chromatin decondensation ("derepression") is believed to be necessary, but not sufficient, for target gene activation. At the third step the HAT complex dissociates, possibly due to acetylation of the coactivator which decreases its ability to interact with the receptor, and the SMCC/DRIP/TRAP complex is assembled through its TRAP220/DRIP205 subunit. The SMCC complex is able to establish contacts with the basal transcription machinery, resulting in transcription initiation. Note however, that it has been proposed that for some receptors p160 coactivator-mediated transactivation can occur without contribution of SMCC factors [55]. This illustration is modified from [66].

→

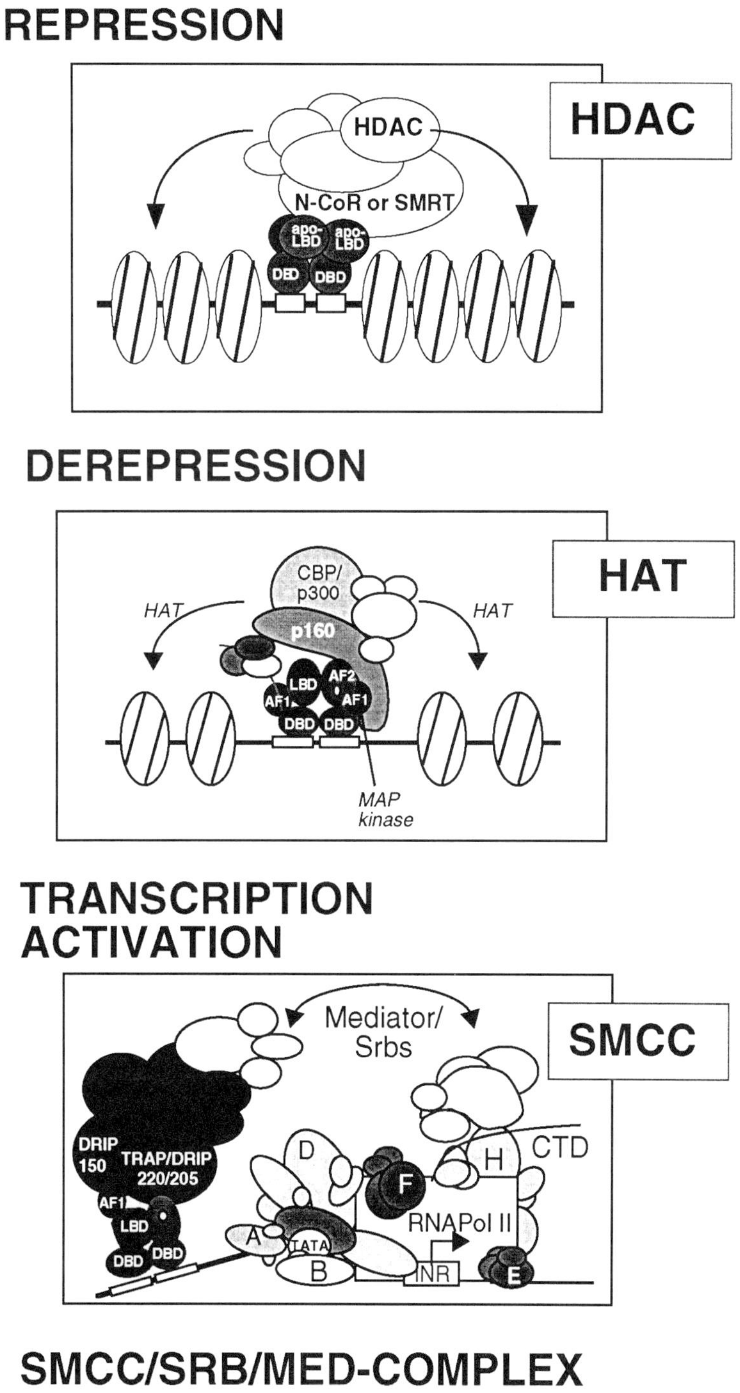
REPRESSION
HDAC
HDAC
N-CoR or SMRT
apo-LBD
apo-LBD
DBD
DBD
DEREPRESSION
HAT
CBP/p300
HAT
HAT
p160
LBD
AF2
AF1
AF1
DBD
DBD
MAP kinase
TRANSCRIPTION ACTIVATION
Mediator/Srbs
SMCC
DRIP 150
TRAP/DRIP 220/205
AF1
LBD
DBD
DBD
D
F
H
CTD
A
TATA
B
RNAPol II
INR
E
SMCC/SRB/MED-COMPLEX

Hereafter this complex is referred to as "SMCC". It is a large multisubunit protein complex that contains several homologues of the yeast mediator complex (RGR1, MED6, MED7, SRB7, SRB10, SRB11, NUT2, SOH1) [65, 67] as well as additional proteins of unknown function. As expected for a mediator complex, SMCC associates with the RNA polymerase II to form holoenzymes [67]. On the other hand, the SMCC complex is able to interact functionally with different transcription factors such as p53 and VP16 [67]. Furthermore, due to its identification as a thyroid hormone and vitamin D receptor-interacting complex it is believed to function as nuclear receptor coactivator. This notion is supported by the demonstration that SMCC can enhance thyroid hormone and vitamin D transcription activation in *in vitro* transcription systems [68, 69]. The subunit of the complex that is responsible for interaction with the agonist-bound ligand-binding domain of nuclear receptors was identified as DRIP205 [70] which is identical to TRAP220 and contains a functional LxxLL NR box motif [71]. Interestingly, another subunit of the SMCC complex, DRIP150, interacts with the N-terminal region of the glucocorticoid receptor, which harbors the activation function AF-1 [72]. The current working hypothesis is that once the chromatin environment at target gene promoters has been decondensed by coactivators complexes containing members of the TIF2 and CBP families, the nuclear receptor recruits RNA polymerase II holoenzymes *via* its association with the TRAP220/DRIP205 subunit of the SMCC. This switch between coactivators and the SMCC complex might be regulated by the acetylation of coactivators within the HAT complex [61] resulting in their dissociation from the nuclear receptor, thus allowing the recruitment of factors such as SMCC *via* the LxxLL motif of the TRAP220/DRIP205 subunit. A recent study tends to support this model and shows that interaction of TR with members of the p160 family occurs rapidly after addition of thyroid hormone followed by the formation of TR-TRAP complexes [73].

Nuclear Receptor Corepressors, Silencing and HDACs

The second class of nuclear receptor coregulators comprises corepressors. Early on, it had been learned that some nuclear receptors can repress transcription in the absence of agonists. This phenomenon had been particularly well established for retinoic acid and thyroid hormone receptors [74-76]. Soon after the identification of ligand-recruited coactivators by two hybrid or direct interactions cloning, similar approaches identified proteins that recognize the ligand-free nuclear receptor. To date several different nuclear receptor corepressors have been identified; by far most studies have been performed with NCoR and SMRT. For these two it was shown recently that a conserved CoRNR box motif interacts with a surface on the ligand-binding domain which is topologically very similar to that recognized by coactivator LxxLL motifs, but does not involve helix H12 [33-35]. Corepressors are believed to reside in, or recruit, high molecular weight complexes that display the opposite activity of coactivator complexes. While coactivator complexes acetylate histones thereby weakening the interaction of the N-terminal histone tails with the nucleosomal DNA, corepressors recruit histone deacetylase activities that reverse this process (illustrated in Fig. 8 as the "repression" step). Deacetylated histones are associated with silent regions of the genome, and it is generally accepted that histone acetylation and deacetylation shuffle nucleosomal targets between a condensed and relaxed chromatin configuration, the latter being requisite for transcriptional activation. An unresolved issue is whether all nuclear receptors are able of active repression.

RXR SUBORDINATION IN HETERODIMERS

While RAR agonists can autonomously activate transcription through RAR-RXR heterodimers, RXR is unable to respond to RXR-selective agonists in the absence of a RAR ligand. This phenomenon is generally referred to a RXR "subordination" or RXR "silencing" by apoRAR. However, in the presence of a RAR agonist or certain RAR antagonists, RXR agonists further stimulate the transcriptional activity of RAR-RXR heterodimers in a synergistic manner ([77] and references therein). RXR subordination solves a potential problem that could arise from the ability of RXR to act as a promiscuous heterodimerization partner for numerous nuclear receptors. This could create a problem for signaling pathway identity, since in the absence of subordination RXR ligands could activate simultaneously multiple heterodimeric receptors, such as RAR-RXR, TR-RXR and VDR-RXR and confuse retinoic acid, thyroid hormone and vitamin D3 signaling, respectively. However, RA-deprived animals do not exhibit abnormalities that could be readily related to impaired thyroid hormone or vitamin D3 signaling. Moreover, RXR-selective ligands on their own could not trigger RXR-RAR heterodimer-mediated RA-induced events in various cell systems. RAR apparently "controls" the activity of RXR-RAR heterodimers in two ways, it (i) induces transcription in response to its own ligand and (ii) silences RXR activity in the absence of an RAR ligand. Consequently, the only way for RXR to affect transactivation in response to its ligand in RXR-RAR heterodimers is through synergy with RAR ligands. This concept of RXR silencing may not apply to all NR partners, as the ligand-induced RXR activity was permissive in heterodimers with NGFI-B, leading even to a synergistic response [78, 79]. However, neither the existence of an endogenous NGFI-B ligand nor a weak constitutive activity of the NGFI-B AF-2 can be excluded; both of these scenarios would readily explain RXR activity and NGFI-B-RXR synergy due to the absence of RXR silencing. Note that the same argument would apply to the important observation of a seemingly RXR ligand-induced activity of RXR-PPAR heterodimers [80] which lead to insulin-sensitization by RXR agonists in an animal model of non insulin-dependent diabetes.

NUCLEAR RECEPTORS: PLATFORMS FOR MULTIPLE SIGNAL INTEGRATION

It has become increasingly well documented in the past few years that nuclear receptor action is not confined to the positive and negative regulation of the expression of cognate target genes. Indeed, these receptors, and most likely also their "downstream" mediators, are targets of other signaling pathways and reciprocally, can modify the activity of such pathways.

Signal Transduction Crosstalk Between Nuclear Receptors and AP1

In addition to transactivation of their own target genes, certain nuclear receptors were shown to crosstalk with other signal transduction pathways. The original observation was made in 1990 when it was observed that GR could inhibit, in a ligand-dependent manner, the ability of AP1 (the heterodimer composed of the proto-oncogene products c-Fos and c-Jun) to transactivate its target gene promoters [81-83]. This transrepression is mutual and requires an unknown state of the receptor which can be induced by both agonists and certain, but not all, antagonists (for reviews see [84, 85]). It is important to point out that the nuclear receptor-AP1 crosstalk does not *per se* imply negative regulation of transcription; several reports show that under certain conditions this crosstalk can lead to positive transcriptional effects [86-88]. AP1 transrepression has been studied intensively because of its likely implication in important events, such as anti-inflammation and cell proliferation. AP1 transrepression is found to be dependent on the promoter structure and on the dimer composition of the AP1 transcription factor [88]. Furthermore, AP1-nuclear receptor crosstalk is also cell type-dependent [86], which probably reflects the involvement of other cellular factors with varying expression as mediators of this mutual transrepression event. In addition to GR, mutual interference has been observed between the transcriptional activities of AP1 and RARs and RXRs, ER, TR/v-ErbA PML-RARα, while MR appears to be insensitive to AP1. Moreover, interference between the transactivation abilities of nuclear receptors and other transcription factors has been reported (reviewed in [84, 85]). The mechanism(s) on nuclear receptor-AP1 crosstalk has remained elusive despite, or because of, several contradictory reports and discrepancies between studies using *in vivo* and *in vitro* approaches.

NF*κ*B and Nuclear Receptor Crosstalk

The second best studied example of transcription factor crosstalk is the mutual interference between glucocorticoid receptors and NFκB proteins. This pathway is again of significant importance, since it may also contribute to the anti-inflammatory, as well as osteoporotic action of glucocorticoids. Albeit some have suggested that glucocorticoid action can be attributed to the increased production of the NFκB inhibitory molecule IκB which in turn would sequester active NFκB in the cell nucleus [89], studies with mutant receptors and "dissociated" glucocorticoids showing IκB-independent repression of NFκR activity are incompatible with a simple IκB-mediated mechanism [90]. The underlying molecular events are still elusive. Other transcription factors that are crossregulated in their activities by nuclear receptors are Oct 2A, RelA (another NFκB family member), STAT5 and Spi-1/PU.1 (for references to original work, see Ref. [84]).

CONCLUSIONS

Nuclear receptors are ligand-regulated transcription factors that have evolved from an ancestral orphan receptor into a highly diverse family present throughout the entire animal kingdom and encompassing receptors for steroid and non-steroid hormones, vitamins, corticoids and metabolic intermediates. These receptors signal through endocrine, paracrine, autocrine and intracrine modes of action to regulate multiple aspects of animal physiology, such as homeostasis, development and reproduction. They regulate target genes, to which they bind directly as either mono-, homo- or heterodimers at cognate response elements, and have the ability to indirectly modulate other gene expression programs ("signal transduction crosstalk"). Through the coordinated expression of genetic programs, nuclear receptors contribute to cell fate determining processes, thereby shaping and sustaining the organism. The signal - binding of the ligand - induces a major allosteric change which is transformed into cascades of protein-protein recognition paradigms involving coregulator and cointegrators proteins.

Direct transcriptional repression in the absence of ligand or the presence of certain antagonists by some nuclear receptors is mediated by corepressor complexes that are associated with the unliganded receptor and condense the chromatin environment at the promoter region through histone deacetylation. Upon ligand binding, the allosteric change in the ligand-binding domain induces corepressor dissociation, and coactivator complexes are recruited.

Bona fide coactivators recognize the active nuclear receptor ligand-binding domain (AF-2) *via* conserved LxxLL NR boxes and often the N-terminal activation function AF-1. The holo-H12 is required for coactivator but incompatible with corepressor binding. Coactivator complexes reverse the repressive effects of chromatin by specific histone acetylation, and allow access of the basal transcription machinery.

Despite their direct actions on the chromatin environment and the transcription machinery, nuclear receptors also regulate transcription by positive and negative interference with other signaling pathways. Different mechanisms for such transcription factor crosstalk have been described, but none is fully accepted and can explain all aspects of the particular crosstalk. The activity of nuclear receptors is regulated by phosphorylation that may serve to fine-tune the signaling or/and to establish a link to other signaling pathways. Finally, the promoter context, and the temporal order of incoming signals on a particular promoter have the likelihood of adjusting the transcriptional potential of nuclear receptors to particular situations. Taken together, nuclear receptors serve as platforms to coordinate cognate signals with those emanating from other signaling pathways, thereby integrating the nuclear receptor signal into the functional context of cellular state and activity.

Acknowledgements

The authors apologize to their colleagues for the many references that have not been cited due to space limitations. Studies in our laboratory are supported by funds from the INSERM, CNRS, Hôpital Universitaire de Strasbourg (HUS), the Fondation pour la Recherche Médicale, the Ministère de la Recherche et de la Technologie and Bristol-Myers-Squibb.

REFERENCES

1. **Gronemeyer H, Laudet V** 1995 Transcription factors 3: nuclear receptors. Protein Profile 2:1173-308
2. **Giguere V** 1999 Orphan nuclear receptors: from gene to function. Endocr Rev 20:689-725
3. **Escriva H, Safi R, Hanni C, Langlois MC, Saumitou-Laprade P, Stehelin D, Capron A, Pierce R, Laudet V** 1997 Ligand binding was acquired during evolution of nuclear receptors. Proc Natl Acad Sci U S A 94:6803-8
4. 1999 A unified nomenclature system for the nuclear receptor superfamily. Cell 97:161-3
5. **Shao D, Lazar MA** 1999 Modulating nuclear receptor function: may the phos be with you. J Clin Invest 103:1617-8
6. **Tremblay A, Tremblay GB, Labrie F, Giguere V** 1999 Ligand-independent recruitment of SRC-1 to estrogen receptor beta through phosphorylation of activation function AF-1. Mol Cell 3:513-9

7. **Bocquel MT, Kumar V, Stricker C, Chambon P, Gronemeyer H** 1989 The contribution of the N- and C-terminal regions of steroid receptors to activation of transcription is both receptor and cell-specific. Nucleic Acids Res 17:2581-95
8. **Berry M, Metzger D, Chambon P** 1990 Role of the two activating domains of the oestrogen receptor in the cell-type and promoter-context dependent agonistic activity of the anti-oestrogen 4-hydroxytamoxifen. Embo J 9:2811-8
9. **Sartorius CA, Melville MY, Hovland AR, Tung L, Takimoto GS, Horwitz KB** 1994 A third transactivation function (AF3) of human progesterone receptors located in the unique N-terminal segment of the B-isoform. Mol Endocrinol 8:1347-60
10. **Alen P, Claessens F, Verhoeven G, Rombauts W, Peeters B** 1999 The androgen receptor amino-terminal domain plays a key role in p160 coactivator-stimulated gene transcription. Mol Cell Biol 19:6085-97
11. **He B, Kemppainen JA, Voegel JJ, Gronemeyer H, Wilson EM** 1999 Activation function 2 in the human androgen receptor ligand binding domain mediates interdomain communication with the NH(2)-terminal domain. J Biol Chem 274:37219-25
12. **Bevan CL, Hoare S, Claessens F, Heery DM, Parker MG** 1999 The AF1 and AF2 domains of the androgen receptor interact with distinct regions of SRC1. Mol Cell Biol 19:8383-92
13. **Benecke A, Chambon P, Gronemeyer H** 2000 Synergy between estrogen receptor a activation functions AF1 and AF2 mediated by transcription intermediary factor TIF2. Embo Reports 1:151-7
14. **Glass CK** 1994 Differential recognition of target genes by nuclear receptor monomers, dimers, and heterodimers. Endocr Rev 15:391-407
15. **Gronemeyer H, Moras D** 1995 Nuclear receptors. How to finger DNA [news; comment]. Nature 375:190-1
16. **Tyagi RK, Amazit L, Lescop P, Milgrom E, Guiochon-Mantel A** 1998 Mechanisms of progesterone receptor export from nuclei: role of nuclear localization signal, nuclear export signal, and ran guanosine triphosphate. Mol Endocrinol 12:1684-95
17. **Baumann CT, Lim CS, Hager GL** 1999 Intracellular localization and trafficking of steroid receptors. Cell Biochem Biophys 31:119-27
18. **Tyagi RK, Lavrovsky Y, Ahn SC, Song CS, Chatterjee B, Roy AK** 2000 Dynamics of intracellular movement and nucleocytoplasmic recycling of the ligand-activated androgen receptor in living cells. Mol Endocrinol 14:1162-74
19. **Bourguet W, Germain P, Gronemeyer H** 2000 Nuclear receptor ligand-binding domains: three-dimensional structures, molecular interactions and pharmacological implications. Trends Pharmacol Sci 21:381-8
20. **Glass CK, Rosenfeld MG** 2000 The coregulator exchange in transcriptional functions of nuclear receptors. Genes Dev 14:121-41
21. **Bourguet W, Ruff M, Chambon P, Gronemeyer H, Moras D** 1995 Crystal structure of the ligand-binding domain of the human nuclear receptor RXR-alpha. Nature 375:377-82
22. **Renaud JP, Rochel N, Ruff M, Vivat V, Chambon P, Gronemeyer H, Moras D** 1995 Crystal structure of the RAR-gamma ligand-binding domain bound to all- trans retinoic acid. Nature 378:681-9
23. **Wagner RL, Apriletti JW, McGrath ME, West BL, Baxter JD, Fletterick RJ** 1995 A structural role for hormone in the thyroid hormone receptor. Nature 378:690-7
24. **Wurtz JM, Bourguet W, Renaud JP, Vivat V, Chambon P, Moras D, Gronemeyer H** 1996 A canonical structure for the ligand-binding domain of nuclear receptors. Nat Struct Biol 3:206
25. **Nolte RT, Wisely GB, Westin S, Cobb JE, Lambert MH, Kurokawa R, Rosenfeld MG, Willson TM, Glass CK, Milburn MV** 1998 Ligand binding and co-activator assembly of the peroxisome proliferator- activated receptor-gamma. Nature 395:137-43

26. **Moras D, Gronemeyer H** 1998 The nuclear receptor ligand-binding domain: structure and function. Curr Opin Cell Biol 10:384-91
27. **Le Douarin B, Nielsen AL, Garnier JM, Ichinose H, Jeanmougin F, Losson R, Chambon P** 1996 A possible involvement of TIF1 alpha and TIF1 beta in the epigenetic control of transcription by nuclear receptors. Embo J 15:6701-15
28. **Heery DM, Kalkhoven E, Hoare S, Parker MG** 1997 A signature motif in transcriptional co-activators mediates binding to nuclear receptor. Nature 387:733-736
29. **Darimont BD, Wagner RL, Apriletti JW, Stallcup MR, Kushner PJ, Baxter JD, Fletterick RJ, Yamamoto KR** 1998 Structure and specificity of nuclear receptor-coactivator interactions. Genes Dev 12:3343-56
30. **Shiau AK, Barstad D, Loria PM, Cheng L, Kushner PJ, Agard DA, Greene GL** 1998 The structural basis of estrogen receptor/coactivator recognition and the antagonism of this interaction by tamoxifen. Cell 95:927-37
31. **Brzozowski AM, Pike AC, Dauter Z, Hubbard RE, Bonn T, Engstrom O, Ohman L, Greene GL, Gustafsson JA, Carlquist M** 1997 Molecular basis of agonism and antagonism in the oestrogen receptor. Nature 389:753-8
32. **Bourguet W, Vivat V, Wurtz JM, Chambon P, Gronemeyer H, Moras D** 2000 Crystal structure of a heterodimeric complex of RAR and RXR ligand- binding domains. Mol Cell 5:289-98
33. **Hu X, Lazar MA** 1999 The CoRNR motif controls the recruitment of corepressors by nuclear hormone receptors. Nature 402:93-6
34. **Nagy L, Kao HY, Love JD, Li C, Banayo E, Gooch JT, Krishna V, Chatterjee K, Evans RM, Schwabe JW** 1999 Mechanism of corepressor binding and release from nuclear hormone receptors. Genes Dev 13:3209-16
35. **Perissi V, Staszewski LM, McInerney EM, Kurokawa R, Krones A, Rose DW, Lambert MH, Milburn MV, Glass CK, Rosenfeld MG** 1999 Molecular determinants of nuclear receptor-corepressor interaction. Genes Dev 13:3198-208
36. **Gampe RT, Jr., Montana VG, Lambert MH, Miller AB, Bledsoe RK, Milburn MV, Kliewer SA, Willson TM, Xu HE** 2000 Asymmetry in the PPARgamma/RXRalpha crystal structure reveals the molecular basis of heterodimerization among nuclear receptors. Mol Cell 5:545-55
37. **Cheskis B, Freedman LP** 1994 Ligand modulates the conversion of DNA-bound vitamin D3 receptor (VDR) homodimers into VDR-retinoid X receptor heterodimers. Mol Cell Biol 14:3329-38
38. **Collingwood TN, Butler A, Tone Y, Clifton-Bligh RJ, Parker MG, Chatterjee VK** 1997 Thyroid hormone-mediated enhancement of heterodimer formation between thyroid hormone receptor beta and retinoid X receptor. J Biol Chem 272:13060-5
39. **Peters GA, Khan SA** 1999 Estrogen receptor domains E and F: role in dimerization and interaction with coactivator RIP-140. Mol Endocrinol 13:286-96
40. **Sladek FM, Ruse MD, Jr., Nepomuceno L, Huang SM, Stallcup MR** 1999 Modulation of transcriptional activation and coactivator interaction by a splicing variation in the F domain of nuclear receptor hepatocyte nuclear factor 4alpha1. Mol Cell Biol 19:6509-22
41. **Montano MM, Muller V, Trobaugh A, Katzenellenbogen BS** 1995 The carboxy-terminal F domain of the human estrogen receptor: role in the transcriptional activity of the receptor and the effectiveness of antiestrogens as estrogen antagonists. Mol Endocrinol 9:814-25
42. **Nichols M, Rientjes JM, Stewart AF** 1998 Different positioning of the ligand-binding domain helix 12 and the F domain of the estrogen receptor accounts for functional differences between agonists and antagonists. Embo J 17:765-73
43. **Chen JD** 2000 Steroid/nuclear receptor coactivators. Vitam Horm 58:391-448

44. **Meyer ME, Gronemeyer H, Turcotte B, Bocquel MT, Tasset D, Chambon P** 1989 Steroid hormone receptors compete for factors that mediate their enhancer function. Cell 57:433-42
45. **Tasset D, Tora L, Fromental C, Scheer E, Chambon P** 1990 Distinct classes of transcriptional activating domains function by different mechanisms. Cell 62:1177-87
46. **Collingwood TN, Urnov FD, Wolffe AP** 1999 Nuclear receptors: coactivators, corepressors and chromatin remodeling in the control of transcription. J Mol Endocrinol 23:255-75
47. **McKenna NJ, Lanz RB, O'Malley BW** 1999 Nuclear receptor coregulators: cellular and molecular biology. Endocr Rev 20:321-44
48. **Torchia J, Rose DW, Inostroza J, Kamei Y, Westin S, Glass CK, Rosenfeld MG** 1997 The transcriptional co-activator p/CIP binds CBP and mediates nuclear- receptor function. Nature 387:677-84
49. **Ding XF, Anderson CM, Ma H, Hong H, Uht RM, Kushner PJ, Stallcup MR** 1998 Nuclear receptor-binding sites of coactivators glucocorticoid receptor interacting protein 1 (GRIP1) and steroid receptor coactivator 1 (SRC- 1): multiple motifs with different binding specificities. Mol Endocrinol 12:302-13
50. **McInerney EM, Rose DW, Flynn SE, Westin S, Mullen TM, Krones A, Inostroza J, Torchia J, Nolte RT, Assa-Munt N, Milburn MV, Glass CK, Rosenfeld MG** 1998 Determinants of coactivator LXXLL motif specificity in nuclear receptor transcriptional activation. Genes Dev 12:3357-68
51. **Xu J, Liao L, Ning G, Yoshida-Komiya H, Deng C, O'Malley BW** 2000 The steroid receptor coactivator SRC-3 (p/CIP/RAC3/AIB1/ACTR/TRAM-1) is required for normal growth, puberty, female reproductive function, and mammary gland development. Proc Natl Acad Sci U S A 97:6379-84
52. **Xu J, Qiu Y, DeMayo FJ, Tsai SY, Tsai MJ, O'Malley BW** 1998 Partial hormone resistance in mice with disruption of the steroid receptor coactivator-1 (SRC-1) gene. Science 279:1922-5
53. **Chen D, Huang SM, Stallcup MR** 2000 Synergistic, p160 coactivator-dependent enhancement of estrogen receptor function by CARM1 and p300. J Biol Chem 275:40810-6
54. **Huang SM, Stallcup MR** 2000 Mouse Zac1, a transcriptional coactivator and repressor for nuclear receptors. Mol Cell Biol 20:1855-67
55. **Shang Y, Hu X, DiRenzo J, Lazar MA, Brown M** 2000 Cofactor dynamics and sufficiency in estrogen receptor-regulated transcription. Cell 103:843-52
56. **Berrevoets CA, Doesburg P, Steketee K, Trapman J, Brinkmann AO** 1998 Functional interactions of the AF-2 activation domain core region of the human androgen receptor with the amino-terminal domain and with the transcriptional coactivator TIF2 (transcriptional intermediary factor2). Mol Endocrinol 12:1172-83
57. **Ma H, Hong H, Huang SM, Irvine RA, Webb P, Kushner PJ, Coetzee GA, Stallcup MR** 1999 Multiple signal input and output domains of the 160-kilodalton nuclear receptor coactivator proteins. Mol Cell Biol 19:6164-73
58. **McInerney EM, Tsai MJ, O'Malley BW, Katzenellenbogen BS** 1996 Analysis of estrogen receptor transcriptional enhancement by a nuclear hormone receptor coactivator. Proc Natl Acad Sci U S A 93:10069-73
59. **Tetel MJ, Giangrande PH, Leonhardt SA, McDonnell DP, Edwards DP** 1999 Hormone-dependent interaction between the amino- and carboxyl-terminal domains of progesterone receptor in vitro and in vivo. Mol Endocrinol 13:910-24
60. **He B, Kemppainen JA, Wilson EM** 2000 FXXLF and WXXLF sequences mediate the NH2-terminal interaction with the ligand binding domain of the androgen receptor. J Biol Chem 275:22986-94

61. **Chen H, Lin RJ, Xie W, Wilpitz D, Evans RM** 1999 Regulation of hormone-induced histone hyperacetylation and gene activation via acetylation of an acetylase. Cell 98:675-86
62. **Imhof A, Yang XJ, Ogryzko VV, Nakatani Y, Wolffe AP, Ge H** 1997 Acetylation of general transcription factors by histone acetyltransferases. Curr Biol 7:689-92
63. **Fondell JD, Ge H, Roeder RG** 1996 Ligand induction of a transcriptionally active thyroid hormone receptor coactivator complex. Proc Natl Acad Sci USA 93:8329-8333
64. **Rachez C, Suldan Z, Ward J, Chang CP, Burakov D, Erdjument-Bromage H, Tempst P, Freedman LP** 1998 A novel protein complex that interacts with the vitamin D3 receptor in a ligand-dependent manner and enhances VDR transactivation in a cell-free system. Genes Dev 12:1787-800
65. **Gu W, Malik S, Ito M, Yuan CX, Fondell JD, Zhang X, Martinez E, Qin J, Roeder RG** 1999 A novel human SRB/MED-containing cofactor complex, SMCC, involved in transcription regulation]. Mol Cell 3:97-108
66. **Freedman LP** 1999 Increasing the complexity of coactivation in nuclear receptor signaling. Cell 97:5-8
67. **Ito M, Yuan CX, Malik S, Gu W, Fondell JD, Yamamura S, Fu ZY, Zhang X, Qin J, Roeder RG** 1999 Identity between TRAP and SMCC complexes indicates novel pathways for the function of nuclear receptors and diverse mammalian activators. Mol Cell 3:361-70
68. **Fondell JD, Guermah M, Malik S, Roeder RG** 1999 Thyroid hormone receptor-associated proteins and general positive cofactors mediate thyroid hormone receptor function in the absence of the TATA box-binding protein-associated factors of TFIID. Proc Natl Acad Sci U S A 96:1959-64
69. **Rachez C, Lemon BD, Suldan Z, Bromleigh V, Gamble M, Naar AM, Erdjument-Bromage H, Tempst P, Freedman LP** 1999 Ligand-dependent transcription activation by nuclear receptors requires the DRIP complex. Nature 398:824-8
70. **Yang W, Freedman LP** 1999 20-Epi analogues of 1,25-dihydroxyvitamin D3 are highly potent inducers of DRIP coactivator complex binding to the vitamin D3 receptor. J Biol Chem 274:16838-45
71. **Yuan CX, Ito M, Fondell JD, Fu ZY, Roeder RG** 1998 The TRAP220 component of a thyroid hormone receptor- associated protein (TRAP) coactivator complex interacts directly with nuclear receptors in a ligand-dependent fashion. Proc Natl Acad Sci U S A 95:7939-44
72. **Hittelman AB, Burakov D, Iniguez-Lluhi JA, Freedman LP, Garabedian MJ** 1999 Differential regulation of glucocorticoid receptor transcriptional activation via AF-1-associated proteins. Embo J 18:5380-8
73. **Sharma D, Fondell JD** 2000 Temporal formation of distinct thyroid hormone receptor coactivator complexes in HeLa cells. Mol Endocrinol 14:2001-9
74. **Baniahmad A, Tsai SY, O'Malley BW, Tsai MJ** 1992 Kindred S thyroid hormone receptor is an active and constitutive silencer and a repressor for thyroid hormone and retinoic acid responses. Proc Natl Acad Sci U S A 89:10633-7
75. **Chen JD, Evans RM** 1995 A transcriptional co-repressor that interacts with nuclear hormone receptors. Nature 377:454-7
76. **Horlein AJ, Naar AM, Heinzel T, Torchia J, Gloss B, Kurokawa R, Ryan A, Kamei Y, Soderstrom M, Glass CK, et al.** 1995 Ligand-independent repression by the thyroid hormone receptor mediated by a nuclear receptor co-repressor. Nature 377:397-404
77. **Chen JY, Clifford J, Zusi C, Starrett J, Tortolani D, Ostrowski J, Reczek PR, Chambon P, Gronemeyer H** 1996 Two distinct actions of retinoid-receptor ligands. Nature 382:819-22

78. **Forman BM, Tontonoz P, Chen J, Brun RP, Spiegelman BM, Evans RM** 1995 15-Deoxy-delta 12, 14-prostaglandin J2 is a ligand for the adipocyte determination factor PPAR gamma. Cell 83:803-12
79. **Perlmann T, Jansson L** 1995 A novel pathway for vitamin A signaling mediated by RXR heterodimerization with NGFI-B and NURR1. Genes Dev 9:769-82
80. **Mukherjee R, Davies PJ, Crombie DL, Bischoff ED, Cesario RM, Jow L, Hamann LG, Boehm MF, Mondon CE, Nadzan AM, Paterniti JR, Jr., Heyman RA** 1997 Sensitization of diabetic and obese mice to insulin by retinoid X receptor agonists. Nature 386:407-10
81. **Jonat C, Rahmsdorf HJ, Park KK, Cato AC, Gebel S, Ponta H, Herrlich P** 1990 Antitumor promotion and antiinflammation: down-modulation of AP-1 (Fos/Jun) activity by glucocorticoid hormone. Cell 62:1189-204
82. **Yang-Yen HF, Chambard JC, Sun YL, Smeal T, Schmidt TJ, Drouin J, Karin M** 1990 Transcriptional interference between c-Jun and the glucocorticoid receptor: mutual inhibition of DNA binding due to direct protein-protein interaction. Cell 62:1205-15
83. **Schule R, Umesono K, Mangelsdorf DJ, Bolado J, Pike JW, Evans RM** 1990 Jun-Fos and receptors for vitamins A and D recognize a common response element in the human osteocalcin gene. Cell 61:497-504
84. **Gottlicher M, Heck S, Herrlich P** 1998 Transcriptional cross-talk, the second mode of steroid hormone receptor action. J Mol Med 76:480-9
85. **Resche-Rigon M, Gronemeyer H** 1998 Therapeutic potential of selective modulators of nuclear receptor action. Curr Opin Chem Biol 2:501-7
86. **Shemshedini L, Knauthe R, Sassone-Corsi P, Pornon A, Gronemeyer H** 1991 Cell-specific inhibitory and stimulatory effects of Fos and Jun on transcription activation by nuclear receptors. Embo J 10:3839-49
87. **Bubulya A, Wise SC, Shen XQ, Burmeister LA, Shemshedini L** 1996 c-Jun can mediate androgen receptor-induced transactivation. J Biol Chem 271:24583-24589
88. **Pearce D, Matsui W, Miner JN, Yamamoto KR** 1998 Glucocorticoid receptor transcriptional activity determined by spacing of receptor and nonreceptor DNA sites. J Biol Chem 273:30081-5
89. **Auphan N, DiDonato JA, Rosette C, Helmberg A, Karin M** 1995 Immunosuppression by glucocorticoids: inhibition of NF-kappa B activity through induction of I kappa B synthesis. Science 270:286-90
90. **Heck S, Bender K, Kullmann M, Gottlicher M, Herrlich P, Cato AC** 1997 I kappaB alpha-independent downregulation of NF-kappaB activity by glucocorticoid receptor. Embo J 16:4698-707

13

Estrogen Receptor Beta:

How Awareness of ERβ Affects Our Understanding of Estrogen Action

Zhang Weihua, Guojun Cheng, Ling Wang, Hideki Sakaguchi, Margaret Warner, Jan-Åke Gustafsson*

Department of Medical Nutrition and Department of Bioscience, Karolinska Institute, NOVUM, Huddinge University Hospital, SE-14186 Huddinge, Sweden

INTRODUCTION

When the discovery of ERβ was reported in 1996 [1], many endocrinologists thought that, since this receptor had gone unnoticed for so long, it must be some sort of vestigial receptor with no function in the endocrinology of estrogen. This idea was quickly dispelled when ERβ$^{-/-}$ mice were created and found to have severely compromised ovarian function and prostatic hyperplasia [2]. Later, two other ERβ$^{-/-}$ mouse lines were created independently in two other laboratories and all three mouse lines were shown to exhibit severe ovarian dysfunction [3-5]. Moreover since the ovarian phenotype in ERβ$^{-/-}$ mice is distinctly different from that in ERα$^{-/-}$ mice [3, 6], it became clear that ERα and ERβ have distinct roles in the body. There is, at present, some argument between different labs concerning other aspects of the ERβ$^{-/-}$ phenotype. If biology was as straight forward as physics, where for each action there is an equal and opposite reaction, understanding knock out phenotypes would be easy. Unfortunately, in endocrinology, it is impossible to make a single alteration without eliciting a cascade of responses.

* jan-ake.gustafsson@mednut.ki.se

Since ERβ is clearly a functional receptor, it is to be expected that in ERβ$^{-/-}$ mice there should be a phenotype in tissues in which this receptor is highly expressed. These tissues include the prostate, lung, central nervous system (CNS), mammary gland, uterus, spleen and bone. Exactly what kind of phenotype would be exhibited in these different tissues in the absence of ERβ, depends on what the functions of ERβ are in the tissue and exactly how the individual tissue responds to loss of ERβ. Even more complex phenotypes are to be expected in tissues such as bone, mammary gland, cardiovascular system, uterus and some regions of the CNS, where both receptors are expressed. Since transcription of many genes is regulated by both ERα and ERβ, but in opposite directions, loss of ERβ should result in exaggerated ERα responses to estrogen in some tissues. In this chapter, we will describe some of the more important discoveries about ERβ which have come from molecular biologists, biochemists and endocrinologists and try to explain how this information helps us to understand the functions of ERβ and the endocrinology of the ERβ$^{-/-}$ mouse.

Although ERα and ERβ are both estrogen receptors which bind estradiol with quite similar affinity and activate transcription of genes which harbor estrogen response elements (EREs) in their promotors, there are very important differences between these receptors, which suggest that they should have distinct functions in the body. The most important of these differences are:

1. Structural differences [7-9] including the presence of functional ERβ splice variants [10-17].
2. Differences in ligand specificity and affinity [18-20].
3. Distinct patterns of tissue distribution [21, 22].
4. Independent and frequently opposing interactions with other transcription pathways (reviewed in Ref. [23]).

REDUCED ACTIVITY OF ACTIVATION FUNCTION-1 (AF-1) SITE IN ERβ

One of the differences between ERα and ERβ which results in clear differences in the behavior of the two receptors, is the weak activity of activation domain AF-1 (see Chapter 12), in ERβ [24]. This is a ligand-independent site which is one surface for interaction of the receptor with other proteins. Both receptors have AF-2 domains, a ligand-dependent site of protein-protein interaction (for a review, see Ref. [9]). The important implication of the difference in AF-1 function is that ERα has the capacity to interact with certain proteins with which ERβ will not be able to interact. One very important pathway in this regard is that of growth factor signaling [25, 26]. MAP kinase, activated by growth factors, phosphorylates the Ser118 residue of the human ERα and this phosphorylation allows coactivators to associate with the AF-1 site and potentiates the transactivation function of the receptor. One can therefore anticipate that the two estrogen receptors will exhibit differences in their capacity to cross-talk with growth factor signaling pathways such as EGF (see Chapter 4) and IGF (see Chapter 5) and that this would lead to differences in growth control of cells which express one or the other of these receptors.

NON-ERE DEPENDENT TRANSCRIPTIONAL REGULATION BY ESTROGEN RECEPTORS

Both estrogen receptors can influence transcription regulated by AP-1 [25-28] and Sp-1 [29] sites in a manner that does not require DNA binding by the receptors themselves. In the presence of estrogen, ERα activates transcription from AP-1 and Sp-1 sites whereas ERβ is silent. In the presence of antiestrogens, however, ERβ can activate transcription from these sites. The promotor of the retinoic acid receptor-alpha-1 (RAR α; see Chapter 12) is one example of a site where ERs act without binding directly to DNA and where ERα and ERβ have opposite actions [29]. In this promotor, the interaction occurs at Sp-1 sites. As with the AP-1 site, ERα activates transcription from the RAR α promotor in the presence of estrogen, but ERβ does not. In the presence of 4-hydroxytamoxifen, ICI-164,384 and ICI-182,780, ERβ does activate this promotor and this activation of ERβ by "antagonists" is blocked by estradiol. What the difference between ERα and ERβ means is that antiestrogens can turn on RARα in ERβ-containing cells. Since RARα in the presence of retinoids can inhibit cellular growth, it appears that tamoxifen in combination with retinoids might be beneficial in breast cancer, even those which do not express ERα. In addition, unless the body compensates for loss of ERβ, there should be marked differences between wild type and ERβ$^{-/-}$ mice in their response to tamoxifen.

REGULATION OF THE ANDROGEN RECEPTOR BY ESTROGEN

In many studies where estradiol and testosterone have similar effects in specific organs, it has been logical to assume that the reason for this is the conversion of testosterone to estradiol *via* the enzyme aromatase within the organ. However, in some systems, non-aromatizable androgens can also mimic estradiol [30, 31] and one possible explanation for similar effects of androgens and estrogens is that ER-AR constitute two sequential steps in a single pathway where ER regulates AR expression. We have found that AR is induced by E_2 in the rodent uterus and this induction is mediated by ERα [32]. This is exactly the opposite of what occurs in the prostate epithelium where the only estrogen receptor is ERβ and where estrogens down-regulate AR [33].

In the immature uterus, ERα colocalizes with AR in the myometrium (Fig. 1) while ERβ is mostly expressed in the stromal cells surrounding the lumen (Fig. 2) and no AR is expressed in these ERβ-containing cells.

Since both flutamide and tamoxifen can block the estrogen induction of IGF-1 and cyclin A (Figs. 3 and 4), it appears that AR mediates some of the proliferative responses to E2 in the uterus, and this implies that this is a pathway where there is sequential activation of ERα followed by AR induction. E_2-induced proliferation of uterine epithelial cells is controlled by growth factors from the stroma and it is ERα in the stroma, but not in the epithelium, which is responsible for this effect [34].

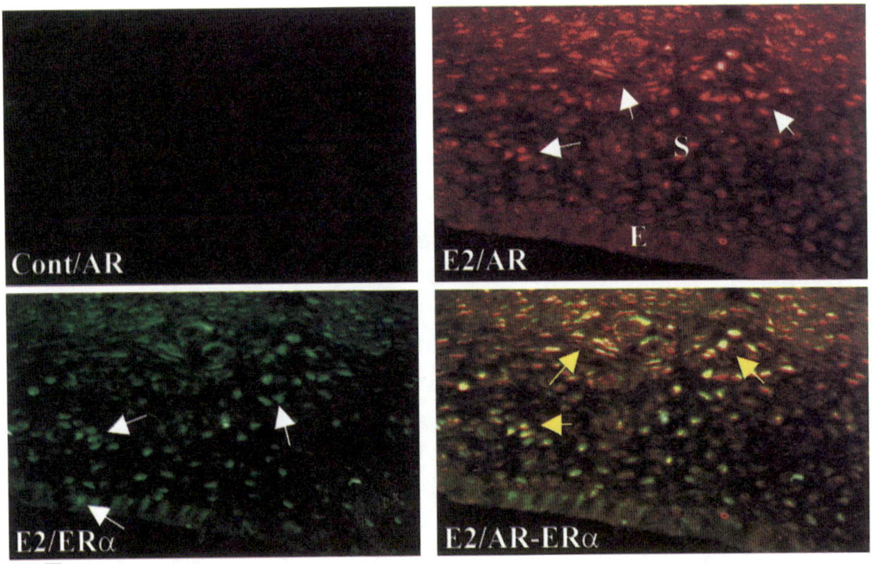

Figure 1. Colocalisation of ERα and AR in the rat uterus by immunohistochemistry with fluorescence detection. FITC-conjugated anti-mouse IgG (green), detects ERα and Cy3-conjugated anti-rabbit IgG (red), detects AR. Cells in which ERα and AR are colocalized are stained yellow or orange (yellow arrows). Symbols: E, epithelium; S, stroma.

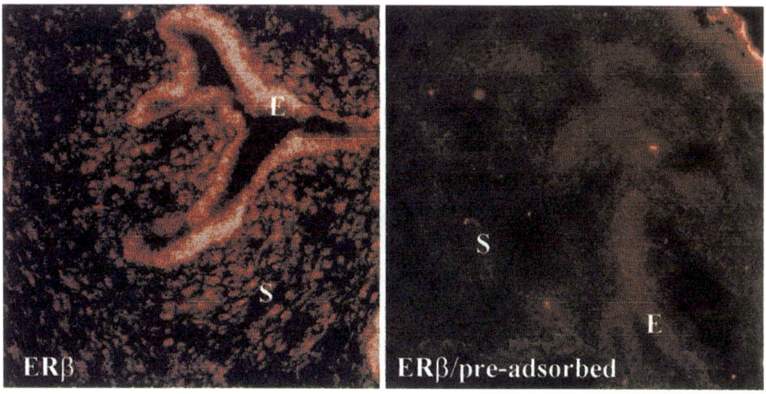

Figure 2. Immunohistochemical localization of ERβ in the immature rat uterus. Uteri were stained with ERβ antibody (ERβ), and pre-adsorbed antibody (ERβ/pre-adsorbed). Symbols: E, epithelium; S, stoma.

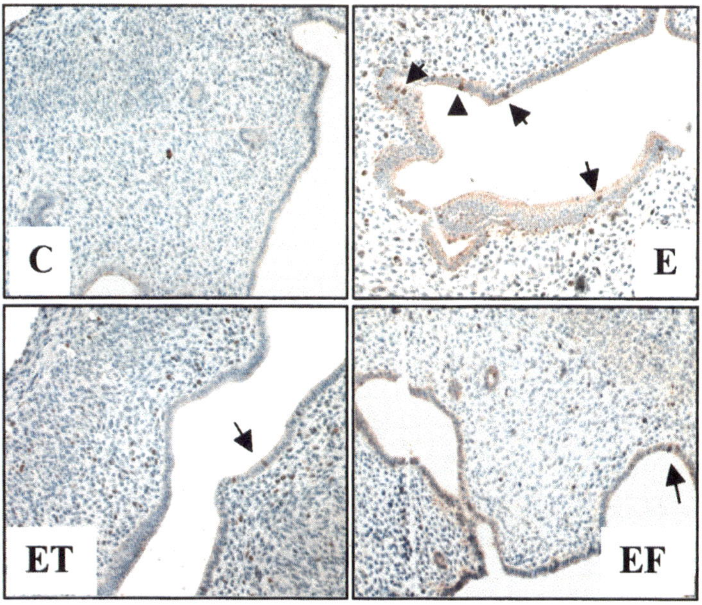

Figure 3. Inhibition of estrogen-induced cyclin A expression by both tamoxifen and flutamide. Cyclin A-positive nuclei are brown (arrows) and the negative cells are counterstained (blue). Cyclin A-positive cells are visible in both the luminal epithelial and stromal cell populations after estrogen treatment of rats (E). There is very little cyclin A expression in the uterus of untreated rats (C). Cyclin A expression is inhibited by tamoxifen in both the epithelium and stroma (ET). Flutamide inhibits estrogen induction of cyclin A in the stroma (EF). Induction is blocked by both antiestrogens and antiandrogens.

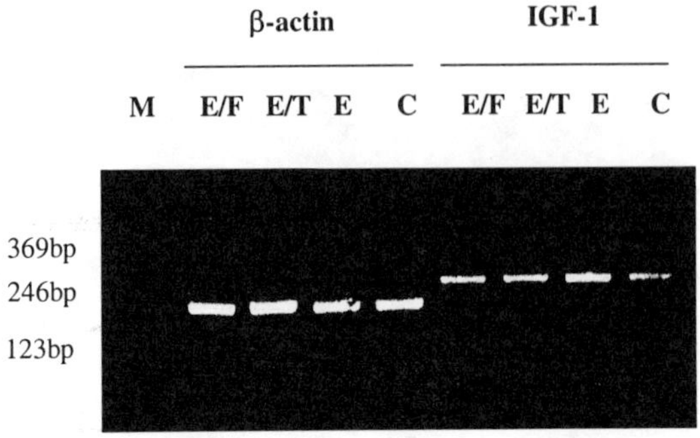

Figure 4. Inhibition of E_2-induced expression of IGF-1 by flutamide. The figure shows RT-PCR detection of IGF-1 mRNA. Rats were given daily subcutaneous injections of estradiol only (E); estradiol plus tamoxifen (E/T); estradiol plus flutamide (E/F); or vehicle only (C). After 35 PCR cycles, products were electrophoresed on agarose gels containing ethidium bromide and visualized under UV light.

IGF-1 is one of the major growth factors involved in stimulation of cell proliferation in the uterus (for a review, see Ref. [35]). In $ER\beta^{-/-}$ mice, expression of AR is increased and this receptor is expressed throughout the uterine stroma, including stromal cells surrounding the lumen. There is also increased expression of IGF-1 and exaggerated cell proliferation. These data indicate that ERβ normally represses AR expression.

The phenotype of the immature $ER\beta^{-/-}$ uterus [32] can now be more easily understood. Loss of ERβ causes over expression of AR, which leads to increased production of IGF-1 and increased cellular proliferation. The two estrogen receptors, thus, have opposite effects on the expression of AR: ERα up-regulates while ERβ down-regulates AR. This is a case where ERα and ERβ have opposite effects in a single organ without a physical interaction between the two receptors. There are cases where ERα and ERβ are co-localized in the same cell and where ERβ can quench the action of ERα. This is particularly true when there is expression of certain splice variants of ERβ.

Interestingly, even though the uterus is considered ERα territory, recent evidence suggests that ERβ plays a role in the timing of the onset of labor [36, 37]. During pregnancy, ERβ is expressed in the myometrium, where it is responsible for repression of labor-associated genes. At the end of pregnancy, ERβ is down-regulated, repression of labor-associated genes is removed, and labor proceeds.

SPLICE VARIANTS OF ERβ

Multiple splice variants of ERβ mRNAs have been reported [10-17]. We have investigated the expression of the proteins corresponding to two of these mRNAs. The original ERβ is now designated as ERβ1 and the variants as ERβins and ERβcx. ERβcx, was discovered by Ogawa and coworkers [12,13]. It is a unique splice variant in which the last exon (exon 8) is not utilized and instead is replaced by an alternative exon. With respect to the protein the last 61 amino acids in ERβ1 are replaced by another peptide of 26 amino acids. This exchange affects the ligand–binding domain and ERβcx does not bind estradiol. When co-expressed with ERα in cells it acts as a dominant negative repressor of estrogen action on EREs [12, 13]. We have made an antibody specific for the C-terminus of ERβcx and found that this protein is expressed in many tissues. One very interesting site of expression is breast cancer cells (unpublished observations), where ERβcx might have some implications for the use of antiestrogen treatment. In this disease, the treatment of choice for tumors which are ERα positive is the anti-estrogen, tamoxifen (see Ref. [38] for a review). Very few patients whose tumors are ERα-negative respond to tamoxifen and a fraction of those who do have ERα-positive tumors do not respond to tamoxifen. One possible explanation for non-responsiveness of ERα positive tumors could be the presence of ERβcx.

At first glance ERβcx, in silencing ERα, would appear to be a natural antiestrogen and having it expressed in breast cancer should provide benefits similar to that achieved by treatment with tamoxifen. This is not necessarily the case. At present, we do not understand the physiological relevance of ERβcx. We do not know whether there are ligands which can activate or inactivate it and, even more important, we do not know whether it silences ERα activity at sites other than at EREs, e.g. AP-1 and Sp-1 sites. What we can say at present is that in breast cancer, co-expression of ERβcx with ERα in the same cells might render those cells insensitive to tamoxifen and this would define one category of ERα-positive non-responders.

The second splice variant which we have studied is ERβins. In this variant, described independently in several labs [10, 11, 16], there is an 18 amino acid insertion in the ligand binding domain. ERβins is widely expressed in many rodent tissues at levels similar to those of ERβ1. It does bind to and is activated by estrogen but only at high concentrations. ERβins is co-localized with ERα in the mammary gland during lactation [39, 40]. Since the mammary gland is insensitive to estrogen during lactation, we have suggested that the co-expression of ERβins with ERα quenches ERα function and leads to unresponsiveness to estrogen [40].

An even more interesting role for ERβins is one which utilizes its capacity to respond to estrogen but only when estrogen levels are high. Such an estrogen response pattern is probably essential in the ovary. Estrogen, produced in granulosa cells of the ovary, is secreted and exerts actions on multiple organs in the body. But, as is clear from the phenotype of the ovary in $ER\beta^{-/-}$ mice [6], estrogen also has functions in the granulosa cells. One of the functions of ERβ in these cells is to down-regulate the AR when follicles are at the antral stage. Inability to down-regulate AR causes follicular arrest, follicular atresia and lack of ovulation [6]. Since an androgenic environment is essential for early follicular development, ERβ activation (and AR down-regulation) should not occur too early in follicular growth. One mechanism through which the granulosa cells could achieve this type of regulation is to express ERβins instead of ERβ1. This would ensure that significant amounts of estrogen are produced before the AR is down-regulated.

One other event in the body where there is a very important response to high levels of estrogen, is the paradoxical LH surge [41]. Normal levels of estrogen down-regulate secretion of LH but as ovarian follicles mature and estrogen levels rise, there is a surge of LH secretion. Without this LH surge, ovulation does not occur. It is well established that ERα is the receptor responsible for down-regulation of LH [42] but the mechanism behind the paradoxical surge has never been elucidated. The presence of ERβins in LHRH neurons would be one mechanism through which this regulation could be achieved. ERβ has been demonstrated in LHRH neurons [43], but the ERβ isoform has not yet been identified. In $ER\beta^{-/-}$ mice, plasma levels of estrogen and LH are normal but ovulation occurs very infrequently. We suggest that in these mice lack of ERβins results in inability to produce an LH surge in response to high levels of estrogen and hence there is no ovulation.

TISSUE AND CELLULAR DISTRIBUTION OF ERβ

There are important differences between ERα and ERβ in their tissue distribution. In general it can be said that ERα has a rather limited tissue distribution while ERβ is expressed in most tissues and is the predominant ER in many tissues in the body [21, 22]. The tissues with the highest levels of ERβ are the epithelium of the ventral prostate, granulosa cells in the ovary and ciliated and non-ciliated secretory cells in the lung. In these sites no expression of ERα can be detected. The detection of ERβ in prostate and lung has removed some of the mystery of estrogen action in these tissues. Estrogen has well documented effects in these tissues but since there is no expression of ERα, the actions of estrogen were difficult to understand. In the lung, estrogen is responsible for sexual dimorphism in both the normal lung physiology and in lung pathology [44-46]. Fetal lung development is more rapid in females and adult females have a greater lung mass and a greater number of alveoli than do males.

Actions of estrogen in the prostate have been well-documented [47] but because of the absence of ERα in the epithelium of the ventral prostate, the action of estrogen was thought to be indirect. ERβ is abundant in the epithelium of the ventral prostate [48-51] and appears to have a role in the control of epithelial cell growth in the ventral prostate [32]. In $ER\beta^{-/-}$ mice, the ventral prostate exhibits multiple foci of epithelial hyperplasia, the androgen receptor is over expressed and most epithelial cells express the proliferation antigen, Ki-67. The type of hyperplasia is quite different from that in TRAMP (transgenic adenocarcinoma of the mouse prostate) mice in which all prostatic epithelial cells over express T antigen and in which there is early development of androgen-independent, adenocarcinoma [52]. It appears that uncontrolled androgen action is one of the main problems in $ER\beta^{-/-}$ mice, but because estrogen actions on the prostate are complex, with distinct actions during development and in the mature gland, the cause of epithelial hyperplasia in $ER\beta^{-/-}$ mice is not yet known. This might be easier to define when conditional $ER\beta^{-/-}$ mice become available and we can distinguish between developmental effects and effects on the mature gland.

High expression of ERβ in the ovarian granulosa cells [1, 6, 53] in which estrogen is synthesized was, at first, a puzzling finding. It was difficult to envisage what function could be served by an estrogen receptor with a dissociation constant in the nM range in cells where the estrogen levels must be in the μM range. As discussed above, the ERβ isoform expressed in granulosa cells is ERβins which has a very low affinity for estradiol and which may be activated only when estrogen levels are high.

In addition to these sites of abundant expression of ERβ, there is exclusive expression of ERβ with no ERα in certain nuclei in the CNS, such as the supraoptic nucleus and the paraventricular nucleus [54-56]. Because of its localization in these neurons it can be anticipated that ERβ mediates the estrogen regulation of neuropeptides such as galanin and vasopressin, and thus plays an important physiological role in the CNS influencing feeding behavior, blood pressure, sexual behavior, social behavior (such as aggressiveness). Behavioral effects of $ER\beta^{-/-}$ mice have been documented [57-59].

It appears that ERβ is necessary for neuronal survival throughout life and might have important influence on the development of degenerative diseases of the central nervous system such as Alzheimer's disease and Parkinson's disease. In the brains of $ER\beta^{-/-}$ mice, as early as at 2 months of age [60], there is marked hypocellularity. This is particularly evident in the somatosensory cortex and the substantia nigra. Sawada and coworkers [61] have shown that estrogen protects against Bleomycin sulfate–induced apoptosis in neurons of the substantia nigra. Apoptosis was also blocked by inhibitors for caspase-3 or c-Jun/AP-1 suggesting that neuroprotection by estradiol is mediated by transcription through AP-1. Interestingly, in these mesencephalic dopaminergic neurons, ERβ is the only estrogen receptor expressed. As discussed above, ERβ in the presence of estrogen has very weak capacity to activate transcription from AP-1 sites. If ERβ and estrogen provide protection against apoptosis, tamoxifen would be expected to do the opposite and increase neurodegeneration. Indeed, the antiapoptotic effect by estradiol on bleomycin-induced apoptosis was blocked by ICI 182,780. It is premature to make predictions about what will be the response of the brain to pharmaceuticals which act on estrogen receptors. The effects of estrogen in the CNS are complex, and involve both estrogen receptors [62-64] as well as rapid non-genomic effects [63] in functions which are much broader than those expected for a "female sex hormone"[66]. To date, no information is available on the morphology of the $ER\alpha^{-/-}$ brain so it is not possible to conclude whether neuronal loss in $ER\beta^{-/-}$ mice is due to loss of ERβ or over activity of ERα.

CONCLUDING REMARKS

Understanding of estrogen action has been changed by the discovery of ERβ. We do not yet understand how the system functions but we never would have been able to do so with a large component of the system missing. We now know that there are actions of estrogen which are controlled by ERβ and those which are controlled by ERα. In addition, the two receptors can oppose each other's actions. In this system, biology uses a principle well known in physics and mathematics, namely, that the most sensitive system of control is not one where there is a single force only, but one where there are two equal and opposing forces.

Acknowledgements

Studies on ERβ were made possible by grants from KaroBio AB and the Swedish Cancer Fund.

REFERENCES

1. **Kuiper GG, Enmark E, Pelto-Huikko M, Nilsson S, Gustafsson J-Å.** 1996 Cloning of a novel receptor expressed in rat prostate and ovary. Proc Natl Acad Sci USA 93: 5925-5930.
2. **Krege JH,. Hodgin JB, Couse JF, Enmark E, Warner M, Mahler JF, Sar M, Korach KS, Gustafsson J-Å, Smithies O.** 1998 Generation and reproductive phenotypes of mice lacking estrogen receptor beta. Proc Natl Acad Sci USA 95:15677-15682.
3. **Couse JF, Curtis Hewitt S, Korach KS.** 2000 Receptor null mice reveal contrasting roles for estrogen receptor α and β in reproductive tissues. J Steroid Biochem Mol Biol. 74:287-296.
4. **Dupont S, Krust A, Gansmuller A, Dierich A, Chambon P, Mark M.** 2000 Effect of single and compound knockouts of estrogen receptors α (ERα) and β (ERβ) on mouse reproductive phenotypes. Development. 127: 4277-4291.
5. **Lyttle RC** 2001 Defining the functions of ERβ: use of knockout mice and selective ligands. In 5 th annual symposium on the long-term effects of estrogen deprivation. Elbow Beach Resort, Bermuda July 8-11.
6. **Cheng G, Mäkinen S, Weihua Z, Mäkelä S, Saji S, Warner M, Gustafsson J-Å, Hovatta O.** 2001 A Role for the Androgen Receptor in Follicular Atresia of Estrogen Receptor β Knockout Mouse Ovary Biol Reprod, in press.
7. **Brzozowski AM, Pike ACW, Dauter Z, Hubbard RE, Bonn T, Engstrom O, Ohman L, Greene G, Gustafsson J-Å, Carlquist M.** 1997 Molecular basis of agonism and antagonism in the oestrogen receptor. Nature 389: 753-758.
8. **Pike ACW, Brzozowski A M, Hubbard R E, Bonn T, Thorsell AG, Engstrom O, Ljunggren J, Gustafsson J-Å, Carlquist M.** 1999 Structure of the ligand-binding domain of oestrogen receptor β in the presence of a partial agonist and a full antagonist. EMBO J 18: 4608-4618.
9. **Nilsson S, Mäkelä S, Treuter E, Tujague M, Andersson G, Enmark E, Pettersson K, Warner M, Gustafsson J-Å.** 2001 Mechanisms of Estrogen Action. Physiol Rev, in press.
10. **Hanstein B, Liu H, Yancisin MC, Brown M.** 1999 Functional analysis of a novel estrogen receptor-β isoform. Mol Endocrinol 13:129-137.
11. **Chu S, Fuller PJ.** 1997 Identification of a splice variant of the rat estrogen receptor β gene. Mol Cell Endocrinol 132:195-199.
12. **Ogawa S, Inoue S, Orimo A, Hosoi T, Ouchi Y, Muramatsu M.** 1998 Cross-inhibition of both estrogen receptor α and β pathways by each dominant negative mutant. FEBS Lett 423:129-132.
13. **Ogawa S, Inoue S, Watanabe T, Orimo A, Hosoi T, Ouchi Y, Muramatsu M.** 1998 Molecular cloning and characterization of human estrogen receptor βcx: a potential inhibitor of estrogen action in human. Nucleic Acids Res 26:3505-3512.
14. **Moore JT, McKee DD, Slentz-Kesler K, Moore LB, Jones SA, Horne EL, Su JL, Kliewer SA, Lehmann JM, Willson TM.** 1998 Cloning and characterization of human estrogen receptor β isoforms. Biochem Biophys Res Commun. 247:75-78.
15. **Maruyama K, Endoh H, Sasaki-Iwaoka H, Kanou H, Shimaya E, Hashimoto S, Kato S, Kawashima H.** 1998 A novel isoform of rat estrogen receptor β with 18 amino acid insertion in the ligand binding domain as a putative dominant negative regular of estrogen action. Biochem Biophys Res Commun 246:142-147.
16. **Petersen DN, Tkalcevic GT, Koza-Taylor PH, Turi TG, Brown TA.** 1998 Identification of estrogen receptor β2, a functional variant of estrogen receptor β expressed in normal rat tissues. Endocrinology 139:1082-1092.

17. **Price RH Jr, Butler CA, Webb P, Uht R, Kushner P, Handa RJ.** 2001 A splice variant of estrogen receptor β missing exon 3 displays altered subnuclear localization and capacity for transcriptional activation. Endocrinology 142:2039-2049.
18. **Kuiper GG,. Carlsson B, Grandien K, Enmark E, Haggblad J, Nilsson S, Gustafsson J-Å.** 1997 Comparison of the ligand binding specificity and transcript tissue distribution of estrogen receptors α and β. Endocrinology 138:863-870.
19. **Barkhem T, Carlsson B, Nilsson Y, Enmark E, Gustafsson J-Å, Nilsson S.** 1998 Differential response of estrogen receptor α and estrogen receptor β to partial estrogen agonists/antagonists. Mol Pharmacl 54:105-112.
20. **Routledge EJ, White R, Parker MG, Sumpter JP.** 2000 Differential effects of xenoestrogens on coactivator recruitment by estrogen receptor (ER) α and ERβ. J Biol Chem 275:35986-35993.
21. **Taylor AH, Al-Azzawi F.** 2000 Immunolocalisation of oestrogen receptor β in human tissues. J Mol Endocrinol. 24:145-155.
22. **Merchenthaler I, Shugrue PJ.** 1999 Estrogen receptor-β: a novel mediator of estrogen action in brain and reproductive tissues. Morphological considerations. J Endocrinol Invest 22:10-12.
23. **Kushner PJ, Agard DA, Greene GL, Scanlan TS. Shiau AK, Uht RM. Webb P.** 2000 Estrogen receptor pathways to AP-1. J Steroid Biochem Mol Biol 74:311-317.
24. **McInerney EM, Weis KE, Sun J, Mosselman S, Katzenellenbogen B S.** 1998 Transcription activation by the human estrogen receptor subtype β (ERβ) studied with ERα and ERβ receptor chimeras. Endocrinology 139: 4513-4522.
25. **Martin MB, Franke TF, Stoica GE, Chambon P, Katzenellenbogen BS, Stoica BA, McLemore MS, Olivo SE, Stoica A.** 2000 A role for Akt in mediating the estrogenic functions of epidermal growth factor and insulin-like growth factor I. Endocrinology 141:4503-4511.
26. **Kato S, Masuhiro Y, Watanabe M, Kobayashi Y, Takeyama KI, Endoh H, Yanagisawa J.** 2000 Molecular mechanism of a cross-talk between oestrogen and growth factor signalling pathways. Genes to Cells 5: 593-601.
27. **Paech K, Webb P, Kuiper GG, Nilsson S, Gustafsson J, Kushner PJ, Scanlan TS.** 1997 Differential ligand activation of estrogen receptors ERα and ERβ at AP1 sites. Science 277:1508-1510.
28. **Webb P, Nguyen P, Valentine C, Lopez GN, Kwok GR, McInerney E, Katzenellenbogen BS, Enmark E. Gustafsson J-Å, Nilsson S, Kushner PJ.** 1999 The estrogen receptor enhances AP-1 activity by two distinct mechanisms with different requirements for receptor transactivation functions. Mol Endocrinol 13:1672-1685.
29. **Zou A, Marschke KB, Arnold KE, Berger EM, Fitzgerald P, Mais DE, Allegretto EA.** 1999 **Estrogen receptor** β activates the human retinoic acid **receptor** α-1 promoter in response to tamoxifen and other **estrogen receptor** antagonists, but not in response to **estrogen**. Mol Endocrinol 13: 418-430.
30. **Armstrong DT, Moon YS, Leung PK.**1976 Uterotrophic effects of testosterone and 5α-dihydrotestosterone in intact and ovariectomized immature female rats. Biol Reprod 15:107-114.
31. **Sahlin L, Wang H, Masironi B, Holmgren A, Eriksson H.** 1999 Regulation of thioredoxin mRNA in the rat uterus by gonadal steroids. J Steroid Biochem Mol Biol 68: 203-209.
32. **Weihua Z, Saji S, Makinen S, Cheng G, Jensen EV, Warner M, Gustafsson J-Å.** 2000 Estrogen receptor (ER) β, a modulator of ERα in the uterus. Proc Natl Acad Sci USA. 97: 5936-5941.

33. **Weihua Z, Mäkelä S, Adersson LC, Salmi S, Saji S, Webster J, Jensen EV, Nilsson S, Warner M, Gustafsson J-Å** 2001 A role for estrogen rceptor β in regulation of growth of the ventral prostate Proc Natl Acad Sci USA, 98:6330-6335.
34. **Kurita T, Lee K, Cooke PS, Taylor JA, Lubahn DB., Cunha GR.** 2000 Paracrine regulation of epithelial progesterone receptor by estradiol in the mouse female reproductive tract. Biol. Reprod 62:821-830.
35. **Richards RG, Walker MP, Sebastian J, DiAugustine RP.**1998 Insulin-like growth factor-1 (IGF-1) receptor-insulin receptor substrate complexes in the uterus. Altered signaling response to estradiol in the IGF-1(m/m) mouse. J Biol Chem 273:11962-1969.
36. **Wu JJ, Geimonen E, Andersen J**. 2000 Increased expression of estrogen receptor β in human uterine smooth muscle at term. Eur J Endocrinol 142:92-99.
37. **Wang H, Stjernholm Y, Ekman G, Eriksson H, Sahlin L.** 2001 Different regulation of oestrogen receptors α and β in the human cervix at term pregnancy. Mol Human Reprod 7: 293-300.
38. **Osborne CK, Fuqua SA** 1994 Mechanisms of tamoxifen resistance. Breast Cancer Res Treatment 32:49-55.
39. **Saji S, Jensen EV, Nilsson S, Rylander T, Warner M, Gustafsson J-Å**. 2000 Estrogen receptors α and β in the rodent mammary gland. Proc Natl Acad Sci USA 97: 337-342.
40. **Saji S, Sakaguchi H, Andersson S, Warner M, Gustafsson J-Å.** 2001 Quantitative Analysis of Estrogen Receptor Proteins in Rat Mammary Gland. Endocrinology 142: 3177-3186.
41. **Tsai HW, Legan SJ.** 2001 Chronic elevation of estradiol in young ovariectomized rats causes aging-like loss of steroid-induced luteinizing hormone surges. Biol Reprod 64:684-688.
42. **Couse JF, Korach KS.**1999 Estrogen receptor null mice: what have we learned and where will they lead us? Endocrine Rev 20: 358-417.
43. **Hrabovszky E, Shughrue PJ, Merchenthaler I, Hajszan T, Carpenter CD, Liposits Z, Petersen SL.** 2000 Detection of estrogen receptor-β messenger ribonucleic acid and 125I-estrogen binding sites in luteinizing hormone-releasing hormone neurons of the rat brain. Endocrinology 141: 3506-3509.
44. **McMillan EM, King GM, Adamson IY.** 1998 Sex hormones influence growth and surfactant production in fetal lung explants. Exp Lung Res 15:167-179.
45. **Adamson IY, Bakowska J, McMillan E, King GM**. 1990 Accelerated fetal lung maturation by estrogen is associated with an epithelial-fibroblast interaction. In Vitro Cellular & Developmental Biology 26:784-790.
46. **Ballard PL**. 1989 Hormonal control of lung maturation. Clin Endocrinol Metab 3:723-753.
47. **Chang WY, Prins GS.** 1999 Estrogen receptor-β: implications for the prostate gland. Prostate 40:115-124.
48. **Royuela M, de Miguel MP, Bethencourt FR, Sanchez-Chapado M, Fraile B, Arenas MI, Paniagua R.** 2001 Estrogen receptors α and β in the normal, hyperplastic and carcinomatous human prostate. J Endocrinol 168: 447-454.
49. **Risbridger G, Wang H, Young P, Kurita T, Wong YZ, Lubahn D, Gustafsson J-Å, Cunha G.** 2001 Evidence that epithelial and mesenchymal estrogen receptor-α mediates effects of estrogen on prostatic epithelium. Developmental Biol 229:432-442.
50. **Makela S, Strauss L, Kuiper G, Valve E, Salmi S, Santti R, Gustafsson J-Å.** 2000 Differential expression of estrogen receptors α and β in adult rat accessory sex glands and lower urinary tract. Mol Cell Endocrinol. 164:109-116.
51. **Sar M, Welsch F.** 2000 Oestrogen receptor α and β in rat prostate and epididymis. Andrologia 32:295-301.

52. **Gingrich JR, Barrios RJ, Kattan MW, Nahm HS, Finegold MJ, Greenberg NM.** 1997 Androgen-independent prostate cancer progression in the TRAMP model. Cancer Res 57: 4687-4691.
53. **Byers M, Kuiper GG, Gustafsson J-Å, Park-Sarge OK.** 1997 Estrogen receptor-β mRNA expression in rat ovary: down-regulation by gonadotropins. Mol Endocrinol 11:172-182.
54. **Hrabovszky E, Kallo I, Hajszan T, Shughrue PJ, Merchenthaler I, Liposits Z.** 1998 Expression of estrogen receptor-β messenger ribonucleic acid in oxytocin and vasopressin neurons of the rat supraoptic and paraventricular nuclei. Endocrinology 139: 2600-2604.
55. **Alves SE, Lopez V, McEwen BS, Weiland NG.** 1999 Differential colocalization of estrogen receptor β (ERβ) with oxytocin and vasopressin in the paraventricular and supraoptic nuclei of the female rat brain: an immunocytochemical study. Proc Natl Acad Sci USA 95: 3281-3286.
56. **Ishunina TA, Kruijver FP, Balesar R, Swaab DF.** 2000 Differential expression of estrogen receptor α and β immunoreactivity in the human supraoptic nucleus in relation to sex and aging. J Clin Endocrinol Metab 85: 3283-3291.
57. **Temple JL, Fugger HN, Li X, Shetty SJ, Gustafsson J-Å, Rissman EF**. 2001 Estrogen receptor β regulates sexually dimorphic neural responses to estradiol. Endocrinology 142: 510-3, 2001.
58. **Ogawa S, Chester AE, Hewitt SC, Walker VR, Gustafsson J-Å, Smithies O, Korach KS, Pfaff DW.** 2000 From the cover: abolition of male sexual behaviors in mice lacking estrogen receptors α and β (α β ERKO). Proc Natl Acad Sci USA 97:14737-14741.
59. **Fugger HN, Foster TC, Gustafsson J-Å, Rissman EF**. 2000 Novel effects of estradiol and estrogen receptor α and β on cognitive function. Brain Res 883:258-264.
60. **Wang L, Warner M, Gustafsson J-Å.** 2001 ERβ: A receptor involved in neuroprotection Proc Natl Acad Sci USA 98: 2792-2796.
61. **Sawada H, Ibi M, Kihara T, Urushitani M, Honda K, Nakanishi M, Akaike A, Shimohama S.** 2000 Mechanisms of antiapoptotic effects of estrogens in nigral dopaminergic neurons. FASEB J 14:1202-1214.
62. **Osterlund M, Kuiper GG, Gustafsson J-Å, Hurd YL.** 1998 Differential distribution and regulation of estrogen receptor-α and -β mRNA within the female rat brain. Brain Res Mol Brain Res. 54:175-180.
63. **Kuiper GG, Shughrue PJ, Merchenthaler I, Gustafsson J-Å.** 1998 The estrogen receptor β subtype: a novel mediator of estrogen action in neuroendocrine systems. Frontiers in Neuroendocrinology 19:253-286.
64. **Dubal DB, Zhu H, Yu J, Rau SW, Shughrue PJ, Merchenthaler I, Kindy MS, Wise PM.** 2001 Estrogen receptor α, not β, is a critical link in estradiol-mediated protection against brain injury. Proc Natl Acad Sci USA 98:1952-1957.
65. **Nadal A, Ropero AB, Laribi O, Maillet M, Fuentes E, Soria B.** 2000 Nongenomic actions of estrogens and xenoestrogens by binding at a plasma membrane receptor unrelated to estrogen receptor α and estrogen receptor β. Proc Natl Acad Sci USA 97:11603-11608.
66. **Shughrue PJ, Merchenthaler I.** 2000 Estrogen is more than just a "sex hormone": novel sites for estrogen action in the hippocampus and cerebral cortex. Frontiers in Neuroendocrinology 21: 95-101.

14

Sensors for Metabolic Control

A Regulatory Network of Nuclear Receptors

Béatrice Desvergne*, Liliane Michalik, and Walter Wahli

Institute of Animal Biology, Faculty of Sciences, University of Lausanne, Bâtiment Biologie, CH- 1015 Lausanne, Switzerland

INTRODUCTION

Nuclear receptors are transcription factors characterized by two important properties: first, they are activated upon the binding of specific ligands for which they have high affinity and low capacity; second, they bind to specific response elements located in the vicinity of the promoter of their target genes (see Chapter 12). Thus, in a simplified view, the effector function of the nuclear receptors in a cell is to adapt the gene expression program according to the signals that they receive in form of specific ligands. Nuclear receptors share a common modular organization. A poorly structured N-terminal domain that may encompass a ligand-independent transactivation domain is followed by the DNA binding domain (DBD) comprised of two zinc fingers which is the hallmark of the nuclear receptor family. A hinge region then links the DNA binding domain to the ligand binding domain (LBD) that has a general fold structured by 12 α helices and 3 β sheets. Nuclear receptors bind to DNA in form of dimers, either homodimers or more often heterodimers with the receptor for 9-cis retinoic acid known as RXR. The DNA response element of nuclear receptors is formed of two sequences corresponding to or closely related to the hexamer AGGTCA. Their organisation in direct repeats or palindromic arrays and the length of the spacing between the two hexamers

---

* beatrice.desvergne@iba.unil.ch

determine the specificity of these response elements towards each dimer of family members. The general scheme for transactivation *via* nuclear receptors is believed to occur in at least two steps. In the absence of ligand, the nuclear receptor dimer binds to a co-repressor protein that inhibits its transactivation properties. In the presence of ligand, or due to an alternative pathway of activation such as phosphorylation, the co-repressor is released and a co-activator is recruited, allowing further contacts to be made with the transcription machinery, eventually resulting in transcription enhancement (also see Chapters 12 and 13).

Since the cloning of the first nuclear receptor cDNA, the glucocorticoid receptor, an ever-increasing number of nuclear receptors that fit this characteristic modular organisation have been described. Considering available information about their respective ligands, they can be organized in three classes (Fig. 1). One class is that of the classic hormone receptors. These specifically recognize one or a few molecules with a high affinity. This is the case for the glucorticoid, thyroid hormone, retinoic acid, oestrogen, vitamin D, as well as progesterone, mineralocorticoid, and androgen receptors. Another class encompasses receptors that possess the structural characteristics of nuclear receptors including a ligand binding domain but for which no ligand has so far been identified. These are the true orphan receptors that are represented for example by Nurr1. Finally, a third class groups the receptors

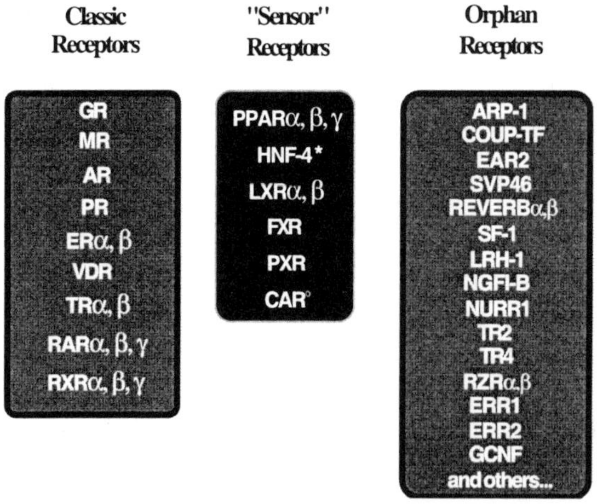

Figure 1. Classification of the nuclear receptors, according to their ligand binding properties. A "classic" receptor recognizes one or few molecules with high specificity. There is no known ligand, to date, for an orphan receptor whereas a "sensor" receptor binds to a broad spectrum of ligands with relatively low affinity. For a phylogenetic classification of these and other superfamily members, see [2].

** HNF4 has been shown to bind acyl CoA thioesters. However, the functional significance of this binding remains to be explored.*

that each binds a broad range of molecules with, as a corollary, a relatively poor affinity. Rather than responding to hormones secreted by endocrine glands with tight feedback controls, these receptors can bind to molecules that belong to metabolic pathways such as substrates, intermediates or end-products. These receptors are sensors of metabolic status, responding to incoming dietary components as well as metabolites generated in the organism, and triggering the required metabolic adaptation at the level of the cell, the organ, and the organism.

Considering this third class of nuclear receptors, there is an emerging scheme that will be described herein which emphasizes the role of these receptors as sensors and adaptators involved in the metabolic pathways of fatty acids, cholesterol, and xenobiotics (Fig. 2).

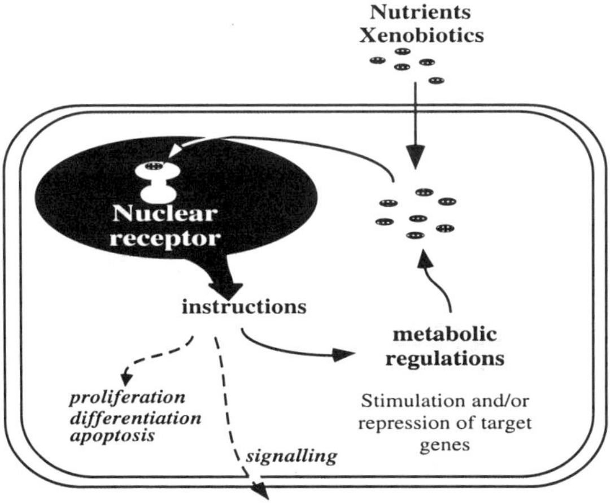

Figure 2. Nutrients and xenobiotics contribute to metabolic regulations via nuclear receptors.

PPARα AND PPARγ : THE SENSORS OF FATTY ACIDS AND KEY REGULATORS OF LIPID METABOLISM

PPARs were the first nuclear receptors identified as a "sensor" rather than a classical hormone receptor (for a review, see Ref. [1]). These are nuclear lipid-activable receptors that control a variety of genes in several pathways of lipid metabolism. Three isotypes of PPAR, PPARα (i.e. NR1C1 according to the unified nomenclature of nuclear hormone receptors [2]), PPARβ (NR1C2, also called δ, NUCI, and FAAR), and PPARγ (NR1C3) have been cloned in *Xenopus*, rodents, and human. Their molecular characteristics fit the modular structure shared by all nuclear receptors. The less conserved N-terminal domain in PPARα and PPARγ possesses a ligand-independent activation domain. The DNA binding domain is extremely well conserved between the three isotypes, whereas more divergence occurs at the level of the ligand binding domain. PPARs must form a heterodimer with RXR (NR2B), in order to bind to DNA. The core of the PPAR response element (PPRE) comprises direct repeats of two hexamers closely related to the sequence AGGTCA, separated by one nucleotide. The five nucleotides that flank the 5' end of this core sequence are also important for determining the efficiency of PPARα:RXR binding [3]. PPAR-mediated transactivation results from the combination of PPAR:RXR binding to a PPRE and ligand activation of this complex.

Ligand Binding Properties of PPARs

The list of ligands that PPARs can recognize is astonishingly long. The first molecules to be recognized as PPARα activators, and later on characterized as ligands, correspond to a group of molecules that have in common the ability to induce peroxisome proliferation in the rodents, thus the name, peroxisome proliferator activated receptor, given to this receptor [4]. This diverse group of substances include for example some plasticizers and herbicides. More interestingly, we and others have shown that various fatty acids, more particularly unsaturated fatty acids, and some eicosanoids derived from arachidonic acid, bind to PPARα, β, and γ with varying affinities (reviewed in Ref. [1]). In addition to being indeed activated by fatty acids, PPARs respond to fibrate hypolipidemic drugs and to insulin sensitisers, demonstrating their potential as drug targets (Table I). The PPAR ligand-mediated activation is considered to be the main pathway of PPAR target gene stimulation. However, other mechanisms include phosphorylation of PPAR, mainly of serines located in the A/B domain, and the activation of the RXR-containing heterodimer by RXR ligands. The relevance of these pathways is presently under study in several laboratories and might lead to interesting new developments.

Table I: Main Compounds or Classes of Compounds Binding to PPARs

	PPARα	PPARβ	PPARγ
Natural ligands	PUFAs (++) 8-HETE LTB4	PUFAs (+)	PUFAs (+) 15dPGJ2 9-HODE 13-HODE
Synthetic ligands	Fibrates	L-165041	Tiazolidinediones

The ability of PPAR to bind such a diverse variety of compounds has been puzzling. Explanation came with the description of the crystal structure of the ligand binding domain of PPARγ [5] and PPARβ [6] (see also Chapter 12). A common structural three-dimensional fold, which consists of an antiparallel α–helical sandwich of 12 helices (Helix 1 to Helix 12) organized in three layers with a central hydrophobic pocket had been previously characterized in classical hormone receptors such as TR, RAR and RXR [7-9]. Upon ligand binding, the swinging of H12 or HAF (activation function helix) closes the ligand binding pocket as a lid, in a so-called "mouse trap model". The ligand binding cavity of PPAR is buried in the bottom half of the LBD and is particularly large, ~1300 Å^3, of which the ligand occupies only about 30 to 40%. This cavity is larger than in other known LBDs (compare with the cavity in TR ~600Å^3, most of this volume, ~530 Å^3, being occupied by the triiodothyronine molecule [9]). Some key hydrogen bonds associated with relatively free non-specific interactions that the hydrophobic part of ligands can develop within the large cavity explain the promiscuous behaviour of PPAR with respect to ligand binding. The other characteristic of this PPAR cavity is its easy accessibility provided by an extra Helix 2' and a particular placement of Helix 2. Overall, PPAR appears to have evolved as a nuclear receptor adapted for binding to multiple natural ligands with relatively low affinity.

The Opposite and Complementary Roles of PPARα and PPARγ in Lipid Metabolism

The patterns of expression of the three PPARs participate in the specificity of their roles. PPARα levels are the highest in brown adipose tissue and in liver, then in heart, kidney and enterocytes. PPARβ expression is ubiquitous, with varying levels in different organs. At the protein level, there are two main PPARγ isoforms, PPARγ1 and PPARγ2. PPARγ1 is mainly expressed in adipose tissues but is also detected in the colon, spleen, retina, hematopoeitic cells and skeletal muscle. PPARγ2 has been mainly found in the adipose tissue ([10]; for a review, see Ref. [1]).

As can be expected from "sensor" molecules, PPARs which can recognize and bind a variety of fatty acids regulate in turn most of the pathways linked to lipid metabolism. The most fascinating is in fact the observation that the two isotypes α and γ allow for balanced regulatory actions between fatty acid oxidation in the liver, *via* PPARα, and fatty acid storage in the adipose tissue, *via* PPARγ (Fig. 3). In contrast, the role of PPARβ remains elusive and will not be discussed.

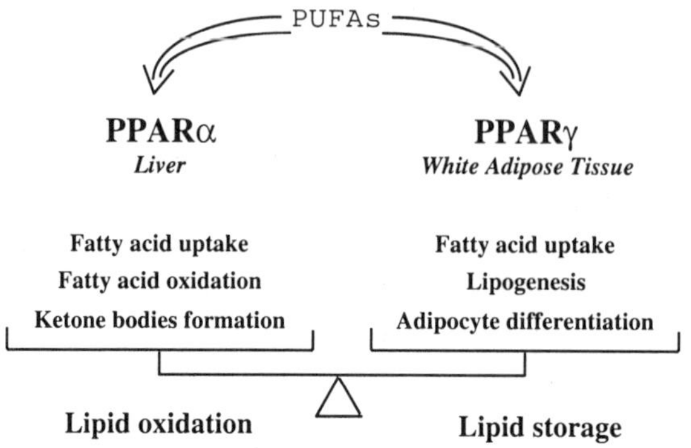

Figure 3. The balanced actions of PPARα and PPARγ in lipid oxidation and lipid storage, respectively. PUFAs: polyunsaturated fatty acids that are PPAR ligands.

In addition to the identification of ligands, there are two main and complementary approaches to understand the function of a nuclear receptor: scrutinising its target genes for the identification of regulated pathways, and generating null allele mutant mice that are then subjected to an adequate challenge. These approaches have contributed to most of our knowledge of PPAR functions as will be exposed below.

PPARα has been mainly studied in the liver where it is highly expressed. PPARα target genes form a comprehensive set of genes that participate in many if not all aspects of lipid catabolism. It includes fatty acid transport, uptake by the cells, intracellular binding, activation *via* acyl-CoA formation, as well as catabolism by β-oxidation in the peroxisomes and mitochondria and ω-oxidation in the microsomes (reviewed in [1]). PPARα KO mice are viable and do not exhibit an obvious phenotype when kept under normal laboratory confinement and diet [11]. However, these mice cannot sustain fasting, a situation that normally results in an enhanced fatty acid load in the liver and oxidation of these fatty acids as the major energy source. Under such a metabolic challenge, PPARα null mice, in contrast, are not capable of enhanced fatty oxidation, which results in hypoketonemia, hypothermia and hypoglycemia [12, 13]. This latter observation links fatty acid and glucose metabolism, a conclusion that is not so surprising if one considers that general energy homeostasis is one of the goals of metabolic adjustment.

In contrast, PPARγ has been clearly linked to the adipocyte differentiation program [14]. It is a late marker of adipocyte differentiation and its forced expression suffices to induce fibroblasts into the adipogenic program. Direct target genes of PPARγ include those coding for the adipocyte fatty acid binding protein (aP2), lipoprotein lipase, acyl-CoA synthase and fatty acid transport protein. PPARγ KO mice are not viable, due to placenta defects [15]. However, a thorough analysis of PPARγ heterozygous mice revealed that they are less prone to developing insulin resistance upon chronic high fat diet [16]. These observations again link glucose and lipid metabolism and deserve further investigations to unveil the molecular mechanism underlying this link.

This short summary of PPARα and PPARγ actions emphasizes the regulatory network that these two receptors establish in the organism, between the liver and the adipose tissue, and between fatty acid and glucose metabolism, allowing for proper adaptation to the everyday succession of fasting and feeding periods [17]. Other transcription factors, such as HNF4, SREBP1, and CEBP are critical for such an adaptation. However, the particular place of PPARs stems from their ability to function as sensors of the metabolic needs of the organism. Any perturbations in these pathways are likely to provoke metabolic alteration such as those seen in diabetes type II, obesity and dyslipidemia, thus the strong interest of the pharmaceutical industry in developing specific PPAR ligands that could be used as drugs aimed at these disorders. In addition, other pathways, noticeably those of inflammation, due to the fact that lipid mediators are ligands for PPAR, are also a major issue for further investigation in the PPAR therapeutic field. However, due to space limitation this aspect will not be developed further.

LXR AND FXR, OXYSTEROL AND BILE ACID SENSORS, RESPECTIVELY: KEY PLAYERS IN CHOLESTEROL HOMEOSTASIS

Cholesterol is an essential component of cell membranes and is critical for growth and viability of cells. It is also the molecule from which steroids are synthesised and which serves as a precursor for bile acid synthesis. The main source of cholesterol is the diet. Its solubilization by bile acids leads to its efficient uptake by the enterocytes. Incorporated into chylomicrons and secreted into the lymph circulation, cholesterol in chylomicron remnants is then delivered to hepatocytes. From there, cholesterol either enters into bile acid synthesis or is redistributed in form of plasma VLDL and LDL particles, which form the primary source of cholesterol for peripheral tissues. The expression of the LDL receptor is thus a main regulatory means for adjusting the import of cholesterol into cells. If cholesterol supply is low, *de novo* synthesis takes place, mainly in the liver and in the intestine. If cholesterol is in excess, it is catabolised to form bile acids that are secreted into the intestine *via* the bile duct. The ability of bile acids to form micelles allows for solubilization of fatty acid and cholesterol, facilitating their uptake by the intestinal mucosa. Thus, cholesterol homeostasis occurs at two levels. Most extra-hepatic cells in the organism can regulate the uptake, as well as the efflux and the synthesis of cholesterol according to their own supply and needs. One major regulator of cellular cholesterol homeostasis has been identified to be SREBP2, a membrane protein that is cleaved by a cholesterol sensitive enzyme. The freed intracellular part locates in the nucleus and affects the expression of key enzymes in cholesterol biosynthesis and uptake [18]. At the level of the organism, most of the regulation occurs through a cross-talk between the liver and the intestine, as described above. SREBP2 again plays an important role at this level but a knot of regulatory pathways involving several nuclear receptors is now being unveiled.

LXR and the Activation of Bile Acid Synthesis

The liver X receptor (LXR, NR1H3) was first described as an orphan receptor, ubiquitously present at low levels but highly expressed in the liver, thus its name [19-22]. Two isotypes LXRα and LXRβ have been identified that share 77% amino acid identity in their DBD and LBD, and are highly conserved between rodents and human. Similarly to PPAR, LXR must form an heterodimer with RXR to bind to its DNA response element, formed of a direct repeat of two hexamers (AGTTCA) separated by four nucleotides. In the absence of known LXR ligands, activation of this heterodimer was shown to be triggered by RXR ligands (reviewed in [23]). However, the discovery that mono-oxidized derivatives of cholesterol were potent ligands of the LXRs caused great excitement. The most potent of these are 22(R)-hydroxycholesterol, 24(S)-hydroxycholesterol, and 24(S),25-epoxycholesterol [24]. Little is known about the sterol hydroxylases that produce these metabolites, but it is assumed that the oxysterol concentration parallels that of cholesterol. Importantly, these compounds are found at μM concentrations in tissues that express high level of LXRα or LXRβ. LXRα null mice are healthy and fertile when bred on a standard mouse diet, but display a severely impaired cholesterol and bile acid metabolism. Fed with a cholesterol-enriched diet, LXRα null mice fail to induce the gene encoding cholesterol 7α-hydroxylase (CYP7A), the rate-limiting enzyme in the neutral bile acid biosynthetic pathway [25]. This results in a dramatic accumulation of cholesteryl ester in the liver and failure to increase bile acid production. No functional compensation by LXRβ is seen in LXRα null mutant mice. LXRβ null mice do not have such phenotype whereas the double mutant mice are more strongly affected than the LXRα null mice [23]. In addition to being the mediator of increased bile acid synthesis in the liver, activation of LXRα in the small intestine leads to the up-regulation of the ABC1 gene whose product promotes efflux of cholesterol from within the epithelial cells to the intestinal lumen [26]. Together, this results in a decrease in the net uptake of cholesterol. These data suggest a rather simple regulatory loop whereby LXR acts as a sensor of cholesterol level *via* its interaction with oxysterols. The activated LXR triggers a decrease in cholesterol availability by positive regulation of cholesterol efflux in the intestine and of cholesterol degradation pathways in the liver (Fig. 4).

FXR and the Inhibition of Bile Acid Synthesis

Next, came the finding that FXR (NR1H4) [27], initially cloned as an RXR interacting protein (RIP14) [20], is a bile acid sensor. Expression of this nuclear receptor is restricted to adrenal cortex, intestine, colon, kidney and liver. FXR forms heterodimers with RXR and can be activated by RXR ligands. FXR:RXR binds to a DNA response element formed of two hexamers (GGGTCA) organized in a palindromic array spaced by 1 nucleotide (also called inverted repeat-1; see Chapter 12). After being initially claimed as a receptor of farnesol metabolites, and thus renamed FXR, it was demonstrated that physiological concentrations of free and conjugated chenodeoxycholic acid, lithocholic acid, and deoxycholic acid activate this nuclear receptor [28-30]. These bile acids are the end-products of the neutral and acidic bile acid biosynthetic pathways, suggesting that FXR may play a role in the regulation of bile acid metabolism. Indeed, FXR null mice, that have no overt phenotype except increased bile acid levels in the blood, cannot sustain a cholic acid enriched diet. They loose weight, develop hepatoxicity and around 30% die after 5 days on that diet [31]. This report and other studies identify the genes encoding the bile salt export pump (BSEP) and the ileal bile acid binding protein as positively regulated by FXR [32]. However, at least as important is the role of FXR in negatively regulating the expression of the *CYP7A* gene, limiting the synthesis of bile acids *via* the neutral pathway, in a negative regulatory feedback loop (Fig. 4).

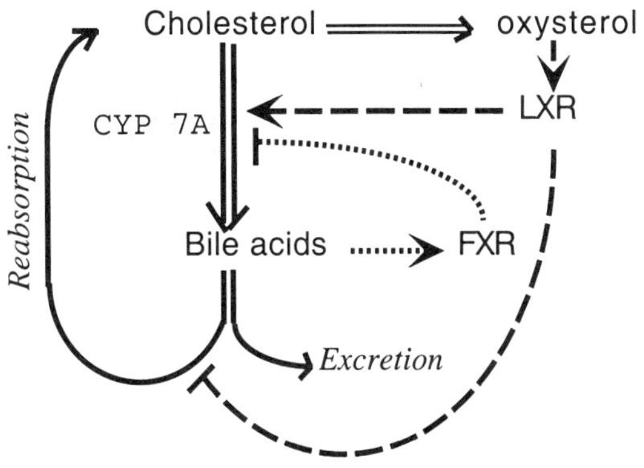

Figure 4. LXR and FXR regulate bile acid synthesis. In the presence of high cholesterol, oxysterols bind and activate LXR. This leads to a stimulation of CYP7A gene expression and increased bile acid production. It also inhibits the net uptake of cholesterol in the gut, by increasing the cholesterol efflux from epithelial cells into the lumen. FXR is activated by high levels of bile acids and inhibits their synthesis through inhibition of CYP7A gene expression.

How FXR negatively regulates *CYP7A* gene expression is a very interesting issue (Fig. 5). The constitutive expression of *CYP7A* in the liver is dependent upon the nuclear factor LRH-1 (NR5A2), which binds to a response element located in the promoter of the *CYP7A* gene. The activity of this so-called "competence factor" for CYP7A expression is also required for its up-regulation by LXR. Activated FXR does not bind to the *CYP7A* gene promoter but up-regulates the expression of the short heterodimerization partner (SHP-1, NR0B2) which binds to LRH-1, forming an inactive heterodimer. This results in down-regulation of CYP7A expression, even in the presence of activated LXR (Fig. 5) [33, 34].

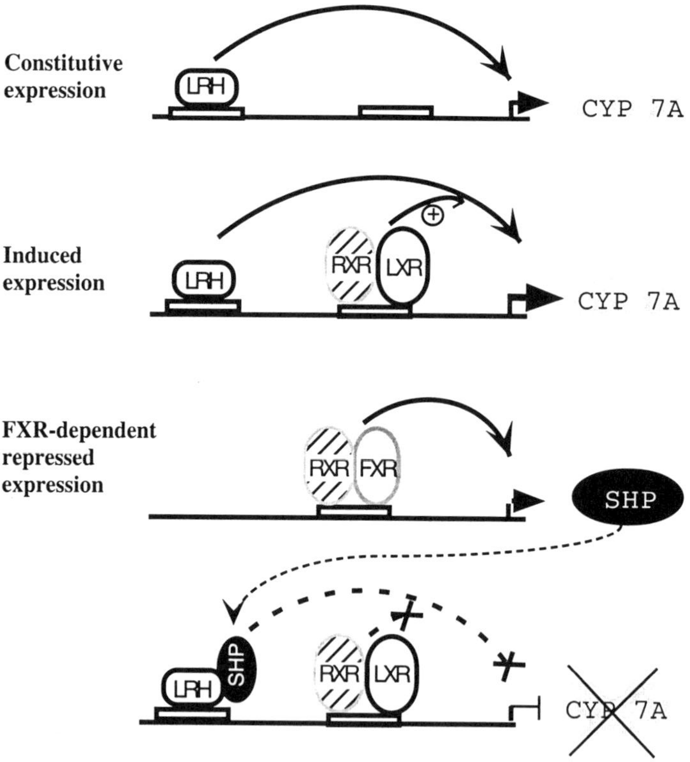

Figure 5. FXR inhibits CYP7A gene expression through stimulation of the synthesis of the inhibitory short heterodimerisation partner (SHP1) of LRH. LRH is a competence factor required for constitutive and induced expression of CYP7A. FXR increases the expression of SHP1 which forms an inactive heterodimer with LRH, thereby inhibiting the expression of CYP7A.

Negative regulation by ligand activated nuclear receptor is not an uncommon feature. However, the modular organization of the receptor with the presence of a positive transactivation domain and the molecular mechanism involving a ligand induced recruitment of a co-activator tend to accredit the hypothesis that most negative regulation by ligand-activated nuclear receptors is achieved *via* an indirect mechanism. The one described for FXR and the *CYP7A* gene expression might reflect a general mechanism for negative regulation by nuclear receptors. Indeed, the same mechanism has been previously described for the inhibitory interaction of Dax-1, closely related to SHP-1, with SF-1, a competence transcription factor for steroidogenic target genes in adrenal gland ([35]; for a review, see Ref. [36]). In that respect, deciphering the FXR negative mode of action might provide the lead for further understanding of other negative regulatory loops that involve nuclear receptors.

How could knowledge of this complex and still incompletely understood network that controls cholesterol homeostasis in an organism be used in a therapeutic approach? Oral treatment with rexinoids, synthetic compounds that activate the RXR moiety of the heterodimers LXR:RXR and FXR:RXR, results in a very efficient inhibition of cholesterol uptake. This rexinoid effect was reported to involve at least two mechanisms, *via* the activation of LXR:RXR in the intestine (increased cholesterol efflux in the lumen), and *via* the activation of FXR:RXR in the liver (inhibition of bile acids synthesis, thus lowering micelle formation and cholesterol uptake in the gut) [26]. As rexinoids are also capable of activating other permissive complexes, such as those formed with PPARα or PPARγ, RXR might not be as specific as originally thought. Hence, these results suggest that FXR and LXR might be interesting alternate targets.

PPAR, PXR AND CAR AS XENOBIOTIC SENSORS: MOLECULAR LINKS TO ENZYMATIC INDUCTION, XENOBIOTIC METABOLISM AND THERAPEUTIC INTERFERENCE

Xenobiotic processing in living organism mainly occurs through two phases of modification. In the first phase, a polar group is added to hydrophobic substrates by hydroxylation and oxidation, *via* the mono-oxygenase system. The second phase involves enzymes that promote the sulfo- or glucuro-conjugation of the polar groups previously added, increasing the hydrophilicity of the compounds and thus their excretion in bile or urine. Phase I enzymes mainly include the cytochrome P450 (CYP) families (reviewed in Ref. [37]). They are haem-containing mono-oxygenases that participate in the oxidative metabolism of a wide variety of substrates. Thirteen out of the seventeen families of these mono-oxygenases are involved in endogenous biosynthetic pathways, such as those used in steroid synthesis, bile acid synthesis (CYP7, as discussed above), and metabolism of other endogenous substrates. They have a selective substrate specificity and, as seen in the case of the CYP7A, are tightly regulated. In contrast, three groups of CYP (CYP 1 to 3), and to a lesser extent the CYP4 family, are essential for the metabolism of hydrophobic exogenous compounds to which the organism is exposed, either voluntarily as drugs, or accidentally such as environmental chemicals and other xenobiotics. In addition, they are also important in the catabolism of steroids. These four groups are mainly expressed in the liver and exhibit a rather broad substrate specificity. Their action and regulation raise two major questions:

i) Depending upon the substrate, CYP-mediated oxidation corresponds either to a first step in the degradation pathway or, in contrast, to the transformation of an inactive substrate into an active molecule. More problematic is the fact that this type of reaction may generate highly reactive electrophilic metabolites that can react with intracellular macromolecules and, in some cases, initiate reactions leading to chemical carcinogenesis and/or cellular toxicity. This is indeed the case for some xenobiotics metabolised by P450s.

ii) Many of the genes belonging to these four families are transcriptionally activated by the substrate that they in turn metabolize, revealing an adaptive mechanism. This mechanism of enzyme induction is considered as the main origin of drug-drug interactions. Indeed, induction of an enzyme by one compound may modify the metabolism, and thus thc effective concentration, of another compound metabolized by the same enzyme. The extent of xenobiotic-induced gene induction, and thus therapeutic interference, cannot always be predicted as sensitivity may be highly variable among human populations.

Recent work has demonstrated that some nuclear receptors are tightly linked to the regulation of xenobiotic metabolism (reviewed in Ref. [38]), making available some new tools for a better understanding and control in drug assessment.

PPARα and the Regulation of the *CYP4A* Genes

The first nuclear receptor discovered as participating in the xenobiotic-induced expression of a member of the CYP family was PPARα. As mentioned above, PPARα is a receptor for a diverse group of compounds referred to as peroxisome proliferators, among which are lipid-lowering drugs of the fibrate class, as well as some plasticizers and herbicides [4]. PPARα, activated by these compounds or by physiologically relevant activators such as unsaturated fatty acids, mediates the induction of the fatty acid hydroxylases of the CYP4A family [39]. The endogenous substrates for this family are fatty acids and prostaglandins which are hydroxylated at the ω position. The importance of fatty acid ω-hydroxylation under normal situations seems to be rather minor, but it might become more important in the context of high load of fatty acids in the liver during starvation or diabetes for example. However, it is so far unclear if the PPARα-mediated induction of some members of the CYP4 family can relate to xenobiotic metabolism and to the carcinogenic potency of the peroxisome proliferators in rodents.

PXR and the Regulation of the *CYP3A* Genes

In contrast, the CYP3A4 enzyme is the most abundant cytochrome P450 expressed in human liver and intestine. The structural variety of its substrates is remarkably broad and altogether, it is estimated that CYP3A4 contributes to the metabolism of around half of the drugs in use today [40]. Thus, variation and regulation of its catalytic activity is of major importance with respect to bio-availability and drug-drug interactions. Importantly, expression of CYP3A4 is strongly regulated by a class of compounds comprising some steroids (dexamethasone at high doses, RU486, and pregnenolone 16α-carbonitrile (PCN)), but also rifampicine and phenobarbital. How this induction, which occurs at the transcriptional level, is achieved by such structurally diverse compounds was still a question two years ago.

PXR (NR1I2) was identified in human and mouse where it is mainly expressed in liver and intestine [41-43]. PXR has the classical modular structure of a nuclear receptor and forms heterodimers with RXR. The heterodimer binds to DNA response elements composed of two hexamers AGTTCA or AGGTCA, organized in a direct repeat with a spacer of 3 nucleotides, or in an everted repeat, spaced by 6 nucleotides. Like other nuclear receptors, PXR positively transactivates target genes *via* the ligand-dependent recruitment of a co-activator, as assessed *in vitro*. PXR was first characterized as a receptor for endogenous steroid metabolites, mainly C21 steroids such as PCN and 5β-pregnane-3,20-dione (thus the name of pregnane X receptor). Interestingly, response elements for PXR:RXR were found in the promoter region of several of the *CYP3A4* genes [43, 44]. The xenobiotics known for activating the expression of this gene family were thus tested and indeed many were found to be able to activate PXR. These include rifampicin, lovastatin, or RU486, an antiprogestin that also acts as a glucocorticoid antagonist. With the identification of the human ortholog also called SXR [42], more systematic analyses of potential activators was carried out. Indeed, PXR from mouse, rat, rabbit and human exhibit significant divergence in their ligand binding domain, which translates into overlapping but not identical subsets of activating compounds [45]. For example, rifampicin is a potent human PXR activator and a rather weak mouse PXR activator. An opposite species-specific potency was found with PCN. The degree of PXR activation remarkably correlates with the potency of the compounds to induce CYP3A in human and rat, respectively. This study [45] and others led to two main conclusions: first, a clear correlation was made with a remarkably diverse group of compounds between their ability to activate human PXR and their ability to induce *CYP3A4* gene expression in human; second, the divergence between human PXR and mouse PXR was found to correlate with species specificity with respect to the compounds that elicit an induction of the expression of *CYP3A4* genes. This conclusion was indeed confirmed by the sensitivity to rifampicine of mice in which the two mouse PXR alleles were deleted but that are transgenic for human PXR [46].

CAR and the Regulation of the *CYP2B10* Gene

Another nuclear receptor, CAR (NR1I3), also plays a major role in these pathways and indeed, as we will see, PXR and CAR might participate in pathways that overlap.

CAR is a former orphan nuclear receptor highly expressed in the liver and also present in the small intestine ([47, 48], reviewed in [49]). As most nuclear receptors in this class, CAR forms a heterodimer with RXR. Surprisingly, CAR:RXR exhibits constitutive activity to a reporter gene containing a retinoic acid response element (a direct repeat of two AGGTCA motifs, spaced by 5 nucleotides). A systematic search for CAR ligands led to the identification of two androstane metabolites (androstanol and androstenol) that inhibit this constitutive CAR activity, thus its name which stands for constitutive androstane receptor [50]. This antagonist activity is associated with the release of a co-activator which binds to CAR:RXR in absence of any known ligands. Interestingly, CAR:RXR was then found to be the complex that binds to the phenobarbital response element present in the promoter of the *CYP2B10* gene, implying that it might be a major relay in the induction of this gene in response to the phenobarbital class of compounds [51]. The most potent compound of this class of xenobiotics, the 1,4-*bis*[2-(3,5-dichlorpyridyloxy)]benzène (TCPOBOP), is indeed a ligand and true agonist of CAR [52]. Its binding to CAR leads to a reinforcement of the co-activator recruitment and further enhances the transcriptional activity of CAR. In contrast, phenobarbital itself does not bind to CAR. Studies using primary hepatocyte culture demonstrate that phenobarbital triggers the nuclear translocation of CAR, *via* a kinase dependent pathway, thereby triggering the induction of the *CYP2B10* gene [53]. Some questions remain unanswered, among them: what is the relevance of androstane in this system? This question arises because sufficient concentration of these metabolites only occurs in tissues that do not express CAR. What is the CAR cellular localisation in an *in vivo* context? This question stems from the observation that CAR is cytoplasmic prior to phenobarbital induction in primary hepatocytes but readily nuclear in hepG2 cells. Are other compounds from the class of phenobarbital also ligands for CAR as is TCPOBOP, or do they act as "translocators" like phenobarbital itself ?

While these questions remain to be answered, the examination of CAR null mice provided the best tool to firmly assess the role of this receptor in xenobiotic induction of drug metabolism. CAR null mice are devoid of any induction of CYP2B10 expression upon phenobarbital or TCPOBOP treatment, resulting in a much higher liver toxicity of these compounds [54]. Pre-treated wild-type mice with phenobarbital are usually resistant to the muscle relaxant Zoxazolamine, due to a marked induction of CYP that degrades this compound. CAR null mice, pre-treated or not, are extremely sensitive and often die following Zoxazolamine administration. In contrast, cocaine, which can cause acute hepatotoxicity when wild type mice are previously exposed to phenobarbital-like substances, does not cause liver damage to pre-treated CAR null mice [54]. All these observations clearly demonstrate that CAR is an important mediator of phenobarbital-like induced gene expression and of drug-drug metabolism interference. However, again, human CAR and mouse CAR also diverge in their ligand binding domains, and thus probably also in their affinity for the various compounds which trigger gene induction *via* this interaction.

The analyses of null mice clearly demonstrated that CAR and PXR, at least in mice, have a specific role that cannot be mutually compensated [46, 54]. However, there is increasing evidence of an overlap in their functions. Studies demonstrated that they share some of their activators and can bind to each other's specific response elements [55, 56]. Thus, the CYP2B family might not be the only target of CAR, as CYP3A4 might not be the only target of PXR. In a simplistic hypothesis, hCAR and hPXR on the one hand and mCAR and mPXR on the other hand together seem to cover most of the known drug metabolism inducers. How these compounds are distributed with respect to their affinity for PXR or CAR within a given species could be hypothesized to be a matter of adaptation and evolution. Finally, the identification of these nuclear receptors is of prime importance with respect to the strategy of leading pharmaceutical industries to promote or retrieve a specific drug. Any molecule capable of activating hPXR or hCAR should be considered as a potential enzymatic inducer and thus stopped from further development [49, 57] (Fig. 6).

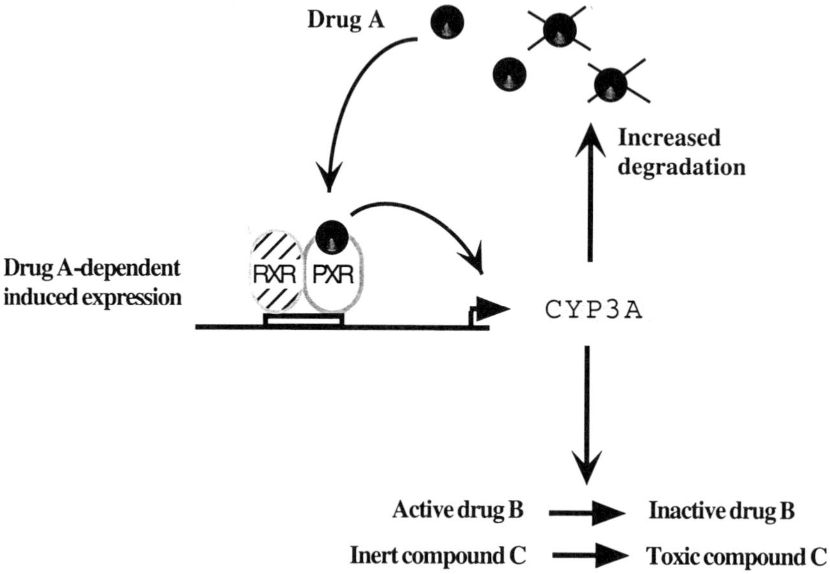

Figure 6. Nuclear receptors mediate xenobiotic-induced enzyme activity, which results in drug-drug interference.

RXR: AT THE CENTER OF THE SENSOR-REGULATED PATHWAYS

Because all the nuclear receptors mentioned above mainly act *via* the formation of heterodimers with RXR, one can wonder if RXR itself may play a role in the general network created by its heterodimerization partners. This question is of major importance if one considers that RXR is a nuclear receptor itself that can be activated by specific ligands, potentially activating the permissive heterodimers previously described.

RXR can be activated by 9-*cis* retinoic acid, an isomer of all-*trans* retinoic acid. While the occurrence of this molecule *in vivo* has been questioned, the identification of two enzymes that participate in the isomerization of all *trans* retinoic acid to form 9-*cis* retinoic acid gives support to its *in vivo* relevance [58, 59]. However, some doubts still linger concerning the nature of the major natural RXR ligand. An oleic acid molecule was found in the RXR ligand binding pocket of a crystal of the heterodimer of RAR and RXR ligand binding domains [60]. This could be considered as an indication that fatty acids might be RXR ligands. Indeed, it has been reported that in the adult brain, RXR can be activated by docosahexaenoic acid, a long chain polyunsaturated fatty acid [61]. However, the physiological relevance of this ligand in organs other than the brain remains to be determined.

There are 3 isotypes of RXR, α, β, and γ, with isoforms for each of them. RXRα seems to be the most critical *in vivo* since RXRα KO mice die at embryonic stages, whereas RXRγ and RXRβ KO animals do not display any severe phenotype [62, 63]. To overcome RXRα embryonic lethality, a tissue specific KO of RXRα has been generated in liver and metabolic studies performed to understand which pathways are the most affected by the absence of this nuclear receptor partner [64]. As could be expected, the PPARα, CAR, LXR, PXR, and FXR pathways were all compromised, at least partially, by the absence of RXRα, an effect that could not be compensated for by RXRβ or RXRγ.

In addition to being a silent heterodimeric partner, activation of RXR *via* ligand binding also activates complexes that it forms with the permissive receptors, i.e. PPAR, LXR, and FXR. Rexinoids have been claimed to be potentially important therapeutic agents due to their ability to trigger PPARγ:RXR activation, resulting in potent antidiabetic effects. However, as mentioned above, rexinoids also trigger LXR:RXR and FXR:RXR pathways to lower cholesterol uptake. Thus, the therapeutic use of rexinoids requires a careful evaluation. Alternatively, the search for and development of heterodimer-specific ligands for RXR might allow a better targeting of the pathway of interest.

CONCLUSIONS AND PERSPECTIVES

In this chapter, we have summarized the main observations that lead to the consideration that a group of nuclear receptors act as metabolite sensors. These sensors exert a feed back regulation upon metabolic pathways allowing for adaptation to the wide variety of nutrients, natural food or xenobiotics to which animals and humans are exposed. The general corollaries that accompany this concept can be summed up as follows.

The definition of a "sensor" for nutrients and their metabolites, as well as for exogenous compounds, requires that these sensors are capable of binding a wide variety of compounds. The structural corollary is that the ligand binding pocket must accommodate this diversity. This has indeed been verified for PPARβ and PPARγ [5, 6]. It will be of interest to see if this applies to the whole group of sensor-nuclear receptors or if PPAR remains the exception.

What humans eat and how they make use of various drugs is quite different from what rabbits, rodents, or other animal species consume or to what they are exposed. Thus, one might expect some divergence among species for the specificity and substrates of these sensors. This was indeed initially discussed for PPARs when it was found that human, mouse, Xenopus and rat PPAR have some overlapping but not superimposed sets of ligands [65]. As discussed above, this divergence has also been well characterized for CAR and PXR. It is believed that this is likely to reflect an adaptive response to different environmental challenges, and will probably be a rule for this class of sensor nuclear receptors.

Another corollary of this divergence across species concerns the use of a rodent model for assessing the usefulness of a new drug, as well as its metabolism and its potential for drug-drug interaction. Screening for hPXR and hCAR activation is a first step. However, it does not replace *in vivo* studies. One possibility is to "humanise" mice for the nuclear receptor of interest. In such a model, the mouse receptor of interest is exchanged by the human receptor. This has been indeed accomplished for PXR, demonstrating that drug metabolism and species specificity can be carried by the receptor and not by the target genes [46]. Upon appropriate controls of the metabolism that is considered, this "humanisation" of mice for the nuclear receptors that act as sensors might become an important tool in drug assessment.

Whether due to a particularly rapid evolution of this category of nuclear receptor-sensors, or to some intrinsic properties of their genes, a high variability is seen among human populations in terms of drug susceptibility, or nutrient use and disposal. The occurrence of polymorphism in the *CYP2D6* and *CYP2C19* genes in a few individuals, called "slow metabolisers", results in a slower metabolism of certain drugs. Variations also exist at the level of *CYP* gene expression. One hypothesis is that this might reflect some polymorphism in the genes encoding the receptors described above. The search for the occurrence of such polymorphism should become a priority both in the diagnostic of metabolic disorders, but also as a predictive tool for individual susceptibility.

Do we know everything, or at least all the main concepts about the network of nuclear receptor-sensors? Certainly not. A mere glance at the long list of remaining orphan receptors is indicative of the pathways that remain to be charted!

Acknowledgement

We thank N. Noy for critical reading of the manuscript and all members of the Institute of Animal Biology for stimulating discussions. Work done in the author's laboratory was supported by the Swiss National Science Foundation and the Etat de Vaud.

REFERENCES

1. **Desvergne B, Wahli W.** 1999 Peroxisome proliferator-activated receptors: nuclear control of metabolism. Endocr Rev 20:649-688.
2. **Nuclear Receptors Nomenclature Committee.** 1999 A unified nomenclature system for the nuclear receptor superfamily. Cell 97:161-163.
3. **Juge-Aubry C, Pernin A, Favez T, Burger AG, Wahli W, Meier CA, Desvergne B.** 1997 DNA binding properties of peroxisome proliferator-activated receptor subtypes on various natural peroxisome proliferator response elements: importance of the 5' flanking region. J Biol Chem 272:25252-25259.
4. **Issemann I, Green S.** 1990 Activation of a member of the steroid hormone receptor superfamily by peroxisome proliferators. Nature 347:645-650.
5. **Nolte RT, Wisely GB, Westin S, Cobb JE, Lambert MH, Kurokawa R, Rosenfeld MG, Willson TM, Glass CK, Millburn MV.** 1998 Ligand binding and co-activator assembly of the peroxisome proliferator-activated receptor-γ. Nature 395:137-143.
6. **Xu HE, Lambert MH, Montana VG, Parks DJ, Blanchard SG, Brown PJ, Sternbach DD, Lehmann JM, Wisely GB, Willson TM, Kliewer SA, Milburn MV.** 1999 Molecular recognition of fatty acids by peroxisome proliferator-activated receptors. Mol Cell 3:397-403.
7. **Bourguet W, Ruff M, Chambon P, Gronemeyer H, Moras D.** 1995 Crystal structure of the ligand-binding domain of the human nuclear receptor RXR-alpha. Nature 375:377-382.
8. **Renaud JP, Rochel N, Ruff M, Vivat V, Chambon P, Gronemeyer H, Moras D.** 1995 Crystal structure of the RAR-gamma ligand-binding domain bound to all-trans retinoic acid. Nature 378:681-689.
9. **Wagner RL, Apriletti JW, McGrath ME, West BL, Baxter JD, Fletterick RJ.** 1995 A structural role for hormone in the thyroid hormone receptor. Nature 378:690-697.
10. **Braissant O, Foufelle F, Scotto C, Dauca M, Wahli W.** 1996 Differential expression of peroxisome proliferator-activated receptors (PPARs): tissue distribution of PPAR-alpha, -beta, and -gamma in the adult rat. Endocrinology 137:354-366.
11. **Lee SS, Pineau T, Drago J, Lee EJ, Owens JW, Kroetz DL, Fernandez-Salguero PM, Westphal H, Gonzalez FJ.** 1995 Targeted disruption of the α isoform of the peroxisome proliferator-activated receptor gene in mice results in abolishment of the pleiotropic effects of peroxisome proliferators. Mol Cell Biol 15:3012-3022.
12. **Kersten S, Seydoux J, Peters JM, Gonzalez FJ, Desvergne B, Wahli W.** 1999 Peroxisome proliferator-activated receptor α mediates the adaptive response to fasting. J Clin Invest 103:1489-1498.
13. **Leone T, Weinheimer C, Kelly D.** 1999 A critical role for the peroxisome proliferator-activated receptor α (PPARα) in the cellular fasting response: the PPARα-null mouse as a model of fatty acid oxidation disorders. Proc Natl Acad Sci USA 96:7473-7478.
14. **Rosen ED, Walkey CJ, Puigserver P, Spiegelman BM.** 2000 Transcriptional regulation of adipogenesis. Genes Dev 14:1293-1307.

15. **Barak Y, Nelson MC, Ong ES, Jones YZ, Ruiz-Lozano P, Chien KR, Koder A, Evans RM.** 1999 PPAR gamma is required for placental, cardiac, and adipose tissue development. Mol Cell 4:585-595.
16. **Kubota N, Terauchi Y, Miki H, Tamemoto H, Yamauchi T, Komeda K, Satoh S, Nakano R, Ishii C, Sugiyama T, Eto K, Tsubamoto Y, Okuno A, Murakami K, Sekihara H, Hasegawa G, Naito M, Toyoshima Y, Tanaka S, Shiota K, Kitamura T, Fujita T, Ezaki O, Aizawa S, Kadowaki T, et al.** 1999 PPAR gamma mediates high-fat diet-induced adipocyte hypertrophy and insulin resistance. Mol Cell 4:597-609.
17. **Kersten S, Desvergne B, Wahli W.** 2000 Roles of PPARs in health and disease. Nature 405:421-424.
18. **Brown MS, Goldstein JL.** 1999 A proteolytic pathway that controls the cholesterol content of membranes, cells, and blood. Proc Natl Acad Sci U S A 96:11041-11048.
19. **Song C, Kokontis JM, Hiipakka RA, Liao S.** 1994 Ubiquitous receptor: a receptor that modulates gene activation by retinoic acid and thyroid hormone receptors. Proc. Natl. Acad. Sci. USA 91:10809-10813.
20. **Seol W, Choi HS, Moore DD.** 1995 Isolation of proteins that interact specifically with the retinoid X receptor: two novel orphan receptors. Mol Endocrinol 9:72-85.
21. **Willy PJ, Umesono K, Ong ES, Evans RM, Heyman RA, Mangelsdorf DJ.** 1995 LXR, a nuclear receptor that defines a distinct retinoid response pathway. Genes Dev 9:1033-1045.
22. **Teboul M, Enmark E, Li Q, Wikstrom AC, Pelto-Huikko M, Gustafsson JA.** 1995 OR-1, a member of the nuclear receptor superfamily that interacts with the 9-cis-retinoic acid receptor. Proc Natl Acad Sci U S A 92:2096-2100.
23. **Repa JJ, Mangelsdorf DJ.** 1999 Nuclear receptor regulation of cholesterol and bile acid metabolism. Curr Opin Biotechnol 10:557-563.
24. **Janowski BA, Willy PJ, Devi TR, Mangelsdorf DJ.** 1996 An oxysterol signalling pathway mediated by the nuclear receptor LXRα. Nature 383:728-731.
25. **Peet DJ, Turley SD, Ma W, Janowski BA, Lobaccaro JM, Hammer RE, Mangelsdorf DJ.** 1998 Cholesterol and bile acid metabolism are impaired in mice lacking the nuclear oxysterol receptor LXR alpha. Cell 93:693-704.
26. **Repa JJ, Turley SD, Lobaccaro JA, Medina J, Li L, Lustig K, Shan B, Heyman RA, Dietschy JM, Mangelsdorf DJ.** 2000 Regulation of absorption and ABC1-mediated efflux of cholesterol by RXR heterodimers. Science 289:1524-1529.
27. **Forman BM, Goode E, Chen J, Oro AE, Bradley DJ, Perlmann T, Noonan DJ, Burka LT, McMorris T, Lamph WW, Evans RM, Weinberger C.** 1995 Identification of a nuclear receptor that is activated by farnesol metabolites. Cell 81:687-693.
28. **Makishima M, Okamoto AY, Repa JJ, Tu H, Learned RM, Luk A, Hull MV, Lustig KD, Mangelsdorf DJ, Shan B.** 1999 Identification of a nuclear receptor for bile acids. Science 284:1362-1365.
29. **Parks DJ, Blanchard SG, Bledsoe RK, Chandra G, Consler TG, Kliewer SA, Stimmel JB, Willson TM, Zavacki AM, Moore DD, Lehmann JM.** 1999 Bile acids: natural ligands for an orphan nuclear receptor. Science 284:1365-1368.
30. **Wang H, Chen J, Hollister K, Sowers LC, Forman BM.** 1999 Endogenous bile acids are ligands for the nuclear receptor FXR/BAR. Mol Cell 3:543-553.
31. **Sinal CJ, Tohkin M, Miyata M, Ward JM, Lambert G, Gonzalez FJ.** 2000 Targeted disruption of the nuclear receptor FXR/BAR impairs bile acid and lipid homeostasis. Cell 102:731-744.
32. **Grober J, Zaghini I, Fujii H, Jones SA, Kliewer SA, Willson TM, Ono T, Besnard P.** 1999 Identification of a bile acid-responsive element in the human ileal bile acid-binding protein gene. Involvement of the farnesoid X receptor/9-cis-retinoic acid receptor heterodimer. J Biol Chem 274:29749-29754.

33. **Goodwin B, Jones SA, Price RR, Watson MA, McKee DD, Moore LB, Galardi C, Wilson JG, Lewis MC, Roth ME, Maloney PR, Willson TM, Kliewer SA.** 2000 A regulatory cascade of the nuclear receptors FXR, SHP-1, and LRH-1 represses bile acid biosynthesis. Mol Cell 6:517-526.
34. **Lu TT, Makishima M, Repa JJ, Schoonjans K, Kerr TA, Auwerx J, Mangelsdorf DJ.** 2000 Molecular basis for feedback regulation of bile acid synthesis by nuclear receptors. Mol Cell 6:507-515.
35. **Seol W, Choi HS, Moore DD.** 1996 An orphan nuclear hormone receptor that lacks a DNA binding domain and heterodimerizes with other receptors. Science 272:1336-1339.
36. **Parker KL, Schedl A, Schimmer BP.** 1999 Gene interactions in gonadal development. Annu Rev Physiol 61:417-433.
37. **Coon MJ, Ding XX, Pernecky SJ, Vaz AD.** 1992 Cytochrome P450: progress and predictions. FASEB J 6:669-673.
38. **Waxman DJ.** 1999 P450 gene induction by structurally diverse xenochemicals: central role of nuclear receptors CAR, PXR, and PPAR. Arch Biochem Biophys 369:11-23.
39. **Muerhoff AS, Griffin KJ, Johnson EF.** 1992 The peroxisome proliferator-activated receptor mediates the induction of CYP4A6, a cytochrome-P450 fatty-acid omega-hydroxylase, by clofibric acid. J Biol Chem 267:19051-19053.
40. **Maurel P.** 1996 The CYP3A family. In: Ioannides C (ed.), Cytochromes P450: Metabolic and Toxicological Aspects, CRC Press Inc., Boca Raton, FL, 241-270.
41. **Kliewer SA, Moore JT, Wade L, Staudinger JL, Watson MA, Jones SA, McKee DD, Oliver BB, Willson TM, Zetterstrom RH, Perlmann T, Lehmann JM.** 1998 An orphan nuclear receptor activated by pregnanes defines a novel steroid signaling pathway. Cell 92:73-82.
42. **Blumberg B, Sabbagh W, Jr., Juguilon H, Bolado J, Jr., van Meter CM, Ong ES, Evans RM.** 1998 SXR, a novel steroid and xenobiotic-sensing nuclear receptor. Genes Dev 12:3195-3205.
43. **Bertilsson G, Heidrich J, Svensson K, Asman M, Jendeberg L, Sydow-Bäckman M, Ohlsson R, Postlind H, Blomquist P, Berkenstam A.** 1998 Identification of a human nuclear receptor defines a new signaling pathway for CYP3A induction. Proc Natl Sci Acad Sci USA 95:12208-12213.
44. **Lehmann JM, McKee DD, Watson MA, Willson TM, Moore JT, Kliewer SA.** 1998 The human orphan nuclear receptor PXR is activated by compounds that regulate CYP3A4 gene expression and cause drug interactions. J Clin Invest 102:1016-1023.
45. **Jones SA, Moore LB, Shenk JL, Wisely GB, Hamilton GA, McKee DD, Tomkinson NC, LeCluyse EL, Lambert MH, Willson TM, Kliewer SA, Moore JT.** 2000 The pregnane X receptor: a promiscuous xenobiotic receptor that has diverged during evolution. Mol Endocrinol 14:27-39.
46. **Xie W, Barwick JL, Downes M, Blumberg B, Simon CM, Nelson MC, Neuschwander-Tetri BA, Brunt EM, Guzelian PS, Evans RM.** 2000 Humanized xenobiotic response in mice expressing nuclear receptor SXR. Nature 406:435-439.
47. **Baes M, Gulick T, Choi HS, Martinoli MG, Simha D, Moore DD.** 1994 A new orphan member of the nuclear hormone receptor superfamily that interacts with a subset of retinoic acid response elements. Mol Cell Biol 14:1544-1552.
48. **Choi HS, Chung M, Tzameli I, Simha D, Lee YK, Seol W, Moore DD.** 1997 Differential transactivation by two isoforms of the orphan nuclear hormone receptor CAR. J Biol Chem 272:23565-23571.
49. **Tzameli I, Moore DD.** 2001 Role reversal: new insights from new ligands for the xenobiotic receptor CAR. Trends Endocrinol Metab 12:7-10.

50. **Forman BM, Tzameli I, Choi HS, Chen J, Simha D, Seol W, Evans RM, Moore DD.** 1998 Androstane metabolites bind to and deactivate the nuclear receptor CAR-beta. Nature 395:612-615.
51. **Sueyoshi T, Kawamoto T, Zelko I, Honkakoski P, Negishi M.** 1999 The repressed nuclear receptor CAR responds to phenobarbital in activating the human CYP2B6 gene. J Biol Chem 274:6043-6046.
52. **Honkakoski P, Zelko I, Sueyoshi T, Negishi M.** 1998 The nuclear orphan receptor CAR-retinoid X receptor heterodimer activates the phenobarbital-responsive enhancer module of the CYP2B gene. Mol Cell Biol 18:5652-5658.
53. **Kawamoto T, Sueyoshi T, Zelko I, Moore R, Washburn K, Negishi M.** 1999 Phenobarbital-responsive nuclear translocation of the receptor CAR in induction of the CYP2B gene. Mol Cell Biol 19:6318-6322.
54. **Wei P, Zhang J, Egan-Hafley M, Liang S, Moore DD.** 2000 The nuclear receptor CAR mediates specific xenobiotic induction of drug metabolism. Nature 407:920-923.
55. **Moore LB, Parks DJ, Jones SA, Bledsoe RK, Consler TG, Stimmel JB, Goodwin B, Liddle C, Blanchard SG, Willson TM, Collins JL, Kliewer SA.** 2000 Orphan nuclear receptors constitutive androstane receptor and pregnane X receptor share xenobiotic and steroid ligands. J Biol Chem 275:15122-15127.
56. **Xie W, Barwick JL, Simon CM, Pierce AM, Safe S, Blumberg B, Guzelian PS, Evans RM.** 2000 Reciprocal activation of xenobiotic response genes by nuclear receptors SXR/PXR and CAR. Genes Dev 14:3014-3023.
57. **Moore JT, Kliewer SA.** 2000 Use of the nuclear receptor PXR to predict drug interactions. Toxicology 153:1-10.
58. **Mertz JR, Shang E, Piantedosi R, Wei S, Wolgemuth DJ, Blaner WS.** 1997 Identification and characterization of a stereospecific human enzyme that catalyzes 9-cis-retinol oxidation. A possible role in 9-cis-retinoic acid formation. J Biol Chem 272:11744-11749.
59. **Romert A, Tuvendal P, Simon A, Dencker L, Eriksson U.** 1998 The identification of a 9-cis retinol dehydrogenase in the mouse embryo reveals a pathway for synthesis of 9-cis retinoic acid. Proc Natl Acad Sci U S A 95:4404-4409.
60. **Bourguet W, Vivat V, Wurtz JM, Chambon P, Gronemeyer H, Moras D.** 2000 Crystal structure of a heterodimeric complex of RAR and RXR ligand- binding domains. Mol Cell 5:289-298.
61. **de Urquiza AM, Liu S, Sjoberg M, Zetterstrom RH, Griffiths W, Sjovall J, Perlmann T.** 2000 Docosahexaenoic acid, a ligand for the retinoid X receptor in mouse brain. Science 290:2140-2144.
62. **Mark M, Ghyselinck NB, Wendling O, Dupe V, Mascrez B, Kastner P, Chambon P.** 1999 A genetic dissection of the retinoid signalling pathway in the mouse. Proc Nutr Soc 58:609-613.
63. **Wendling O, Chambon P, Mark M.** 1999 Retinoid X receptors are essential for early mouse development and placentogenesis. Proc Natl Acad Sci U S A 96:547-551.
64. **Wan YJ, An D, Cai Y, Repa JJ, Hung-Po Chen T, Flores M, Postic C, Magnuson MA, Chen J, Chien KR, French S, Mangelsdorf DJ, Sucov HM.** 2000 Hepatocyte-specific mutation establishes retinoid X receptor alpha as a heterodimeric integrator of multiple physiological processes in the liver. Mol Cell Biol 20:4436-4444.
65. **Keller H, Devchand PR, Perroud M, Wahli W.** 1997 PPAR alpha structure-function relationships derived from species-specific differences in responsiveness to hypolipidemic agents. Biol Chem 378:651-655.

INDEX

Numbers in **bold** indicate chapters wherein the corresponding item is specifically detailed

****